DEVELOPMENTAL BIOLOGY OF FERN GAMETOPHYTES

V. RAGHAVAN
Department of Botany
The Ohio State University

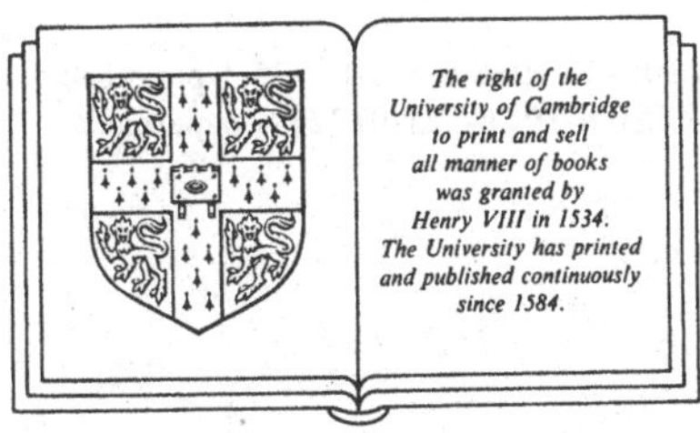

CAMBRIDGE UNIVERSITY PRESS
Cambridge
New York Port Chester
Melbourne Sydney

CAMBRIDGE UNIVERSITY PRESS
Cambridge, New York, Melbourne, Madrid, Cape Town, Singapore, São Paulo

Cambridge University Press
The Edinburgh Building, Cambridge CB2 2RU, UK

Published in the United States of America by Cambridge University Press, New York

www.cambridge.org
Information on this title: www.cambridge.org/9780521330220

First published 1989
This digitally printed first paperback version 2005

A catalogue record for this publication is available from the British Library

Library of Congress Cataloguing in Publication data

Raghavan, V. (Valayamghat), 1931–
Developmental biology of fern gametophytes / V. Raghavan.
p. cm. – (Developmental and cell biology series)
Bibliography: p.
Includes indexes.
ISBN 0 521 33022 X
1. Fern gamephotypes–Development. I. Title. II. Series.
QK521.R34 1989
587′.31043–dc 19 88-22935 CIP

ISBN-13 978-0-521-33022-0 hardback
ISBN-10 0-521-33022-X hardback

ISBN-13 978-0-521-01725-1 paperback
ISBN-10 0-521-01725-4 paperback

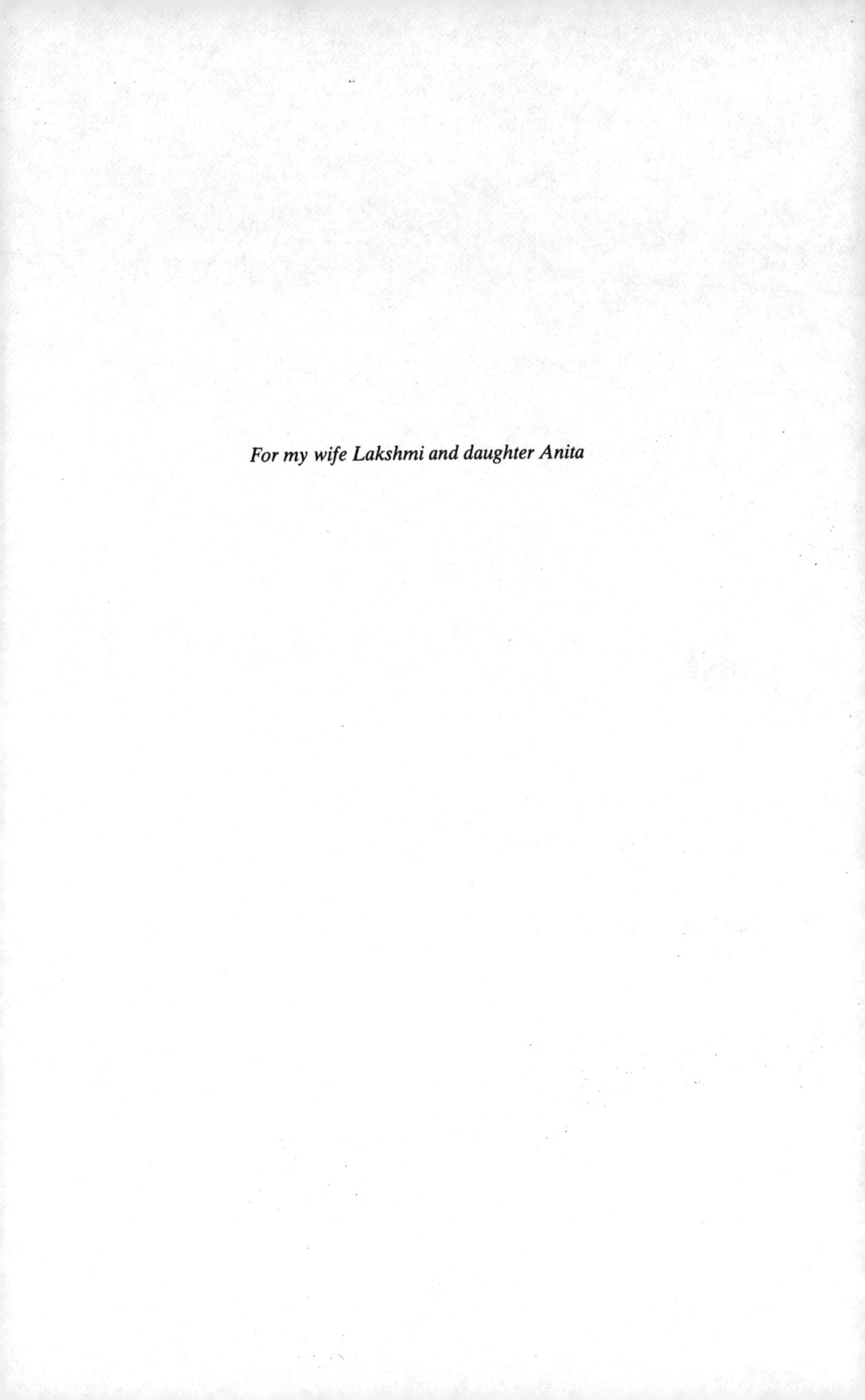

For my wife Lakshmi and daughter Anita

Contents

Preface

As the dominant phase in the life cycle of ferns and other pteridophytes, the sporophyte – the leafy fern plant with its characteristic foliage and growth habit – has received and continues to receive much attention in morphological, phylogenetic and evolutionary considerations, as well as in books and monographs. As a departure from this trend, the present book attempts to focus on the other phase of the fern life cycle – the unobtrusive gametophyte. Although slow to begin with, there has been a growing realization of the potentialities of the fern gametophyte for experimental studies, providing impetus for significant new interdisciplinary research on its growth, differentiation and sexuality. Since no unified summary of these investigations comprising the entire gametophytic phase exists, it is the purpose of this book to present a brief, but comprehensive account of the developmental biology of fern gametophytes. This work is also an inevitable outcome of my own interest in the study of gametophytes for nearly a quarter of a century.

The overall philosophy of the book has been to present the story of the gametophyte in a familiar developmental sequence beginning with the single-celled progenitor, going to the multicellular state and ending with the initiation of the sporophytic phase. The content of each chapter is focused around the physiological, cytological and biochemical background of the topic under discussion. The morphology of the gametophyte is integrated throughout the text and is related, where appropriate, to the developmental biology of the system. Each chapter begins by introducing the basic ideas related to its theme and develops them with reference to specific experiments. This has meant drawing examples from a large number of experimental organisms. The subject matter of some chapters is wide-ranging, with several aspects of research serving as focal points for discussion. In writing these chapters, I have felt that the full force of the insight developed by these researches can become evident only by putting the pieces of information together, at the risk of increasing the bulk of the chapters.

I envision the book as an introduction, to beginners in botany, of the potentialities of this fascinating experimental system and as a major reference work to those who are already working in the field. It is also my hope that this volume will be used as a supplementary text in courses in developmental botany dealing with model systems. In order to keep the book within reasonable bounds, no encyclopedic treatment has been attempted; indeed, most of the references cited represent papers published since 1950. In these days, when the data published in research papers are also found in the contributions to multi-authored volumes and symposia proceedings, I have relied on these latter sources only when the relevant information is not available in regular journal articles.

The binomials used in the book were referred to Drs John T. Mickel (New York Botanical Garden, New York) and Alan R. Smith (University of California, Berkeley) who were very generous with their time in checking the plant names and relating them to names now recognized as the valid ones. I am very appreciative of this help. Several other investigators have aided me in the preparation of this work by acceding to my request for original prints of photographs from their published papers. It is a pleasure to thank them for this help and the various authors and their publishers for granting me permission to use their illustrations in the book. I should also like to acknowledge the assistance of Drs Paul Green and Peter Barlow, two editors of the Developmental & Cell Biology Series, who read the first draft of the book and provided valuable comments for its scientific and stylistic improvement. Finally, I wish to pay tribute to my wife, Lakshmi, and daughter, Anita, for their encouragement and understanding, which prompted me to complete this book sooner than I expected.

V. Raghavan October, 1988

Abbreviations

ABA	Abscisic acid
AMP	Adenosine monophosphate
ATP	Adenosine triphosphate
ATPase	Adenosine triphosphatase
BUDR	5-Bromo-2-deoxyuridine
CCC	2-Chloroethylmethyl ammonium chloride
CEPA	2-Chloroethylphosphonic acid
CIPC	Isopropyl-N-3-chlorophenylcarbamate
DNA	Deoxyribonucleic acid
2,4-D	2,4-Dichlorophenoxyacetic acid
EGTA	Ethylene glycol-bis(β-amino-ethyl ether)-N,N,N′,N-tetraacetic acid
ER	Endoplasmic reticulum
FLU	5-Fluorouracil
GA	Gibberellic acid
IAA	Indoleacetic acid
IBA	Indolebutyric acid
ICL	Isocitrate lyase
MS	Malate synthase
NAA	Naphthaleneacetic acid
NOA	2-Naphthoxyacetic acid
PCIB	*p*-Chlorophenoxyisobutyric acid
P_r	Red-absorbing form of phytochrome
P_{fr}	Far-red absorbing form of phytochrome
$P_{b\text{-}nuv}$	Pigment absorbing blue and near UV light
Poly(A)	Polyadenylic acid
Poly(A) + RNA	Poly(A)-containing RNA
RNA	Ribonucleic acid
hnRNA	Heterogenous nuclear RNA
mRNA	Messenger RNA
rRNA	Ribosomal RNA
tRNA	Transfer RNA

RNase	Ribonuclease
SHAM	Salicyl-hydroxamic acid

1

Introduction

Extant ferns comprise a group of about 12 000 species of plants widely distributed throughout the world in many habitats and niches. As wide as their distribution is their range in size, with extremes such as the small water ferns with leaves less than 1 cm long and the giant tree ferns which attain heights of almost 25 m and bear crowns of leaves 30 cm or more in diameter. In many contemporary systems of classification with which developmental botanists will feel comfortable, ferns are assigned to the group Pteropsida or Filicopsida. Members of this group along with those of Psilopsida, Lycopsida and Sphenopsida constitute a major division of the plant kingdom known as Pteridophyta (pteridophytes). A distinctive anatomical feature of pteridophytes, which they share with gymnosperms and angiosperms, is the presence of a vascular system in the plant body, but pteridophytes differ from the latter two divisions in lacking the seed habit (hence the name, seedless vascular plants, for the division). During their evolutionary past, pteridophytes have stabilized and almost perfected the vascular system for a seedless plant so much so that they are also designated as vascular cryptogams. Most pteridophytes, including ferns, are trapped into a life cycle in which they are constrained by some primitive features such as the production of motile sperm and the requirement for free water for fertilization.

The life cycle of a fern, like that of other sexually reproducing plants, involves an alternation between a sporophytic and a gametophytic generation. Both the sporophyte and gametophyte are free-living multicellular plants, the former representing the asexual and the latter the sexual phase. The macroscopic fern plant is a sporophyte with a diploid number of chromosomes in each cell. It has its origin in the zygote, the product of fertilization of the egg by a sperm. The sporophyte possesses an underground stem or rhizome from which leaves and adventitious roots are produced at regular intervals. The most prominent feature of the sporophyte is its leaf, which is known as the frond, both in science and in common parlance. Ferns are unique among vascular cryptogams in possessing large leaves or megaphylls. The hallmarks of a megaphyll are its branched venation system and the frequent association of the leaf trace with one or more leaf gaps in the vascular cylinder of the stem. The fern frond may be a simple expanded lamina with a petiole or stipe or, more commonly, the lamina is cut up into leaflets or pinnae to give it the appearance of a compound leaf. Owing to the absence of an expanded blade on the pinnate leaf, the petiole continues as the main axis or rachis of the frond bearing the pinnae. In most ferns, the meristematic young leaves are rolled into tight spirals (fiddleheads) in a bud typifying circinate vernation. Coiling of the young leaf and its subsequent unwinding result from differential growth on its upper and lower surfaces and are mediated by changes in the concentration of endogenous auxin. The external morphological features of the typical fern leaf are not seen in the leaves of water ferns, which are in a class by themselves.

The reproductive structure of the sporophyte is the sporangium, which is a multicellular container for the spores. The fruiting dots on the margins and undersurfaces of fern leaves are indeed familiar to the casual observer; each dot represents a cluster of sporangia called the sorus, which is usually overlaid by a protective covering, the indusium. However, the sporangia are not restricted to leaves. In some ferns, they occur on highly differentiated leaf-like structures whose only function is to bear sporangia; in still others, some leaves become differentiated into sporangia, while the rest remain green and photosynthetic.

The young sporangium is filled with sporogenous cells which give rise to sporocytes (spore mother cells). The most fundamental process initiating the gametophytic generation occurs in the sporangium by meiosis in the sporocytes, yielding spores with the haploid or gametic number of chromosomes. From the moment it is liberated from the restraining wall of the tetrad, the spore generates new levels of organizational complexity as it prepares for sexual recombination by way of formation of a multicellular gametophyte, sex organs and gametes. Genetic information is decoded and utilized as the spore unfolds its developmental program. Thus, a spore born out of reduction division of the sporocyte is the single-celled progenitor of the gametophyte and all essential properties of interactions occurring in the adult gametophyte emanate from the information pool of this cell. (The word 'gametophyte' is used in this book loosely to designate any stage of the haploid generation, although, strictly speaking, it means a gamete-producing plant.)

This book is about how the spore achieves the basic body plan of the gametophyte, how the gametophyte grows and reproduces and, finally, how it is set on track to initiate the sporophytic phase. The objective is to provide an account of those aspects of the developmental biology of the gametophyte, beginning with the single-celled spore, that impinge on the morphological, cytological, physiological, biochemical and molecular changes in the cell population. It is important to realize that the progressive evolution of form and function of the gametophyte is specified by a series of developmental decisions. To understand how the spore discharges its genetically determined functions through the developmental landscape requires a study of the mechanism of these decisions.

For convenience, the book is divided into four parts. Chapters 2 to 5 deal with the structure, cytology and physiology of spores and their germination. The primary thrust of the discussion in these chapters is on the changing biochemical potential of the spore from a dormant cell to one undergoing metabolic fluxes as it germinates to form cells destined for different fates and functions. The next four chapters (Chapters 6 to 9) review the mechanisms of growth of the protonema initial of the germinated spore. This includes its growth as a filamentous structure, attainment of planar morphology and growth of the planar gametophyte. Chapters 10 and 11 ask what sort of mechanisms have evolved to account for the induction of sexuality in the gametophytes and to preserve genetic diversity in the population. The third chapter of this part (Chapter 12) examines the cellular and subcellular aspects of gametogenesis and the events leading to fertilization and initiation of the sporophytic phase. In the final part of the book, two chapters (Chapters 13 and 14) examine the alternative developmental programs of the sporophyte and gametophyte, resulting, respectively, in the formation of the gametophyte without meiosis and sporophyte in the absence of fertilization. Most of the information that is presented in the various chapters concerns ferns, which is as it should be in a book whose title specifies this group of plants. The excuse for occasional references to works on pteridophytes other than ferns is that they are of historical importance in a particular context or that they illuminate the concepts under discussion further.

Classification of ferns and other pteridophytes

There is a great diversity of opinion on the taxonomic arrangement of pteridophytes, especially of the ferns. Since 1940, more than 10 separate schemes or modifications of existing schemes of classification of ferns have been proposed and even to this day there is no consensus on the phylogenetic position of certain genera and families which have provoked severe disagreement in the past. Although the subject matter of this book is not even remotely connected to phylogeny, it is important that the genera discussed be identified in a phylogenetic framework to reflect relationships. In reviewing the different schemes of classification of ferns and other pteridophytes, I have found that the one proposed by Crabbe, Jermy and Mickel (1975) has incorporated many features that make it useful to the systematic pteridologist as well as to the developmental botanist. These investigators have divided the pteridophyte genera into seven assemblages, identifying families and subfamilies within each assemblage, as follows (subfamilies omitted here):

1. Fern allies (includes Psilopsida, Lycopsida and Sphenopsida)
   - Lycopodiaceae
   - Selaginellaceae
   - Isoetaceae
   - Equisetaceae
2. Eusporangiate ferns and Plagiogyriaceae
   - Ophioglossaceae
   - Marattiaceae
   - Osmundaceae
   - Plagiogyriaceae
3. Schizaeoid ferns to adiantoid ferns
   - Schizaeaceae
   - Parkeriaceae
   - Platyzomataceae
   - Adiantaceae
4. Filmy ferns and related groups
   - Loxosomaceae
   - Hymenophyllaceae
   - Hymenophyllopsidaceae
5. Gleichenioid ferns to polypodioids and grammatid ferns, by way of *Matonia, Cheiropleuria* and *Dipteris*
   - Stromatopteridaceae
   - Gleicheniaceae
   - Matoniaceae
   - Cheiropleuriaceae
   - Dipteridaceae
   - Polypodiaceae
   - Grammitidaceae
6. Protocyatheoid and cyatheoid ferns to dennstaedtioid, to thelypterid, to asplenioid ferns, the 'Aspidiales' and the blechnoid ferns
   - Metaxyaceae
   - Lophosoriaceae
   - Cyatheaceae
   - Thrysopteridaceae
   - Dennstaedtiaceae
   - Thelypteridaceae
   - Aspleniaceae
   - Davalliaceae
   - Blechnaceae
7. Hydropterides (water ferns)
   - Marsileaceae

Salviniaceae
Azollaceae

In an older classification scheme of extant ferns proposed by Bower (1923–28), the genera of Osmundaceae, Plagiogyriaceae and of assemblages 3 to 7 delimited by Crabbe *et al.* (1975) were incorporated into an order known as Filicales. Basic families recognized in this order are Osmundaceae, Schizaeaceae, Gleicheniaceae, Matoniaceae, Hymenophyllaceae, Loxosomaceae, Dicksoniaceae, Plagiogyriaceae, Protocyatheaceae, Cyatheaceae, Dipteridaceae, Polypodiaceae, Masileaceae, and Salviniaceae. Later, Copeland (1947) introduced a system of classification in which several natural families were carved out of Bower's Polypodiaceae. The order Filicales proposed by Copeland includes Osmundaceae, Schizaeaceae, Gleicheniaceae, Loxosomaceae, Hymenophyllaceae, Pteridaceae, Parkeriaceae, Hymenophyllopsidaceae, Davalliaceae, Plagiogyriaceae, Cyatheaceae, Aspidiaceae, Blechnaceae, Aspleniaceae, Matoniaceae, Polypodiaceae, Vittariaceae, Marsileaceae, and Salviniaceae. Occasional references made in this book to filicalean ferns are to Filicales of Copeland's classification.

The following convention is adopted in the book in respect to plant names. With rare exceptions, plants have been identified only by their scientific names, even though a common or a vernacular name might exist. The first occasion a genus or species is mentioned in the book, its familial position in the classification of Crabbe *et al.* (1975) is indicated in parenthesis. In all instances, in the text as well as in the tables, binomials as used by the original author are given, even though they are not recognized now as the valid names; in such cases, the genus or species in which it is felt that a particular plant belongs, preceded by the sign =, is given in parenthesis. It is safe to say that, for a period of time, a lack of consensus on the phylogenetic position of certain families and genera of ferns was matched to an equal degree by ambiguity about the generic and specific status of some ferns. Fortunately, recent morphological, cytological, and taxonomic studies have helped to bring a much-needed order in the matter of identification of ferns, although a lot more remains to be done. As mentioned earlier, controversy about the phylogeny of ferns continues with no end in sight.

Part I

The beginning

2

The spore – beginning of the gametophytic phase

It was noted in the introductory chapter that, in ferns, the genesis of a spore marks the beginning of the developmental process that produces a gametophyte. From the standpoint of understanding the evolution of the gametophyte in multicellularity, sexuality and functional competence, an understanding of the formation of the spore and of its structure is important. Therefore, in this chapter we focus on sporogenesis in ferns and on the normal course of morphogenesis of the spore to attain maturity. For the most part we will be concerned with description of events, with little emphasis on the underlying mechanisms. As we shall see later, some features of sporogenesis are surprisingly similar across the entire group of ferns, although the final products generated vary somewhat in their morphology.

Sporogenesis

Existing evidence suggests that the types of spores produced by ferns are of great significance in the evolution of the group. Although evolutionary considerations are not emphasized in this book, we shall nonetheless begin this part with an account of the two major spore types found in ferns.

Homospory and heterospory

One of the most important concepts generated from studies on sporogenesis in ferns is the recognition of homosporous and heterosporous plants (Fig. 2.1). Homosporous types produce only one kind of sporangium and just one kind of spore, as is characteristic of the vast majority of extant ferns. Exosporic gametophytes, that is, gametophytes that are free-living and not restricted within a spore wall, are common to homosporous ferns. In contrast, some ferns generate two kinds of sporangia and two kinds of spores, the large megaspores, produced in the megasporangium giving rise to female gametophytes, and the small microspores, encased in the microsporangium, yielding male types. This

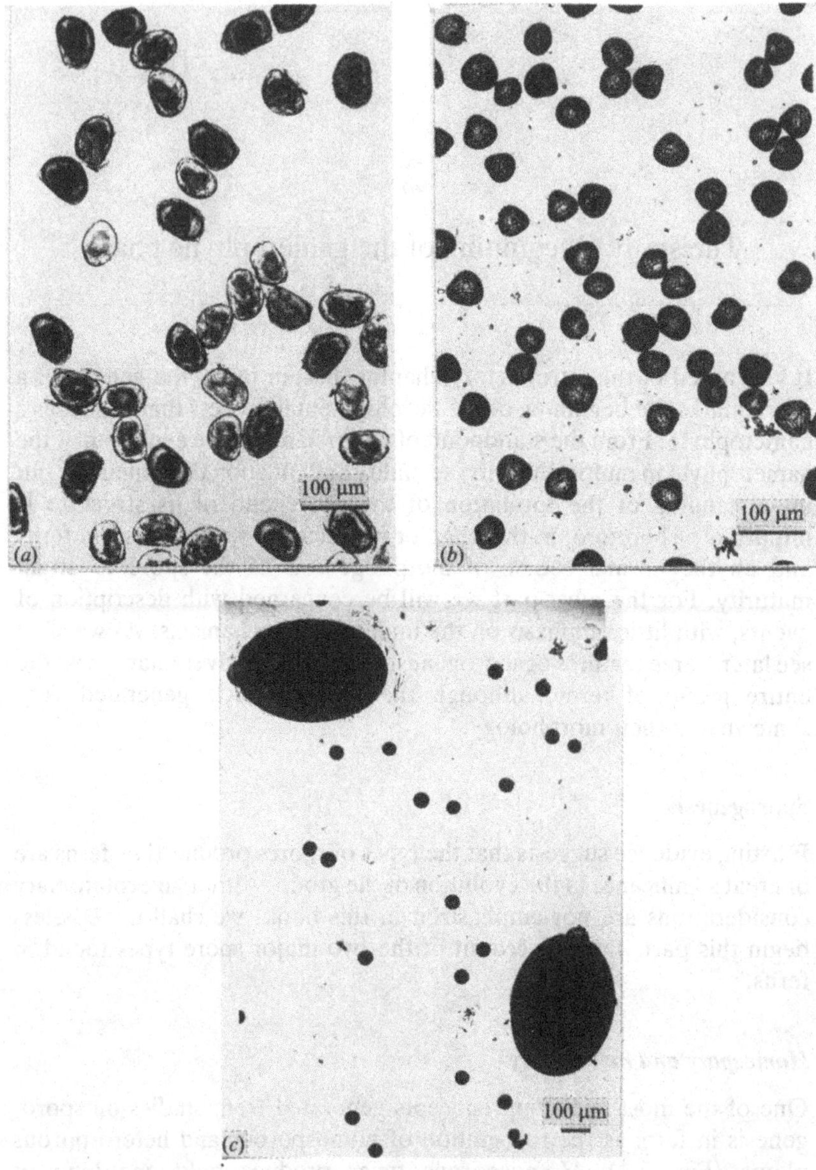

Figure 2.1. Whole mounts of spores of homosporous and heterosporous ferns. (*a*) *Onoclea sensibilis*. (*b*) *Ceratopteris thalictroides*. (*c*) *Marsilea vestita*. In *c* the small dark structures are microspores and the large ones are megaspores.

is the heterosporous condition, well known in Marsileaceae, Azollaceae and Salviniaceae, among ferns and in Selaginellaceae and Isoetaceae, among other pteridophytes. Both macro- and microgametophytes are nonphotosynthetic and, being enclosed within the original spore wall for the most part of their life, are endosporic. A unique situation, characterized as 'incipient heterospory', is encountered in *Platyzoma microphyllum* (Platyzomataceae), which produces a limited number of small and large spores in morphologically different types of sporangia. The small spores are committed to maleness and the large spores yield females which subsequently turn bisexual. The deviations from typical heterospory are concerned with the photosynthetic and exosporic nature of gametophytes of both kinds (Tryon, 1964).

It is generally assumed that plants that produce spores of two kinds with different potentials are better adapted to a terrestrial existence than those that produce a single type of spore. Accordingly, heterospory has been interpreted as a major landmark in the evolution of vascular plants. From our perspective, an important question is: what causes heterospory? It is to be emphasized here that, while heterospory is often reflected in size differences in spores, it is actually defined on the basis of the sexual function of spores. In a thoughtful essay, Sussex (1966) has argued that spore dualism is an expression of the sex-determining process of the plant. This argument relegates the great majority of homosporous ferns to a group in which sex determination is delayed until spore germination and gametophyte development. In contrast, in heterosporous ferns sex determination is thought to occur somewhat early, at a specific stage in the sporophytic phase of the life cycle. On the whole, it appears from this that the strategy of ferns is one of flexibility, probably depending upon the selective activation and repression of genes concerned with the functioning of sex determinants.

Sporangial structure and ontogeny in homosporous ferns

Studies on sporogenesis in homosporous ferns have emphasized evolutionary comparisons and the number of initial cells that go into the formation of the sporangium. Typically, a mature fern sporangium is borne on a definite stalk and contains one or more spores enclosed within a protective wall. The wall becomes multilayered and massive, as in members of Ophioglossaceae and Marattiaceae, or is reduced to a single layer of cells, as in most filicalean ferns. The development of the wall has been widely studied as an adaptation for dehiscence of the sporangium. Undoubtedly, the most specialized and familiar dehiscence mechanism of the sporangium of many filicalean ferns is one implemented by a special layer of unevenly thick-walled cells, known as the annulus, and a few thin-walled cells constituting the stomium. On the other hand, in *Angiopteris* (Marattiaceae), an annulus as such is absent and the dehiscence mechanism of the sporangium is made up of a band of enlarged lignified cells. In most homosporous ferns, the sporangial wall is separated from the sporogenous tissue by one or two layers of thin-walled, densely cytoplasmic cells known as the tapetum. This is a short-lived tissue whose cells break down before or just after meiosis in the sporocytes. Despite the scanty evidence, it is widely presumed that the cytoplasm of tapetal cells is appropriated as nourishment by the developing sporocytes and spores (Bierhorst, 1971).

Ontogeny of the sporangium. Typically, as seen in *Phlebodium aureum* (Polypodiaceae), sporangial development is initiated by an oblique division of a single superficial initial cell (Wilson, 1958). The inner of the two cells formed is subsequently partitioned transversely to separate a basal cell from an upper pair of cells constituting the sporangium primordium. The basal cell virtually remains undivided as the sporangium is fabricated from the primordium. Three oblique divisions ensue in quick succession in the outer cell of the sporangium primordium to form a tetrahedral mother cell surrounded by three cells. These latter cells by further divisions contribute to the wall of the sporangium and part of the subtending stalk. The other part of the stalk is generated by divisions of the inner cell of the sporangium primordium. A critical stage in the development of the sporangium is the occurrence of one transverse and three oblique divisions, in that order, in the tetrahedral cell. The result is the formation of a central cell enclosed by four cells which separate it from the sporangial wall. Each of these four cells divides by anticlinal and periclinal walls to produce a two-layered tapetum. Concomitantly, the central cell divides in various planes to form the sporogenous tissue which in turn yields sporocytes.

As far as is known, details of the major division sequences leading to the establishment of the sporangium are very similar in the various genera of homosporous ferns, but the origin of the sporangium itself has been a point of contention. This has given rise to a widely accepted division of extant ferns into two unequal groups. In the smaller of the two groups, known as the eusporangiate type, the sporangium originates from several initial cells. Other characteristics which go with the eusporangiate ferns are the relatively large size of the sporangium, production of numerous spores and the development of a several-layered wall. In the second group, known as the leptosporangiate type, the sporangium arises from a single parent cell or initial cell and has a wall composed of but a single layer of cells. The number of spores produced per sporangium is also limited and is usually a multiple of two. The most common configuration is one in which a group of 16 sporocytes yield, by meiosis, a total of 64 spores. Ophioglossaceae and Marattiaceae are eusporangiate ferns while most filicalean ferns are leptosporangiate. Later ontogenetic studies have shown that none of the criteria used to distinguish the eusporangium from the leptosporangium are consistent and dependable and that intermediate types of sporangial development are common. Although this has led to the suggestion that the terms eusporangium and leptosporangium should be erased from the fern literature (Bierhorst, 1971), they continue to be used.

Meiosis. Maturation of the sporangium involves meiotic division of the sporocytes to form tetrads of spores. Spores remain encased in the original wall of the tetrad for a short period of time before they are released into the sporangial cavity by dissolution of the wall. Although the cytoplasmic events that occur during meiosis are fundamental to the genetic make-up of the spores, there have been few detailed studies on the ultrastructural cytology of spore formation in ferns. An unusual pattern of organelle behavior seen during sporogenesis in several ferns is the aggregation of mitochondria in the form of a granular disc between the diad nuclei of the first meiotic (heterotypic) division (Marengo, 1954, 1962, 1977). This apparently represents the morphogenetic phase of the organelle which precedes a redistribution phase at the end of the first or second meiotic (homotypic) division of the sporocyte and stands in sharp contrast to the uniform distribution of mitochondria in the young sporocyte.

Sporogenesis in *Pteridium aquilinum* (bracken fern; Dennstaedtiaceae) is accompanied by the selective dedifferentiation and redifferentiation of mitochondria and plastids during meiosis (Sheffield and Bell, 1979). The mitochondrial profile changes from the extensively ramified type seen in the premeiotic cells to rod-shaped ones with finger-like cristae in the preleptotene cells and circular type of cristae in the diplotene–diakinesis period. Typical pleomorphic mitochondria reappear by telophase II. During meiosis, plastids pass through phases characterized by the absence of an envelope and increase in surface area and show regression of the internal lamellar system. A striking aspect of sporogenesis is a decrease in frequency of ribosomes during meiosis, reaching a low value at metaphase I, and their replenishment at the diad stage. Since somewhat similar cytological events are observed during sporogenesis in the apogamous fern *Dryopteris borreri* (=*D. affinis*; Aspleniaceae), it seems that mechanisms that induce and maintain the morphology of organelles in their correct spatial organization set the stage for the transition from sporophytic to gametophytic growth, irrespective of a simultaneous change-over from diplophase to haplophase (Sheffield, Laird and Bell, 1983). Unfortunately, we know very little about these mechanisms.

Control of sporangium development. Because of the range of diversity between fertile fronds and vegetative fronds of ferns, factors that control the initiation and development of sporangia are of particular interest. Some evidence of involvement of hormonal factors in sporangium development has come from the observation that exogenous application of indoleacetic acid (IAA), 2,4-dichlorophenoxyacetic acid (2,4-D) or 2-naphthoxyacetic acid (NOA) to potential fertile fronds of intact plants of *Anemia collina* (Schizaeaceae) inhibits the outgrowth of sporangia (Labouriau, 1952). This stands in contrast to the finding that gibberellic acid (GA) treatment enhances the fertility of leaves of *Ceratopteris thalictroides* (Parkeriaceae; Stein, 1971). Other studies by *in vivo* defoliation experiments and photoperiodic treatments have emphasized a positive relationship between the carbohydrate status of the frond and induction of fertility (Steeves and Wetmore, 1953; Labouriau, 1958; Wardlaw and Sharma, 1963). Support for the role of carbohydrates in inducing sporangium development on fronds has also come from tissue culture experiments. Sussex and Steeves (1958) found that the production of sporangia on excised, cultured fronds of *Leptopteris hymenophylloides, Todea barbara* and *Osmunda cinnamomea* (all Osmundaceae) is promoted by high concentrations of sucrose in the medium, although sporangial development does not proceed beyond the premeiotic stage. In attempts to identify the factors that control meiosis in the sporocytes, it was found that culture of pinnae of *O. cinnamomea* at the premeiotic stage in a medium supplemented with 3% sucrose leads to the development of mature sporangia enclosing viable spores. The only deviations from normal development were the lack of synchrony in sporogenesis and failure of exine differentiation on the spore wall (Clutter and Sussex, 1965). In an improvement of this method, Harvey and Caponetti (1972, 1973) obtained sporangial differentiation and production of viable spores from uncommitted leaf primordia of *O. cinnamomea* cultured in the dark in a medium containing 6% sucrose or 4% glucose. Functional spores have also been obtained on the fronds of *in vitro* developed plants of *T. barbara* (DeMaggio, 1968) and *C. thalictroides* (Rashid, 1972) cultured on relatively simple media. The import of these observations is that whatever the mechanisms are that promote sporangial development on fronds, some elements, possibly carbohydrates, must come from the sporophyte plant for completion of meiosis.

Sporogenesis in heterosporous ferns

In heterosporous ferns, sporangia are enclosed in hard, nut-like reproductive structures known as sporocarps. The classic example of a sporocarp is that of the widely distributed *Marsilea* (Marsileaceae). Although the morphological nature of the sporocarp of *Marsilea* remains unresolved, one view based on anatomical and ontogenetic evidence is that it is a modified fertile pinna folded along the margins. A more radical interpretation views the sporocarp as a new organ phylogenetically derived from a pinna through major genetical upsets. With regard to the distribution of sori, each half of the sporocarp bears rows of elongate sori in an apparent internal position. Each sorus consists of a ridge-like receptacle bearing a megasporangium at its apex and several microsporangia on its flanks and is covered by a delicate indusium. Large microsporangiate and small megasporangiate sporocarps are present in *Azolla* (Azollaceae), although both types begin their existence as megasporangiate structures. During further development of the megasporangium, microsporangial initials appear on its stalk. This is the cut-off point in ontogeny when developmental programs of the sporocarp diverge. If the megasporangium degenerates, as often happens, the microsporangial initials spring forth and grow and a microsporocarp is formed. But, if the megasporangium survives, microsporangial degeneration sets in and a megasporocarp with a single megasporangium results. Thus, each sporocarp passes through an interesting ontogenetic sequence of megasporangiate and then both mega- and microsporangiate, and ends up as either mega- or microsporangiate type (Bierhorst, 1971).

General features of megasporogenesis in *Marsilea* and *Azolla* are well documented at the light microscopic level (Feller, 1953; Demalsy, 1953; Boterberg, 1956; Bonnet, 1957; Konar and Kapoor, 1974; Bilderback, 1978*a*). The megasporangium has its origin in a single initial cell which follows division sequences of the leptosporangiate type of development. During megasporogenesis, walls of the tapetal cells are eroded and their contents intermingle to form a periplasmodium. Later, the megasporocytes are deprived of their walls and the rounded cells become ringed by a layer of the periplasmodial tapetum. Following an undistinguished meiosis, 32 megaspores in eight tetrads are produced in each sporangium. However, survival record of the megaspores is dismal, since three megaspores in each tetrad regress rapidly and of the eight viable megaspores only one, which bloats enormously and fills the whole sporangium, becomes functional. We have no information about the underlying basis of megaspore functioning. According to Bell (1981), there are no consistent differences in size or in the complement of cytoplasmic components between the four megaspores of a tetrad of *M. vestita* to account for their differential survival. In *Azolla*, granular materials of the periplasmodium on the top of the megaspore are partitioned among three membrane-bound chambers. Later, as the megaspore matures, the cytoplasmic materials harden to form a honeycomb-like structure known as the float (Konar and Kapoor, 1974; Lucas and Duckett, 1980; Herd, Cutter and Watanabe, 1986).

Detailed studies on microsporogenesis in heterosporous ferns include those on *Azolla* by Demalsy (1953), Bonnet (1957), Konar and Kapoor (1974), Calvert, Perkins and Peters (1983), and Herd *et al.* (1985). Following its origin in a single-celled initial, a microsporangium containing several sporogenous cells surrounded by a single-layered tapetum and a microsporangial wall is formed. The sporogenous cells gradually enlarge to form the microsporocytes, while the tapetal cells become periplasmodial. Following meiosis, all the 64 microspores derived from 16 spore mother cells are functional. During maturation of the sporangium, the microspores are loosely embedded in a meshwork of membranous material originating from the tapetum and transformed into a cushion-like structure known as the massulae. A characteristic feature of both mega- and microsporangia of *Azolla* is the presence of akinetes of the blue green algal symbiont *Anabaena azollae* which occupy a space between the top of the sporangium and the indusium.

Release of spores from sporocarps of heterosporous ferns relies on the breakage of the hard coat. It is probable that, in nature, microbial action accounts for the weakening of the sporocarp coat and dissemination of the spores. In the laboratory, these processes are accelerated by scarifying the sporocarp and immersing it in water. Within a short time after imbibition of water, a long gelatinous structure bearing sori, known as the sorophore, emerges from within the sporocarp. According to Bilderback (1978*b*), the sorophore of *M. vestita* is transformed into a long column as a result of water uptake by the hygroscopic polysaccharides in its cells. After several hours in water, the sporangial walls become gelatinous, releasing the microspores and megaspores into the soral cavity covered by the indusium, and finally they are set free in water by the dissolution of the latter. Thus, not only is the sporocarp of *Marsilea* a morphological enigma, it is also a paradigm of physiological ingenuity.

Morphogenesis of spore maturation

Among homosporous ferns, depending upon their arrangement in the tetrad, the overall form of mature spores is either tetrahedral (or trilete) or bilateral (or monolete). The former results from the arrangement of spores in a tetrahedrally symmetrical fashion. Upon their release from the tetrad, tetrahedral spores are identified by a characteristic triradiate mark on their proximal face, representing the point of attachment of the spore to other members of the quartet. (By convention, the face of the spore opposite to the proximal face is the distal face; the equatorial axis of the spore runs perpendicular to the proximal–distal axis.) Bilateral spores, which have only a single scar at their proximal face, are arranged in two pairs, with the long axes of one pair oriented at right angles to the long axes of the other pair. What determines the arrangement of spores in a tetrad remains one of the vexing questions facing cytologists. One idea is that it is related to the manner of cytokinesis during meiosis, especially to the changes in configuration of the mitochondrial plate described earlier. According to this view, bilateral spores are formed when granules composing the mitochondrial plate are partitioned into four equal groups during the homotypic division, as seen in *Onoclea sensibilis* (sensitive fern; Aspleniaceae) (Marengo, 1949). In *Osmunda regalis* (Marengo, 1954), and *Adiantum hispidulum* (Adiantaceae; Marengo, 1962), which have tetrahedral spores, it is assumed that the mitochondrial plate is folded into a six-partite granular disc marking the limits of the internal walls of the newly formed spores in the tetrad. Thus, the method used by sporocytes to ensure the orderly arrangement of spores they generate is to modify the movement of cytoplasmic inclusions during meiosis. How this is accomplished, however, remains a mystery.

Spore wall morphogenesis

One of the most differentiated and permanent parts of the fern spore is the wall enclosing the protoplasm bequeathed at meiosis. The wall is generally constituted of two layers, an inner endospore or intine and an outer exospore or exine. The intine is seen as a thin covering over the plasma membrane and is putatively cellulosic in nature. The exine is variously ornamented and its surface configurations, as seen in the scanning electron microscope (SEM), are increasingly being used as correlative data in taxonomic characterization of homosporous ferns. In a few cases, a third layer of comparable stability, the perine or perispore, surrounds the exine as a sac-like covering. The most resistant spore wall is the exine, whose main constituent is an acetolysis-resistant, carotenoid-containing polymer known as sporopollenin. A multifaceted research program including light microscopy (see Nayar and Kaur, 1971 for references), transmission electron microscopy (Lugardon, 1974), and SEM (Gastony, 1979; Gastony and Tryon, 1976) has contributed a wealth of information on the wall structure of spores of diverse homosporous ferns; the details of this large subject lie outside the scope of this book.

Among heterosporous ferns, wall structure of megaspores and microspores of *Marsilea* (Feller, 1953; Boterberg, 1956; Pettitt, 1966, 1979*a*) and *Azolla* (Konar and Kapoor, 1974; Lucas and Duckett, 1980; Herd *et al.*, 1985, 1986) has been extensively studied and the presence of sporopollenin in acetolysis-resistant materials of megaspores and microspores of *A. mexicana* has been established by infra-red spectroscopy (Toia *et al.*, 1985). Despite the fact that different species have figured in the different investigations, there is a surprising degree of harmony in the accounts of spore wall ontogeny. In both megaspores and microspores, the picture that emerges is that of walls clearly demarcated into an inner intine, middle exine and outer perine. The perine is enveloped by a polysaccharide layer which swells to form a mucilaginous sheath upon hydration of spores. Also worth mentioning is the fact that the megaspore perine is differentiated into an outer prismatic layer and an inner reticulate layer while the corresponding wall of the microspore consists of a single pleated layer. Although the perine covers almost the entire surface of both types of spores, toward their proximal face it diminishes in thickness. It is in the proximal region of the megaspore that the female gametophyte is formed. These structural features of spores of heterosporous ferns are of adaptive value because they make it possible for spores to survive desiccation and pave the way for their germination upon hydration.

Major problems concerning morphogenesis of the spore wall include the need to synthesize precursors of sporopollenin; the polymerization of precursors; the orderly accumulation of the finished product to form the exine, and the creation of an ornamentation pattern where there was none before. Although some aspects of the ultrastructure of spore wall development have been investigated in *Marsilea drummondii* (Pettitt, 1966), *Botrychium lunaria* (Ophioglossaceae; Pettitt, 1979*b*), *Anemia phyllitidis* (Surova, 1981; Schraudolf, 1984) and *Azolla microphylla* (Herd *et al.*, 1986), based on static electron micrographs alone, it is impossible to provide anything but a partial view of the morphogenetic mechanisms involved in the process. However, close analogy exists between the structure of the fern spore wall and the wall of the more widely investigated angiosperm pollen grains, and ideas and concepts on wall differentiation in the pollen grain are useful in elucidating the origin of the fern spore wall. Pettitt (1979*b*) has consolidated into a useful review much of the available information on spore wall morphogenesis in ferns and other pteridophytes in the context of development of the wall layers in pollen grains.

In *B. lunaria*, preparations for wall formation are launched in the cytoplasm of the sporocyte even before it begins to divide. The first step toward elaboration of the wall is the formation of a surface coat on the newly liberated spore. This is often foreshadowed by the presence of stacks of dictyosomes in the cytoplasm of the pre- and post-meiotic cells. It is believed that these vesicles are guided by microtubules to the cell surface where they fuse with the plasma membrane to form an initial layer of exine. Two key changes that occur at this time that profoundly affect the surface architecture of the spore are the retraction of the plasma membrane and the incipient exine from the surface coat on the distal pole of the spore to form pockets and the deposition of electron-opaque aggregates of sporopollenin in them. Circumstantial evidence implies that these aggregates are polymerized precursor molecules which have their origin outside the spore in the tapetal periplasmodium. Although this interpretation is suggested *a priori* by the conceptual difficulty of explaining the origin of the aggregates in the spore cytoplasm and their subsequent passage through the spore coat, the enormous potential of the transient tapetum as a supplier of spore wall components should not be overlooked. After this initial sporopollenin condensation at the distal pole of the spore is under way, a similar change occurs at the proximal pole with the difference that sporopollenin is deposited as plates rather than as aggregates. This suggests the possibility that sporopollenin precursors could be specifically targeted to the distal and proximal poles of the spore and thus differentially affect the final exine configurations at the spore surfaces. During the terminal phase of differentiation of the wall, its thickness is augmented by the addition of vesicles from the spore cytoplasm and of sporopollenin from the periplasmodium (Pettitt, 1979*b*).

From an ultrastructural study of spores of *Anemia phyllitidis*, there is also good reason to suppose that the periplasmodial tapetum fulfils some role in the growth and ornamentation of the exine. The cells of the tapetum display characteristics of a secretory tissue with well-developed rough and smooth endoplasmic reticulum (ER), annulate lamellae and active dictyosomes. At the end of meiosis in the sporocytes, the tapetum apparently breaks down and the periplasmodium engulfs the juvenile spores, but there is no direct evidence of actual transfer of wall material from the tapetum to the growing exine (Schraudolf, 1984). A microtubule meshwork and coated vesicles are dominant components of the tapetal zone closest to the developing megaspore of *Azolla microphylla* and it is believed that granular contents of the vesicles contribute to the thickness of exine (Herd *et al.*, 1986).

Finally, there remains the question of interpreting what causes the form and ornamentation of the exine. This is a difficult task because the most rigorous investigation of this question requires a way to identify the sporophytic or gametophytic cells of the sporangium in which exine precursors are synthesized and demonstrate gene expression for their synthesis. Despite the large volume of work done on pollen grains, no consensus has been reached on this issue, so we can speak here only in very general terms. One view is that differences in the composition and configuration of macromolecules in the initial spore coat determine the form in the spore wall (Pettitt, 1979*b*). According to another view, a prepattern of microtubule associations in the tapetum provides the blueprint for the design of the spore wall (Schraudolf, 1984). Very little evidence exists for either of these mechanisms and thus there is no solid basis to choose between them. As alluded to earlier, exine morphology of fern spores is a stable character worthy of taxonomic considerations and hence, is probably gene-regulated. Evidence reviewed here indicates that sporophytic rather than gametophytic genes determine the final form of the exine.

Perine formation. Perine is differentiated after exine takes shape and, as shown in the microspores of *Marsilea drummondii*, evidence for mobilization of materials resident in the tapetal periplasmodium in perine fabrication rests on a strong foundation (Pettitt, 1979*b*). Under the normal sequence of events, perine development begins with the formation of deep infolds on the tapetal periplasmodium facing the microspore. As precursor molecules migrate to the surface of the folds, an external surface coat is formed on the membrane of the periplasmodium. It seems that the perine arises by continued deposition of discrete droplets of sporopollenin from the tapetum on the surface coat which thus defines the outer limit of this wall layer. Structurally, the perine covering megaspores of *Marsilea* and other genera of Marsileaceae is more intricate than that found on microspores; some evidence also suggests that the perine of megaspores is formed from granular contents of small vesicles and vacuoles of the tapetal periplasmodium (Pettitt, 1979*a*; Herd *et al.*, 1986).

On the whole, these investigations show that, in ferns, tapetal cells are not only sensitive to the state of differentiation of spores but they also function as the major contributor to spore differentiation. This probably involves the operation of a highly specific biochemical system of cell–cell recognition. One of the handicaps in relating precisely the changes in tapetal metabolism to spore wall morphology is the highly selective nature of the participating cells and their entrenched location in the sporangium.

Internal structure of the spore

Spores freshly released from the tetrad are immature and will not germinate. They need to complete a series of cytological and biochemical changes which are necessary for maturation and for potentiation of germination. It is something of an irony that as the wall is framed around the spore protoplasm, it bars the entry of conventional fixatives employed in light and electron microscopy necessary to monitor the structural evolution of the protoplasm. This has resulted in a paucity of information about the cytoplasmic events associated with spore maturation in ferns, as the following account will show.

Cytological changes during spore maturation ensure the synthesis and orderly rearrangement of organelles in the cytoplasm. Developing spores of *Onoclea sensibilis* have been used to follow organelle rearrangement because the nucleus and chloroplasts are easily identified under the light microscope and their simplicity is highly diagrammatic without any loss of important concepts. The newly liberated spore has a central nucleus suspended in a moderately vacuolate cytoplasm containing lipid droplets, amyloplasts and mitochondria. However, soon after its liberation into the sporangial cavity, the spore enlarges, accompanied by the formation of a large vacuole almost filling the entire cell. It seems that the increased size would provide an enormous volume for storage of reserve materials. During continued enlargement of the spore, the nucleus initially migrates equatorially before it comes to rest at the proximal pole of the spore. At this stage of development, the spore reveals the presence of proplastids as small granules in the cytoplasm surrounding the nucleus. The proplastids follow a characteristic polarity as they multiply and radiate from the nucleus to other parts of the spore. During the final stage of maturity, the nucleus executes another movement back to the center of the spore. This sequence of events, illustrated in Fig. 2.2, culminates in the formation of a spore endowed with a generous supply of chloroplasts surrounding a central nucleus (Marengo, 1949; Marengo and Badalamente, 1978).

At the biochemical level, maturation of the spore is probably due to the endogenous synthesis of a great variety of molecules utilizing the breakdown products of the tapetum and cytoplasmic materials of the sporangial cavity. Light and electron microscopic investigations of mature spores of several homosporous ferns have revealed organelles and structures not inconsistent with the idea of massive synthesis of new materials. For example, the spore of *Matteuccia struthiopteris* (ostrich fern; Aspleniaceae) is an interesting package that includes a large number of protein bodies in the cytoplasm already enriched with chloroplasts, mitochondria and ribosomes, and a central nucleus that gives the spore legitimacy as a cell (Gantt and Arnott, 1965; Marengo, 1973). A recent observation (Templeman, DeMaggio and Stetler, 1987) that the protein bodies contain globulins that are homologous to seed storage proteins has raised some questions about the evolution of storage proteins in phylogenetically distinct groups of plants. In the spore of *O. sensibilis* the cytoplasmic contents surrounding the nucleus present a different configuration; here, a zone predominantly rich in chloroplasts is sandwiched between narrow layers containing both lipid droplets and protein granules (Bassel, Kuehnert and Miller, 1981). In nongreen spores of *Blechnum spicant* (Blechnaceae; Gullvåg, 1969), *Anemia phyllitidis* (Raghavan, 1976), *Pteris vittata* (Adiantaceae; Raghavan, 1977*b*), and *Lygodium japonicum* (Schizaeaceae; Rutter and Raghavan, 1978), fixation and sectioning difficulties have made it impossible to identify anything more than storage protein granules and lipid droplets surrounding a central nucleus. Fig. 2.3 shows sections of spores representing examples of homosporous ferns.

Quantitative analyses have shown that lipid content of spores varies

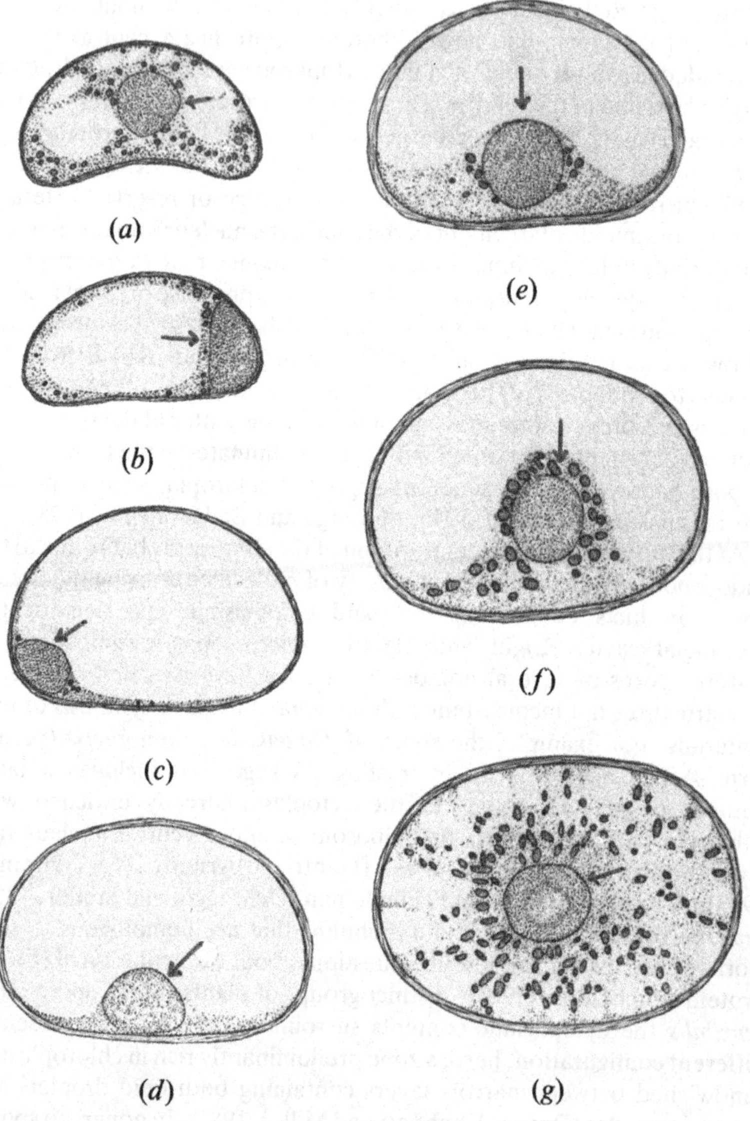

Figure 2.2. Cytological changes during maturation of the spore of *Onoclea sensibilis*. (*a*) Beginning of vacuolation during spore enlargement. (*b*) Nucleus at one end of the spore at the completion of vacuolation. (*c*) Spore enlargement accompanied by a decrease in cytoplasm. (*d*) A stage prior to the development of proplastids showing enlargement of the nucleus. (*e*) First appearance of proplastids around the nucleus. (*f*) Continued development of plastids and their centripetal migration. (*g*) Mature spore with a central nucleus and numerous chloroplasts. Arrows point to the nucleus. (From Marengo, 1949.)

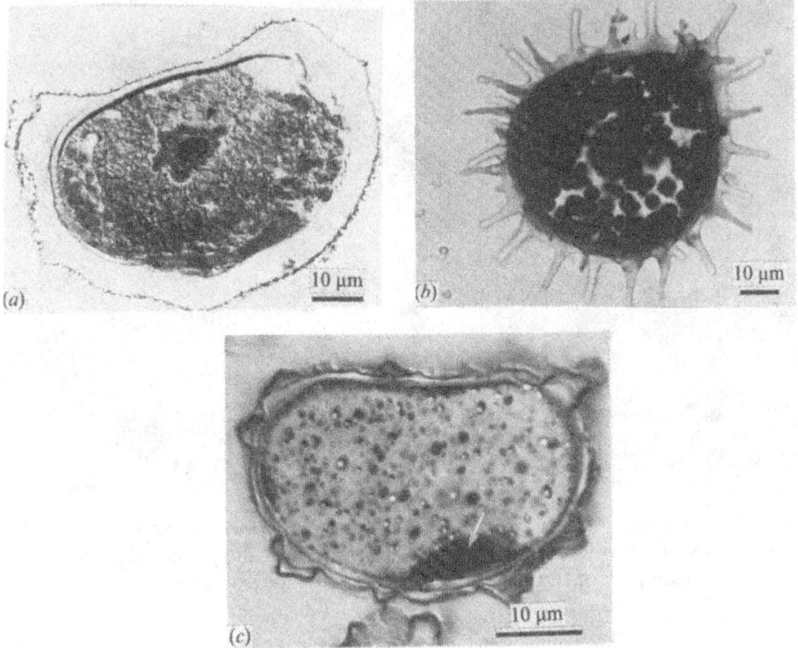

Figure 2.3 Sections of dry spores of (*a*) *Onoclea sensibilis*, (*b*) *Anemia phyllitidis* and (*c*) *Asplenium nidus* (photographed with Nomarski optics) showing internal structure. Arrows point to the nucleus. A nucleus is not clearly seen in the spore of *A. phyllitidis*, which is stained with aniline blue–black for proteins. In (*a*) and (*c*), the proximal face of the spore is toward the top of the page.

from 4% of the weight in *Ceratopteris thalictroides* to 79% in *Polypodium meyenianum* (=*Aglaomorpha meyeniana*; Polypodiaceae) (Gemmrich, 1977*b*, 1980; Seilheimer, 1978; DeMaggio and Stetler, 1980; Esteves, Felippe and Melhem, 1985). The predominant components of the lipids of spores of *A. phyllitidis* are galactolipids of the chloroplast membrane complex (Kraiss and Gemmrich, 1986). Other substances identified in spores of homosporous ferns include sugars and amino acids (Courbet, 1963; Courbet and Metche, 1971), plant hormones such as GA (Weinberg and Voeller, 1969b) and abscisic acid (ABA) (Cheng and Schraudolf, 1974; Bürcky, 1977*c*; Yamane *et al.*, 1980), minerals such as K, Mg, Na, and Ca (Wayne and Hepler, 1985*b*), and specific proteins such as calmodulin (Föhr, Enssle and Schraudolf, 1987).

Mature megaspores of members of Marsileaceae, Salviniaceae and Azollaceae are the products of complex series of developmental processes. To begin with, the newly formed megaspore of *Marsilea vestita* is highly vacuolate and is embedded in a cytoplasmic cavity surrounded by

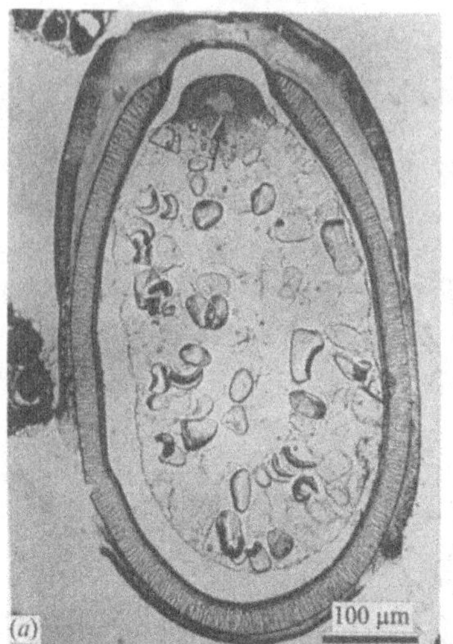

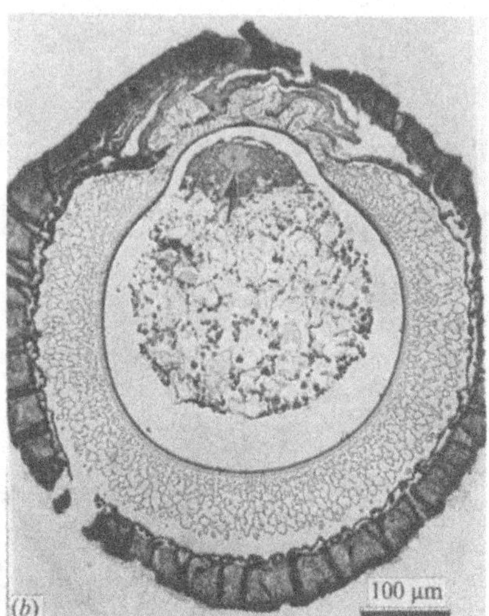

Figure 2.4. Sections of mature megaspores of (*a*) *Marsilea vestita* and (*b*) *Regnellidium diphyllum*. Arrows point to the nucleus at the proximal part of the spore. The distal part of the spores is filled with storage granules.

the tapetum. The nucleus is confined to the proximal part of the spore which grows by expansion of the vacuole distal to the nucleus. The production of numerous tubular extensions of the nucleus which traverse in the cytoplasm among the plastids and mitochondria, is suggestive of intense nucleocytoplasmic interactions (Bell, 1985). Later, the megaspore becomes fully engorged with cytoplasm and accumulates an acervate complex of storage products such as lipids, starch and protein bodies (Fig. 2.4).

Developing megaspores of *Azolla pinnata* (Konar and Kapoor, 1974), *A. filiculoides* (Lucas and Duckett, 1980) and *A. microphylla* (Herd *et al.*, 1986) share with *M. vestita* such features as the presence of a large vacuole, an incipient cytoplasm and abundant storage products. Much less is known about the maturation of microspores of heterosporous ferns, but it seems to follow the same route as megaspores. In *A. microphylla*, the newly formed microspore has a lightly staining cytoplasm with poorly defined organelles. During later stages of development, as the microspore increases in size, a thin layer of peripheral cytoplasm, surrounding a large central vacuole, is all that remains (Herd *et al.*, 1985). The fully mature microspore consists mostly of storage lipids embedded in the cytoplasm (Fig. 2.5).

The phase of sporogenesis in ferns is overtly terminated with the maturation of the spore. At this stage, the spore is equipped with all the essential information to produce a gametophyte when it is supplied with the conditions for germination.

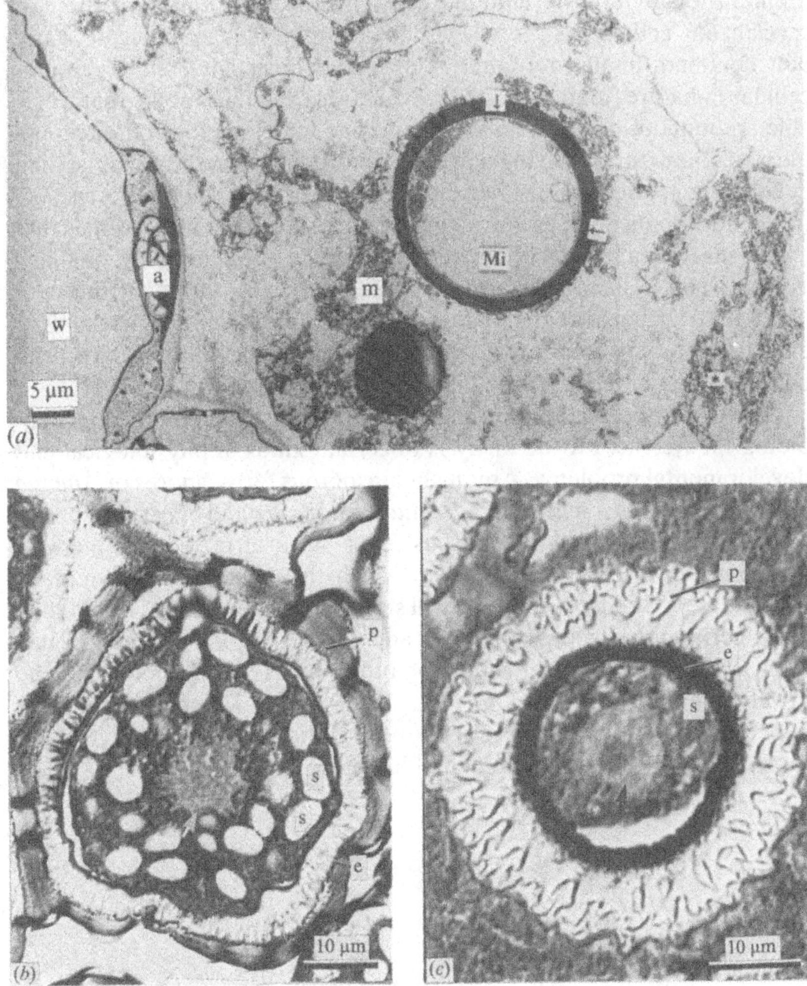

Figure 2.5. (*a*) Electron micrograph of part of the microsporangium of *Azolla microphylla* showing a vacuolate microspore (Mi) surrounded by membranous aggregates (m). Arrows point to the exine. Cells of the microsporangial wall (w) contain amyloplasts (a). (From Herd *et al.*, 1985). Light microscopic views (photographed with Nomarski optics) of sections of mature microspores of (*b*) *Marsilea vestita* and (*c*) *Regnellidium diphyllum*. Arrows point to the nucleus. e, exine; p, perine and the mucilaginous layer; s, storage granules.

General comments

The brief overview of sporogenesis in ferns presented in this chapter has focused on major landmarks in the transition of a diploid sporocyte to a haploid spore. Included in the progression of differentiation are mitotic proliferation of undifferentiated cells, meiotic division, formation of specialized cell types, synthesis of complex macromolecules and polymers and, finally, termination of metabolic activity. Although unicellular in nature, mature fern spores are large and thus require considerable amounts of proteins and other storage materials for housekeeping purposes and for surviving adverse conditions. Whether spores, like developing cotyledons or endosperm tissues of angiosperms, use selective gene amplification as a strategy to produce proteins in quantities that are required, is a subject of interest for future investigation. The similarities between pollen wall formation and spore wall formation make it likely that further research will result in the integration of mechanistic and structural information to determine how the complex architectural pattern of the outer spore wall is achieved.

As we will see in the next three chapters, fern spores are excellent biological systems for the analysis of certain kinds of physiological and developmental problems. Among homosporous ferns, spores of *Anemia phyllitidis, Onoclea sensibilis, Pteridium aquilinum* and of certain species of *Marsilea*, among heterosporous ferns, have attracted considerable attention because of the ease with which they can be experimentally manipulated. Still later chapters will show that gametophytes originating from germinating spores of these and other species have established themselves as worthy of study of other kinds of physiological and developmental problems in their own right. Taken together, as experimental systems, the fern spore and the fern gametophyte have much to offer to the developmental biologist.

3

Physiology of spore germination

Homosporous ferns produce an immense quantity of spores. Fern spores are relatively resistant to extremes of environmental conditions and persist in a viable, yet metabolically inactive state for long periods of time. Although no one can explain all that is going on in the spore during this period of enforced rest, the hiatus may well be valuable to the spore to plot a survival strategy and form the gametophyte when conditions are most propitious. While the spore is the first cell of the gametophytic phase, development of the gametophyte itself begins with the germination of the spore. It is therefore appropriate that attention is paid to the various factors that keep spores alive for prolonged periods and potentiate their germination. Accordingly, the present chapter is concerned with the physiological aspects of viability, dormancy, and germination of spores; the morphogenetic, metabolic, and biochemical changes during germination are treated in the next two chapters. Understanding of the physiological mechanisms of dormancy and germination of fern spores has benefited immeasurably in recent years from developments in the field of seed germination. However, references will be made in this chapter to the inescapable parallels that exist between seed germination and spore germination only when it is considered necessary to explain certain concepts further.

Germination of spores of homosporous ferns is heralded by an asymmetric division of the spore. This division is the culmination of events of a preceding cell cycle; as a result, the developmental potential of the spore is parceled out to two cells which pursue divergent differentiative pathways. A small lens-shaped cell (the rhizoid initial) that is formed elongates into a narrow, colorless rhizoid, specialized as an anchoring and absorbing organ. The large cell, on the other hand, divides again to produce an isodiametric cell (the protonema initial) endowed with a munificient supply of chloroplasts. This cell is the forerunner of the mature gametophyte. As the rhizoid initial begins to elongate, the exine is ruptured; the appearance of the tip of the rhizoid outside it is the first visible sign of germination. In this book, the term germination is used loosely as an ensemble of one or more of the processes involved such as breakage of the exine, nuclear and cell division and outgrowth of the daughter cells.

Viability and other physiological attributes of spores

The length of time that spores retain their ability to gérminate, known as viability, and the rapidity of germination and subsequent growth of gametophytes are basic to the successful competition of ferns in a new habitat. In general, viability of spores varies enormously among ferns and periods ranging from a few days to a few years have been reported in the literature (reviewed by Okada, 1929; Miller, 1968*a*). Among homosporous ferns, spores of several species of *Osmunda, Todea* and *Leptopteris* (Stokey and Atkinson, 1956*a*; Lloyd and Klekowski, 1970), as well as members of Gleicheniaceae, Grammitidaceae and Hymenophyllaceae (Page, 1979*b*) are known to lose their viability within a few days after harvest. More commonly, viability of spores is reckoned in months or years. The longest life span reported is for spores collected from a 50-year-old herbarium specimen of *Pellaea truncata* (Adiantaceæ; Windham, Wolf and Ranker, 1986). This is a remarkable achievement for a single cell. Among heterosporous ferns, *Marsilea* is the classic example where spores retain their viability for an unbelievable length of time. The newest record is for the production of viable megaspores and microspores from 99- to 100-year-old sporocarps of *M. oligospora* (Johnson, 1985). Although there are other reports on the ability of *Marsilea* sporocarps to 'germinate' after long periods in storage, these should not be confused with viability of spores. As indicated in Chapter 2, release of spores from the sporocarp is a purely physical phenomenon which requires the contact of a properly scarified sporocarp with water even if none of the enclosed spores are viable. In *Marsilea, Regnellidium* and *Pilularia* (all of Marsileaceae), germination of viable megaspores is accompanied by the development, at the proximal end, of a rhizoid-bearing green cushion of cells harboring the archegonium, while nonviable megaspores produce neither a cushion nor rhizoids (Mahlberg and Baldwin, 1975). Likewise, the unmistakable sign of viability of microspores is the production of sperm.

What determines longevity of spores is a moot question, but it has become clear in recent years that, apart from the capricious fluctuations of the environment, the genetic and certain physiological attributes of spores have to be considered in formulating a general hypothesis to explain longevity. To be sure, ferns and other pteridophytes have evolved into cytotypes with chromosome numbers over and above the diploid number. We have all kinds of evidence testifying to a pervading effect of gene dosage on various aspects of development of ferns. That the genetic status of the sporophyte might play a role in determining spore viability is suggested by the observation that the percentage of viable spores produced on a tetraploid race of *Polypodium virginianum* is higher than that produced on the diploid race (Kott and Peterson, 1974). A similar correlation appears to exist between ploidy levels of different species of *Isoetes* (Isoetaceae) and megaspore viability (Kott and Britton, 1982). At the physiological level, Lloyd and Klekowski (1970) have made comparisons between the viability of chlorophyllous and nonchlorophyllous spores of several species of ferns. Their results show that under ordinary storage conditions, green spores succumb early, averaging a life-span of only 48 days, while nongreen spores live on an average up to 2.8 years. It has been reasoned that spores might have evolved the chlorophyllous state primarily as a safeguard against expenditure of energy required to lapse into dormancy. Since chlorophyllous spores have a high respiratory activity, it was naturally thought that they metabolize stored reserves in an apparently shorter period of time than nongreen spores (Okada, 1929). In a few determinations made, it has also been established that short-lived green spores have a higher water content than long-lived nongreen spores suggesting that uncontrolled loss of water may have some relevance to the life span of spores (Okada, 1929). Since the presence of an additional spore coat might partially ameliorate the desiccation of cell contents, it comes as no surprise that out of about 50 different species of chlorophyllous spores tested, only the perine-covered spores of *Onoclea, Onocleopsis* (Aspleniaceae) and *Matteuccia* possess the longest life-span (Lloyd and Klekowski, 1970). Shirt-lived spores of certain members of the Osmundaceae can be kept in a viable state for a long stretch of time by storage under conditions which minimize desiccation, such as a low temperature (Stokey, 1951*b*; von Aderkas and Cutter, 1983*a*).

Even under favorable conditions of storage, increasing spore age leads to a decline in viability. Such a loss of viability is not to be mistaken for a sudden or abrupt failure of all spores in a population to germinate. Although spores of *Pteridium aquilinum* remain viable for more than ten years, germination decreases by as much as 30% by the fourth month after collection and by 50% by the end of the first year (Conway, 1949). In an analysis of the viability of an aging population of spores of *Polypodium vulgare* conducted over a period of seven years, Smith and Robinson (1975) obtained a sigmoid type of spore survival curve (Fig. 3.1). This work also showed that increasing age of the spore delays the appearance of the first morphological signs of germination and leads to the formation of a disproportionate number of abnormal gametophytes in culture. As commonly noted under laboratory storage conditions, delayed germination with increasing spore age is a feature that *P. vulgare* shares with spores of several other ferns.

With regard to the length of time that elapses between sowing and

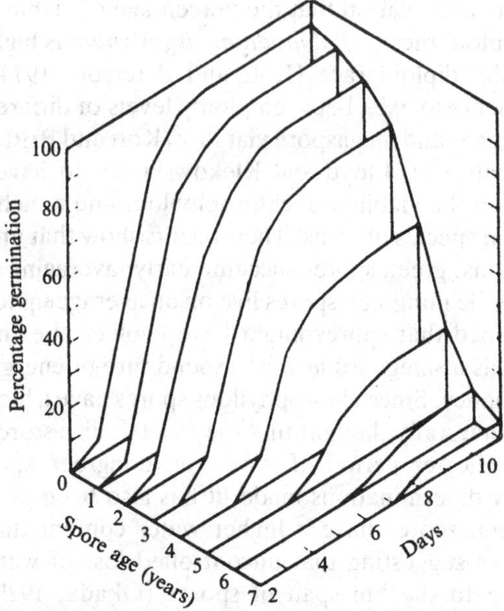

Figure 3.1. Effect of increasing age of spores of *Polypodium vulgare* on the percentage of germination and viability. The line connecting germination percentages on day 11 is the viability curve. (From Smith and Robinson, 1975.)

actual germination, comparisons are also germane between green and nongreen spores. The general tendency is for chlorophyllous spores to germinate more rapidly than spores impoverished of chloroplasts, the former germinating on an average in 1.5 days after sowing, as compared to 9.5 days required by the latter (Lloyd and Klekowski, 1970). The rapidity of germination of green spores can perhaps be attributed to their physiological state at the time of culture; endowed as they are with photosynthetically active pigments, green spores pass into a growth phase with relative ease when challenged by suitable conditions for germination. In nongreen spores, hydrolysis of storage granules to simpler molecules is a necessary first step to provide energy for the synthetic activities of germination.

Quiescence and dormancy of spores

As applied to fern spores, quiescence and dormancy are not sharply separated states and so, some comments are in order concerning them. The assumption made here is that quiescent spores are those that germinate readily in an environment supplied with adequate moisture, a range of temperature favorable for plant growth and the normal composition of the atmosphere. In contrast, dormancy may be visualized as a developmental block resulting in the failure of perfectly viable spores to germinate when supplied with conditions that promote germination of quiescent spores. This block is probably the result of subtle chemical changes during maturation of spores so that germination occurs only under conditions that reverse the changes. Fern spores may remain dormant for a variable period of time and in some cases dormancy may continue almost indefinitely until a special condition is met. The enormous quantity of spores produced by a fern provides an effective method of their dispersal in space, while an enforced period rest ensures their dispersal in time. It is also certain that dormancy mechanisms have evolved in fern spores in response to competitive selective pressures. To support this statement we have to look at spores of Ophioglossaceae which produce subterranean, tuberous gametophytes. When spores germinate in the soil, they invariably establish a symbiotic relationship with an endophytic fungus. It is likely that the prolonged dormancy displayed by spores of Ophioglossaceae could increase their chance to sift slowly down in the soil and become buried in close proximity to a fungal mycelium. However, the immediate relevance of our knowledge of dormancy mechanisms in spores to the survival strategy of the vast majority of ferns is not obvious since there are no comparative studies on the evolutionary consequences of dispersal of species with dormant and nondormant spores.

Spores of several ferns including those of Hymenophyllaceae (Stokey, 1940; Atkinson, 1960) and Grammitidaceae (Stokey and Atkinson, 1958) scarcely exhibit any dormancy, since they are known to germinate within the sporangium. At the other extreme are spores of members of Schizaeaceae and Adiantaceae which are decidedly dormant; among the signals that have been identified in the breakage of spore dormancy, light and hormones are the principal ones. Admittedly, apart from the primary need to hydrate the spores, in many cases the mechanisms triggering their germination are not understood.

Light and spore germination

Since the mid-nineteenth century, it has been known that light is an important factor in the germination of spores of several ferns. A tabulation by Miller (1968*a*) is particularly instructive in revealing that out of more than 80 species tested, spores of only seven species germinate in substantial numbers in complete darkness. Recognition of the importance of light in germination came, despite the fact that some of the early studies were bedeviled by conflicting claims from different laboratories on the light requirements of spores of one and the same species. Attempts were also made by some investigators in the early part of this century to study the effects of different wavelengths of light on spore germination. It is worth noting that until precision-built interference filter monochrometer systems or refined plastic filters which isolate relatively narrow wavelengths of light became available, studies on the effects of light quality on spore germination were undertaken using either dyed gelatin filters or liquid filters such as alkali-dichromate solution for the red, and copper sulfate for the blue parts of the spectrum. Since these filters give only crude separation of the spectrum and introduce contaminating wavelengths, the interpretation of the data obtained is open to question. Another problem ignored by early researchers was the necessity of providing equal fluences of light at different wavelengths. In spite of the inherent deficiencies of these studies, they provided a broad picture of promotion of germination of light-sensitive spores in the yellow–red region of the spectrum and delay of germination in the blue part of the spectrum. There was also suggestive evidence in these studies for the existence of two different photoreceptors in fern spores, one absorbing in the red and the other perhaps in the blue part of the spectrum. Brief summaries of these investigations are given in the reviews of Sussman (1965), Miller (1968*a*), Raghavan (1980), and Furuya (1978, 1983).

Effects of light quality on germination

Although we are concerned here principally with visible light, a few observations on the effects of ionizing radiation on germination should be mentioned first. In general, exposure of spores to α-rays (Zirkle, 1932), fast neutrons, X-rays, (Zirkle, 1932, 1936; Howard and Haigh, 1970; Palta and Mehra, 1973) and ultraviolet radiation (Charlton, 1938) leads to decrease in germination and delay in the display of germination symptoms in the surviving spores, accompanied by anarchic cell divisions. Sensitivity of spores of *Osmunda regalis* to X-rays is found to increase during the S phase of the mitotic cycle, indicating the intriguing possibility that inhibition of germination occurs because some event linked to S→M progression in spores is blocked after a lethal dose of radiation (Howard and Haigh, 1970). Spores of some ferns are also known to tolerate ionizing radiation and germinate normally or even at an accelerated rate (Döpp, 1937; Rottmann, 1939; Knudson, 1940; Maly, 1951; Kawasaki, 1954*b*; Kato, 1964*a*).

The last three decades have witnessed a remarkable broadening of our understanding of the control of germination of dormant fern spores by visible light. The primary stimulus for this has been the startling new discoveries in the field of seed germination and the impressive technological progress achieved in the fabrication of filter systems which permit isolation of relatively narrow wavelengths of light. Bünning and Mohr (1955; see also Mohr, 1956*a*) determined the action spectrum for promotion and inhibition of germination of fully imbibed spores of *Dryopteris* *filix-mas* (male fern) after 24 hours of continuous irradiation with equal fluences of monochromatic light isolated with interference filters. Evaluation of the results seven days after the beginning of irradiation showed that red light (650–670 nm) is the most effective region of the spectrum to induce germination; subsequent exposure of red-induced spores to far-red light (733–750 nm) and to some extent to blue light inhibits their germination. One other critical point to emerge from this work was that the effects of red and far-red light are mutually reversible. In other words, germination of spores potentiated by red light can be inhibited by far-red light; if far-red exposed spores are further irradiated with red light, the spores behave as if they have not seen any far-red light at all, and germinate. These results are similar to those reported for seed germination, control of flowering in photoperiodically sensitive plants, unfolding of the plumular hook of etiolated plants and a variety of other photomorphogenetic processes to suggest that germination of fern spores is mediated by the pigment, phytochrome.

From physiological studies on seed germination and other photomorphogenetic systems, it has been deduced that phytochrome exists in two photoreversible forms, P_r (red-absorbing form) and P_{fr} (far-red absorbing form). The function of phytochrome in spore germination is fulfilled when red light presumably converts the inactive P_r phytochrome to the active P_{fr} form which is now capable of absorbing far-red light. When the pigment is in P_{fr} form, spore germination is favored, while in P_r state it is prevented.

With varying degrees of complexity, involvement of phytochrome in the germination of spores of other ferns has been reported. Some of these cases will be briefly examined here. In spores of *Osmunda claytoniana* and *O. cinnamomea*, where breakage of the exine occurs after exposure to brief periods of red light and emergence of the rhizoid initial requires a long period (4 days) of red or blue light in an atmosphere enriched with CO_2, explanations along the lines of phytochrome control of the former and photosynthetic control of the latter have been sought (Mohr, Meyer and Hartmann, 1964). From the sustained investigations of Sugai and Furuya (1967, 1968), a general picture of involvement of at least three photoreactions, two of which are phytochrome-controlled, in the germination of spores of *Pteris vittata* has emerged. The first phytochrome-mediated photoreaction is the promotion of germination which is a typical low-fluence, repeatedly reversible, red/far-red light effect (Fig. 3.2). Since red-potentiated germination is also inhibited by blue light with peaks at 380 nm and 440 nm and by UV-light with a peak at 260 nm, there is reasonable evidence for the existence of a photoreaction in which a blue and near UV-light absorbing pigment ($P_{b\text{-}nuv}$) participates (Sugai, 1971; Sugai, Tomizawa, Watanabe and Furuya, 1984). The other phytochrome-mediated reaction is in the recovery of spores from blue light

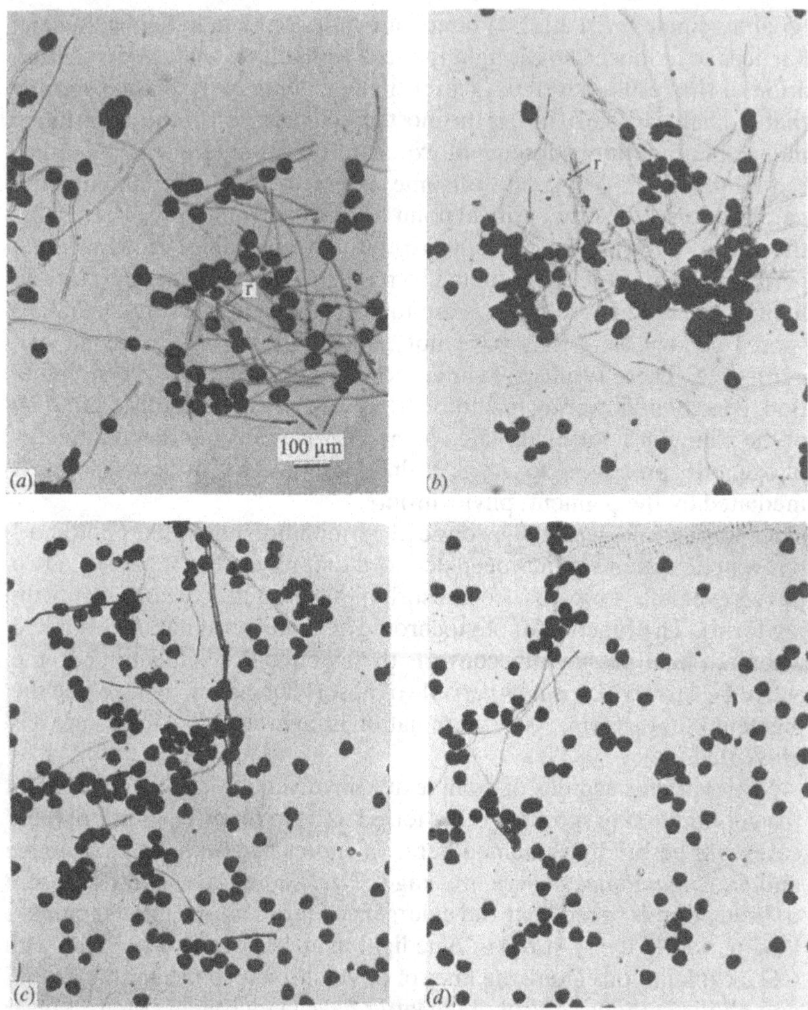

Figure 3.2. Effect of red (R, 1 min) and far-red (FR, 4 min) irradiations given in sequence on the germination of spores of *Pteris vittata*. Outgrowth of the rhizoid (r) indicates germination. Spores were photographed four days after irradiation. (*a*) R. (*b*) R/FR/R. (*c*) R/FR. (*d*) R/FR/R/FR. Scale bar in (*a*) applies to all.

inhibition of germination. As seen in Table 3.1, this is accelerated by red light given intermittently during the first eight hours after blue light and annulled by far-red light administered after each red exposure (Sugai and Furuya, 1968). An additional fact of interest is that the presence of a low concentration of ethanol in the medium rescues spores from blue light inhibition of germination (Sugai, 1970). While many details remain to be filled in, the picture that appears is that blue light is absorbed by

Table 3.1 *Effects of red and far-red light on the recovery from blue light-induced inhibition of germination of spores of* Pteris vittata[a]

Schedule of light treatment (Hours after blue irradiation)									Average percentage of germination
0	1	2	3	4	5	6	7	8	
R		R		R		R		R[b]	62.0
B/F		F		F		F		F	1.8
B/R		R		R		R		R	57.5
B/F/R		F/R		F/R		F/R		F/R	56.3
B/R/F		R/F		R/F		R/F		R/F	6.0
B/R/F/R		R/F/R		R/F/R		R/F/R		R/F/R	61.3

[a]From Sugai and Furuya (1968).
[b]R, red light 5 min (800 ergs/cm^2/s); B, blue light 5 min (800 ergs/cm^2/s); F, far-red light 5 min (18 kiloergs/cm^2/s). Spores were incubated in the dark at 26 °C between treatments.

carotenoids which generate an inhibitor of germination. The synthesis of this compound is apparently prevented by ethanol which acts as a reducing agent. How far this conclusion will fit in with a later observation (Sugai, 1982) that anaerobiosis and respiratory inhibitors such as azide and cyanide also impede the blue light effect on spore germination, is at present conjectural. Although phytochrome control and/or blue-near UV light interaction have also been established in the germination of spores of *Asplenium nidus* (Aspleniaceae; Raghavan, 1971*a*), *Cheilanthes farinosa* (Adiantaceae; Raghavan, 1973*a*), *Polypodium* (= *Phlebodium*) *aureum* (Spiess and Krouk, 1977), *P. vulgare* (Agnew, McCabe and Smith, 1984), *Lygodium japonicum* (Sugai, Takeno and Furuya, 1977), *Thelypteris kunthii* (Thelypteridaceae; Huckaby and Raghavan, 1981*a*), *Onoclea sensibilis* (Huckaby, Kalantari and Miller, 1982), *Mohria caffrorum* (Schizaeaceae; Reynolds and Raghavan, 1982), *Adiantum capillus-veneris* (Sugai and Furuya, 1985), *Dryopteris paleacea* (= *D. wallichiana*), *Polystichum minutum* (= *P. munitum*) (Aspleniaceae; Haupt, 1985), *Schizaea pusilla* (Schizaeaceae; Guiragossian and Koning, 1986), and *Ceratopteris richardii* (Cooke, Racusen, Hickok and Warne, 1987), probably no other system has been the target of as much experimental investigation as *P. vittata*. No convincing evidence for the existence of other pigments controlling spore germination has also been offered in these studies. In the germination of spores of *Matteuccia struthiopteris*, the action spectrum with two major peaks at 550 nm and 625 nm and a minor peak at 450 nm is so unusual that it does not show even remote resemblance to the absorption spectrum of any known plant pigments (Jarvis and Wilkins, 1973).

While germination of spores of some ferns clearly exhibits a low-fluence, red/far-red light effect, there are also cases in which such a reaction is disguised in subtle ways. To this category belong spores of *Anemia phyllitidis* collected from plants reared under red light during sporogenesis. They germinate in complete darkness while those collected from plants grown under ordinary greenhouse conditions require red light for germination (Schraudolf, 1987*b*). A critical assumption underlying this observation is that red light exposure during sporogenesis causes an increase in P_{fr} level of spores to potentiate germination without additional doses of red light. Phytochrome action is also masked in spores which are typically induced to germinate after prolonged exposure to red light. For instance, fully imbibed spores of *Asplenium nidus* go through the complete germination program and form rhizoid and protonema initials after exposure to continuous red light of moderate intensity for 72 hours followed by 8 days in the dark. If spores are held in darkness after 12 hours in red light, they form only the rhizoid initial. Irradiation of spores with far-red light immediately after red light reverses the effect of the latter and leads to significant decrease in the percentage of spores with a rhizoid initial. Spores with a rhizoid initial form a protonema initial when exposed to an additional period of red light for 12 hours. The effect of the second red light is also annulled by a subsequent irradiation with far-red light. As shown in Fig. 3.3, promotion and inhibition of growth of the rhizoid and protonema initials by red and far-red exposures are repeatedly reversible. These results are consist with the notion that in *A. nidus* spores, P_{fr} made by a single saturating exposure to red light does not persist long enough to permit full germination and that the outgrowth of rhizoid and protonema initials is controlled by separate phytochrome reactions perhaps with different P_{fr} requirements (Raghavan, 1971*a*). A similar effect of pretreatment with a long period of red light on a subsequent enhancement by pulses of red light has been found in the germination of spores of *Dryopteris filix-mas* (Haupt and Filler, 1986).

Although it is common parlance to speak of phytochrome control of germination based on red/far-red light reversibility, the proof lies in the spectrophotometric demonstration of the pigment and its phototransformation from one form to another in the spore. Thus far this has been accomplished only in spores of *Lygodium japonicum*. Consistent with the physiological data, Tomizawa, Manabe and Sugai (1982) showed that the native phytochrome of *L. japonicum* spores is in P_r form which becomes spectrophotometrically detectable after three days of dark imbibition. Is this pigment a hold-over from sporogenesis, or is it synthesized *de novo* upon hydration of the spore? An observation that gabaculine, presumed to be an inhibitor of chlorophyll and phytochrome synthesis, prevents red light-induced germination of spores lends credence to the first alternative (Manabe *et al.*, 1987), but the opposite effects of this drug on the germination of spores of *Anemia phyllitidis* make the second alternative also plausible (Schraudolf, 1987*a*). Direct measurements and estimates

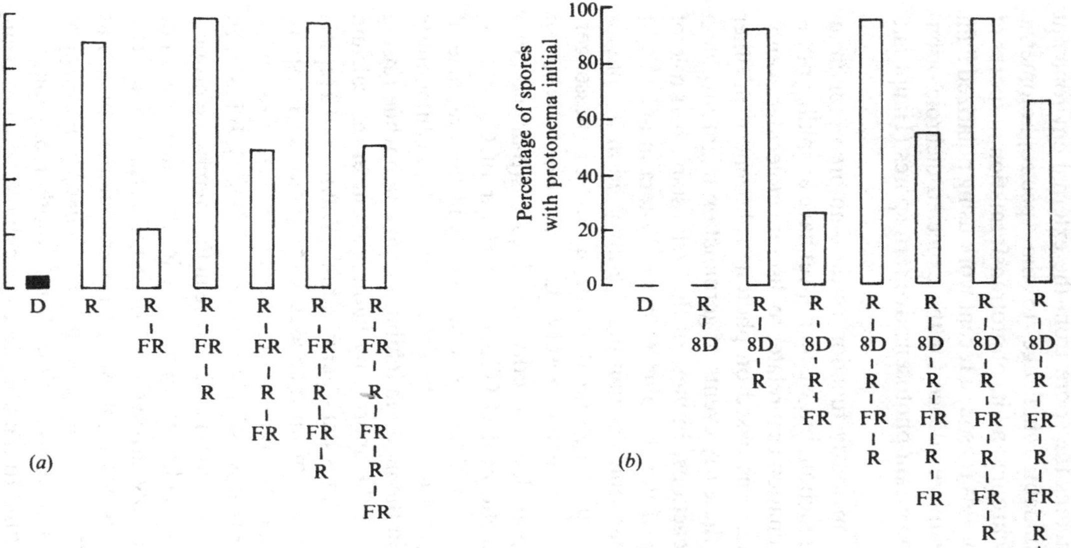

Figure 3.3. Control of initiation of the rhizoid and protonema initials during germination of spores of *Asplenium nidus* after exposure to red (R) and far-red (FR) light in sequence. (*a*) Rhizoid initiation during germination. Initial exposures to R and FR were for 12 hours and 48 hours, respectively, and subsequent exposures were for 10 min and 8 hours, respectively. Spores were examined 8 days after the last light exposure. (*b*) Protonema initiation during germination. Spores were exposed to 12 hours R and then held in darkness for 8 days (8 D). They were then exposed initially to R and FR in sequence for 12 hours and 48 hours, respectively; subsequent exposure to R and FR were for 10 min and 24 hours, respectively. Spores were examined 4 days after the last light exposure. D, dark control. (From Raghavan, 1971*a*.)

of P_{fr} in photoinduced spores of *L. japonicum* (Tomizawa, Sugai and Manabe, 1983) and *Pteris vittata* (Furuya, Kadota and Uematsu-Kaneda, 1982), respectively, have shown that germination is closely related to the percentage of active phytochrome available in the spore. The presence of a thick exine which insulates the spore from the external environment seems to pose a formidable challenge to the spectrophotometric demonstration of phytochrome and its phototransformation in spores of other ferns. Probably for this reason, a recent work using polarized light has failed to obtain positive evidence for the expected dichroic orientation of the pigment in certain photosensitive fern spores (Haupt and Björn, 1987).

Phytochrome should obviously function as the photoreceptor in an ensuing photochemical reaction, but we do not yet have an inkling of the role of the latter in germination as it relates to discrete molecular events. However, from all indications based on photomorphogenesis in other systems it appears that the early events of germination might involve a chain of independent reactions. In view of the well-established role of calcium in the coupling of the stimulus–response system in animals, it comes as no surprise that ionic calcium may be involved in the phytochrome-mediated fern spore germination as a second messenger. Evidence for this comes from the observation that when spores of *Onoclea sensibilis* are percolated in ethylene glycol-bis(β-amino-ethyl ether)-N,N,N′,N-tetraacetic acid (EGTA), a chelator of Ca^{2+}, their subsequent germination by red light is secured only in the presence of external Ca^{2+} (Fig. 3.4). Moreover, the presence of calcium antagonists such as La^{3+} and Co^{3+} in the medium inhibits germination while raising the internal Ca^{2+} level in the spores, by bathing them in the calcium ionophore A23187, potentiates dark germination (Wayne and Hepler, 1984). Direct demonstration of an increased Ca^{2+} influx in the spore following exposure to a saturating dose of red light has been possible by atomic absorption spectroscopy (Wayne and Hepler, 1985*a*). It has been reasoned that a phytochrome-mediated change in the membrane properties of the dormant spore leading to an increase in the concentration of free Ca^{2+} may be somehow involved in triggering germination. The question is, how. A recent work (Wayne, Rice and Hepler, 1986) has discounted the possibility that an increase in intracellular pH of the spore may be a link in the signal transduction chain. Another possibility is that free calcium might bind to the ubiquitous calcium-dependent regulator, calmodulin. The observation (Föhr *et al.*, 1987) that calmodulin content of spores of *A. phyllitidis* increases during the early phase of photoinduced germination is consistent with this suggestion. A new twist has been added to this problem by the discovery that, besides Ca^{2+}, an external supply of K^{+} is also essential for germination of ion-depleted spores of *O. sensibilis* (Miller and Wagner, 1987). The sum of these

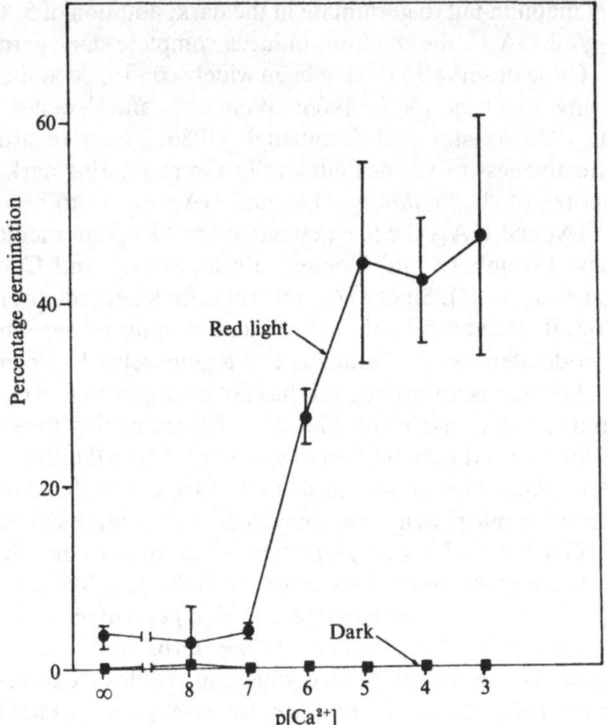

Figure 3.4. Effect of externally applied Ca^{2+} on dark and red light-induced germination of spores of *Onoclea sensibilis*. Spores were sown in Ca^{2+}-EGTA-buffered medium in the presence of increasing concentrations of free Ca^{2+}. Red light was applied at a fluence rate of 2.4 $Jm^{-2}s^{-1}$ for 1 min. ($P[Ca^{2+}] = -\log$ [free Ca^{2+}]). (From Wayne and Hepler, 1984.)

results is that we have to keep open the possibility that small molecules, including ions, may have regulatory functions, even at the level of the gene.

Hormonal substitution of light requirement

Stimulation of dark germination of spores of *A. phyllitidis* by GA reported by Schraudolf (1962) was the first clear case of a specific chemical circumvention of light requirement in fern spore germination. More than any other work on the physiology of fern spore germination, this report, coming in the wake of the discovery of GA stimulation of germination of the achenes of lettuce (lettuce seeds), served to popularise the use of fern spores as the single-celled model for the multicellular angiosperm seed, to investigate the mechanism of action of light and hormones in germination. Schraudolf showed that whereas spores sown in the basal medium fail to germinate in the dark, addition of 5×10^{-6} to 5×10^{-5} g/ml GA to the medium induces complete dark germination (Fig. 3.5). These observations have been widely confirmed and extended by other investigators (Näf, 1966; Weinberg and Voeller, 1969*a*; Raghavan, 1976; Nester and Coolbaugh, 1986). With regard to the relative effectiveness of various gibberellins in promoting dark germination of spores of *A. phyllitidis*, GA_4 and GA_7 appear to be the most effective, GA_5 and GA_{13} the least effective and GA_3 intermediate in its effectiveness (Weinberg and Voeller, 1969*a*; Nester and Coolbaugh, 1986; Sugai *et al.*, 1987). Since GA substitutes for a light requirement for germination, it was naturally thought that photoinduced germination of spores is modulated by the production of a gibberellin-like compound. However, this issue is unsettled, as it has not been possible to confirm an earlier claim of isolation of a GA-like substance from the culture medium used for light-induced germination of spores of *A. phyllitidis* (Weinberg and Voeller, 1969*b*; Nester and Coolbaugh, 1986). It might be that there is a common chemical denominator which tantalizingly controls both light- and GA-induced spore germination. In view of this it will be instructive to compare these observations with the germination of spores of *Lygodium japonicum* which respond in high percentages alike to GA and light (Sugai *et al.*, 1987). Spores of other ferns unrelated to *Anemia* and *Lygodium* are insensitive to GA while those of close relatives such as *Mohria caffrorum* and *Schizaea pusilla* germinate sparsely or not all in the dark in the presence of GA (Näf, 1966; Weinberg and Voeller, 1969*a*; Guiragossian and Koning, 1986).

Another source of compounds which substitute for light requirement for germination of spores is the culture filtrate of gametophytes. Näf (1966) found that a liquid culture medium of seven-week old gametophytes of *A. phyllitidis (Anemia* medium) completely cancels the light requirement for germination of spores of *L. japonicum*. However, *Lygodium* medium which induces a weak germination response in *Lygodium* spores is not reciprocally active against *Anemia* spores. Similarly, an ability to induce dark-germination of spores of *A. mexicana* is inherent in *A. mexicana* medium, but the latter is ineffective against spores of *A. phyllitidis* and *Ceratopteris thalictroides* (Nester and Schedlbauer, 1982). Factors inducing dark-germination of spores are also present in the culture filtrates of spores of *Pityrogramma calomelanos* (Adiantaceae; Dubey and Roy, 1985) and *Pteris vittata* (Gemmrich, 1986*b*). According to Endo *et al.* (1972), a substance in the *Anemia* medium which promotes dark germination of spores of *A. phyllitidis* is identical with a species-specific hormone that controls antheridia formation on the gametophytes of this species (see Chapter 10). The hormone belongs to the class of diterpenes with a carbon skeleton structurally

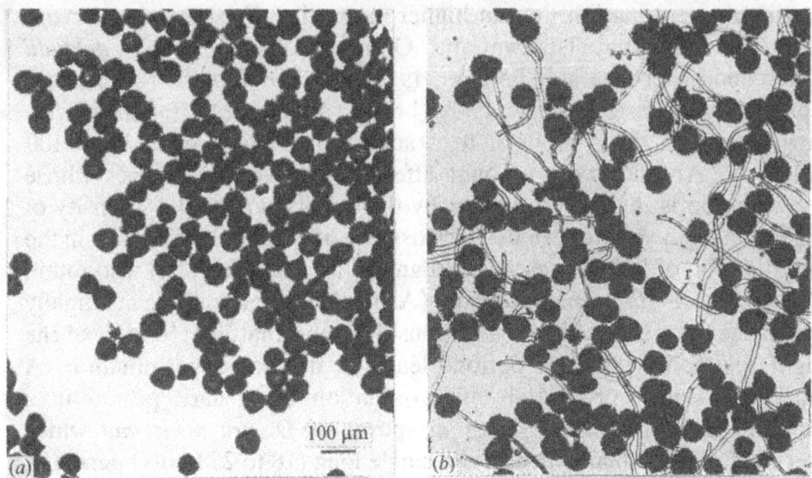

Figure 3.5. Induction of germination of spores of *Anemia phyllitidis* by GA. (*a*) Dark control. (*b*) Spores imbibed in GA in the dark. Outgrowth of the rhizoid (r) indicates germination. Spores photographed 3 days after GA treatment. Scale bar in (*a*) applies to both.

related to that of gibberellins (Nakanishi, Endo, Näf and Johnson, 1971).

How GA induces spore germination is unknown. Since the first sign of change seen in spores of *A. phyllitidis* as early as 12 hours after GA application is the hydrolysis of storage granules, one guess is that the hormone might induce the synthesis of hydrolytic enzymes (Raghavan, 1977*a*). This presupposes that messenger RNA (mRNA) is present in the dry spore – a fact confirmed recently (Fechner and Schraudolf, 1982), but it is capable of directing translational function only in the presence of GA. This, of course, is one of the hypotheses that may be formulated to explain the mode of action of GA in fern spore germination. Whatever mechanism eventually turns out to be the valid one, the simplified system of the fern spore will make the hypothesis useful to explain the action of GA in seed germination also.

Effects of light intensity and photoperiodism

Germination of spores of some ferns is dependent not so much on the quality of light as on the intensity of light. Although no systematic investigations have been undertaken on the light intensity factor in the germination of fern spores, its importance has been mentioned in passing in the literature. Generally speaking, at lower light intensities or when light is limiting, it takes much longer for a given lot of spores to attain maximum germination than at higher intensities. For spores of *Dryopteris crassirhizoma* (Isikawa and Oohusa, 1954), *Cyathea delgadii* (Marcondes-Ferreira and Felippe, 1984), and *Trichipteris* (=*Cyathea*) *corcovadensis* (Esteves *et al.*, 1985) (both Cyatheaceae), irradiation with low intensity light is superior to irradiation with light of × 10 or 100 intensity. Another type of light effect is exhibited by spores whose germination is influenced neither by the quality nor by the intensity of light, but by the photoperiod. The existence of photoperiodic effect in the germination of fern spores was recognized by Isikawa (1954) who found that spores of *Athyrium niponicum* (Aspleniaceae) germinate maximally when they are exposed to continuous light and that interruption of the light regimen with dark periods leads to decreased germination. A striking case of promotion of germination by a dark period in a photoperiodic regimen is seen in spores of *D. crassirhizoma* which germinate best when exposed to a single long (18 to 25 hours) period of illumination. The total duration of the light period required for maximum germination is reduced when a dark period of 15 to 20 hours is inserted between light periods of 3 hours each (Isikawa and Oohusa, 1956).

Although there is a requirement for high intensity light for germination of spores of *Matteuccia struthiopteris*, the situation is complicated by the involvement of a photoperiodic effect. This is reflected in a much lower percentage of germination when spores are exposed to a short period of illumination at high intensity than when light is given for a long period (Pietrykowska, 1962*a*). Similar results have been reported in the germination of spores of *Acrostichum aureum* (Loxosomaceae; Eakle, 1975).

Variations in the germination response of spores collected during different seasons of the year to the same stimulus have been reported in *Polypodium* (=*Phlebodium*) *aureum* (Spiess and Krouk, 1977). Only 24% of spores collected during October–January germinate after a saturating exposure to red light, but 79% germination is noted in spores collected during January–April. Since germination of *P. aureum* spores is under phytochrome control, it has been suggested that the amount of P_{fr} formed in spores may be regulated by the duration of light incident upon plants during spore maturation. As shown by Sugai *et al.* (1977), depending upon the localities of collection, timing of harvest and storage conditions, spores of *Lygodium japonicum* exhibit great heterogeneity in germination in red light. From an ecological perspective, this will assure that, in spite of the different selective pressures which spores face after they leave the sporophyte, they germinate only under conditions advantageous to the survival of the species.

Light inhibition of spore germination

In contrast to the amount of work done on light promotion of spore germination, light inhibition has received only scant attention. According to Whittier (1973), when axenically cultured spores of *Botrychium dissectum* are exposed to as low as 5 lux fluorescent light for 12 hours daily, no germination occurs; successful germination of spores requires culture in relatively long periods of complete darkness. Later studies showed that the dormancy of spores of certain other species of *Botrychium*, *Ophioglossum vulgatum*, and *Helminthostachys zeylanica* (Ophioglossaceae; Gifford and Brandon, 1978; Whittier, 1981, 1987) is also broken by culturing them in the dark. The inhibitory effect of light on spore germination is of considerable survival value for ferns whose prothalli are mycorrhizal in nature and subterranean in habitat. Presumably both these requirements are met in nature by spores which lie buried in the soil where light would not normally permeate and where there is an increased chance of encounter with a fungal partner. Although spores of a few additional ferns are known to germinate in complete darkness (see Table 1 in Miller, 1968*a*), there is no clear evidence to show that light inhibits their germination.

Other factors controlling germination

As mentioned in the early part of this chapter, when quiescent fern spores are provided with conditions for normal metabolic activity, they germinate without further provocation. There are some differences between spores of diverse ferns in their ability to germinate under optimum conditions of culture; changes in the culture milieu also modify the germination potential of spores. In order to cover these investigations in a logical manner, the various findings will be described for each culture condition in which both quiescent and dormant spores are used. It is assumed that when dormant spores are used, dormancy is broken by the appropriate pretreatments.

Water and medium composition

Imbibition of water is the initial step in the germination process leading to the rehydration of the spore contents, in particular of the chromatin and storage granules. It is likely that apart from changes in the physical state of the spore, some metabolic changes that prime it for germination also occur during imbibition. For example, soaking spores of *Pteris vittata* at 4 °C does not make them photosensitive, indicating that phytochrome or other components are probably synthesized during imbibition under favorable temperature conditions (Zilberstein, Arzee and Gressel,

1984). Entry of water into the spore is determined by the thickness of the outer coats and by the state of hydration of the enclosed protoplasm and for this reason, variable periods of imbibition are necessary for spores of different species before they initiate synthetic activities associated with germination. In dormant spores which require an exposure to light for germination, photosensitivity increases with time after imbibition and remains stable thereafter. This is seen in Fig. 3.6 where the response of spores of *P. vittata* to a saturating dose of red light following imbibition for different periods is plotted (Sugai and Furuya, 1967). For chlorophyllous spores, presoaking times required for attaining photosensitivity are relatively short – for example, three hours for *Osmunda cinnamomea* (Mohr *et al.*, 1964) and six hours for *Onoclea sensibilis* (Towill and Ikuma, 1973).

In spore germination experiments, water for imbibition is generally provided as the major component of a mineral salt medium, of the type usually used in plant tissue culture. Nevertheless, spores of several ferns possess the peculiar virtue of being able to germinate in distilled water or on the surface of agar solidified in water. However, when comparisons are made between the effects of a mineral salt medium and distilled water, the former is found to be superior in enhancing germination of spores, probably by sensitizing them to light (Haupt, 1985). In this context, it is important to note that certain heavy metal ions and constituents of air pollutants are toxic to spores and inhibit their germination; included in the former category are divalent cations of Hg, Cd and Co (Petersen, Arnold, Lynch and Price, 1980; Petersen and Francis, 1980), while sulfite belongs to the latter group (Wada *et al.*, 1986; Wada, Shimizu and Kondo, 1987). Sulfite-induced inhibition of germination is apparently due to decreased utilization of reserve lipids and proteins, accompanied by decreased synthesis of insoluble glucans and low amino

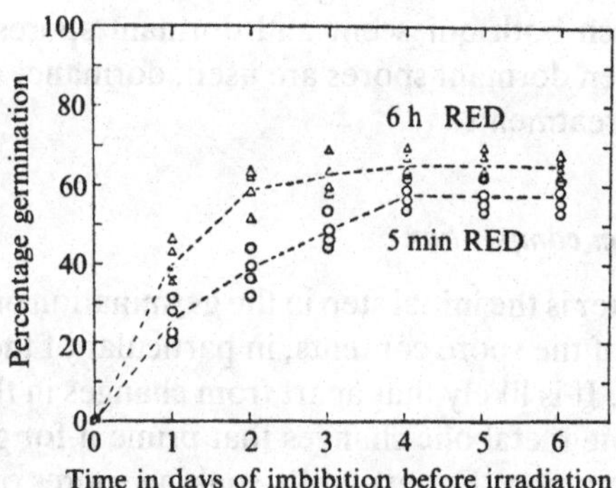

Figure 3.6. Germination of spores of *Pteris vittata* exposed to red light for 5 min (○) and 6 hours (△) after different periods of imbibition in the dark. (From Sugai and Furuya, 1967.)

acid pool size in spores (Minamikawa, Masuda, Kadota and Wada, 1987).

The physical state of the medium is also important in the germination of spores because it determines the rate of imbibition and the survival rate of germinated spores. Although both solid and liquid media have been successfully employed, the latter is apparently superior for spore germination. It is likely that a solid medium affects spore germination by preventing the breakage of spore coats and the outgrowth of rhizoid and protonema initials.

Temperature

The outcome of early investigations in which spores of a number of ferns were found to germinate quite readily at a range of laboratory temperatures thwarted attempts to study germination as a function of temperature. In general, the tolerated range of temperatures for optimum germination of fern spores is 20–25 °C. The precise sensitivity of spores to temperature varies with the species and is invariably related to the temperature requirements for subsequent growth of germinated spores. The ability of spores of certain species of *Notholaena, Pellaea* (=*Notholaena*) *limitanea* (both Adiantaceae), *Cheilanthes eatoni* and *C. lindheimeri* to withstand temperatures as high as 40–50 °C and of spores of *Matteuccia struthiopteris* to resist a temperature as low as 11 °C during germination has been attributed to their natural growth in xeric and temperate environments, respectively (Hevly, 1963; Warne and Lloyd, 1980). The lowest temperature tolerances reported for germination of spores are 5 °C for *Cyathea bonensimensis* (Kawasaki, 1954*a*) and 1–2 °C for *Pteridium aquilinum* (Conway, 1949). According to Whittier (1981), for germination of spores of *Ophioglossum engelmannii*, a sequence of temperature treatments including three months at 24 °C, followed by three months at 3 °C and a second period at 24 °C for six months, all in the dark, is necessary.

The effects of temperature on the germination of spores of some ferns appear somewhat complex because of interactions with other environmental cues, especially light or its lack thereof. The response to red light of spores of *Dryopteris filix-mas*, which germinate optimally at 20 °C, is typical of this kind of interaction between light and temperature. Mohr (1956*a*) found that when spores are incubated at temperature regimens up to 36 °C for 24 hours following irradiation with a saturating dose of red light and then kept at 20 °C for completion of germination, there is a progressive decrease in germination. However, as the interval between the end of light treatment and the beginning of high temperature treatment becomes longer, spores escape from inhibition so that in about 32 hours after light exposure, high temperature does not inhibit germination. Since germination of *D. filix-mas* spores is controlled by phytochrome, the simplest explanation of these results is that high temperature reinstates P_{fr}, formed in the photoinduced spores, back to its inactive form; when the administration of the temperature shock is delayed, the germination-promoting action of P_{fr} is completed before the effect of temperature becomes pervasive.

According to Hartt (1925), spores of *Onoclea sensibilis* germinate optimally in intense diffuse light or in absolute darkness; in the dark, however, there is a sharp cut-off point so that no germination occurs at temperatures below 27 °C or above 29 °C. Following this early report, yet more details of the temperature-dependent germination of *O. sensibilis* spores have emerged in recent years. For example, Towill and Ikuma (1975*a*) showed that the photoinduction of germination of spores is independent of temperature and can proceed even at 4 °C. This is not surprising in view of the photochemical nature of the reaction. Another study (Towill, 1978) showed that incubation of spores at 30 °C, immediately after sowing, induces a high percentage of dark germination (Fig. 3.7). Although imbibition of spores at 25 °C for 12 hours or more during the preinduction phase results in a subsequent loss of their thermosensitivity, a similar treatment does not affect their photosensitivity. It would

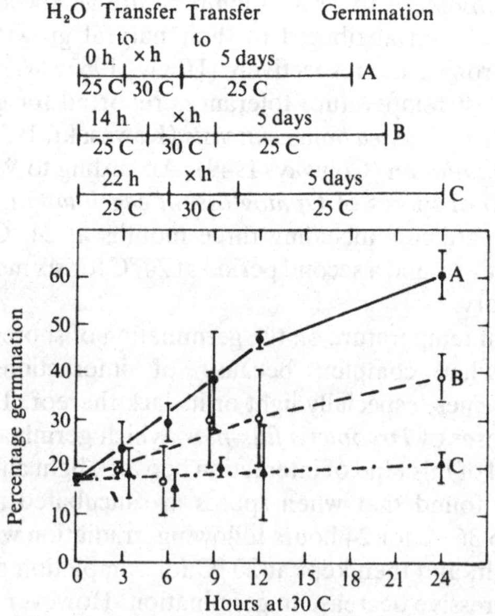

Figure 3.7. Effect of incubation for different periods at 25 °C on the subsequent dark germination at 30 °C of spores of *Onoclea sensibilis*. Experimental protocol is shown at the top of the graph. (From Towill, 1978.)

appear from these results that the mechanism of thermoinduction of germination is different from that of photoinduction. However, as shown by Chen and Ikuma (1979), photosensitivity developed by spores of *O. sensibilis* during dark incubation at 25 °C can be abolished by incubation of spores at 40 °C for eight hours or more, but normal germination is restored by a subsequent dark incubation at 25 °C. The post-induction phase which heralds germination is also blocked by an elevated temperature, the most sensitive period for the high temperature shock being confined to the first three hours after photoinduction. Admittedly, a variety of targets including the spore membranes is physiologically altered by temperature changes, but we are very much in the dark about the underlying basis for temperature sensitivity of *O. sensibilis* spores. In many ways, these experiments have made us realize that in the future, a simplistic picture of temperature effects on fern spore germination is not likely to be a satisfying one.

Effects of pH

There are some early reports of significant increases in the germination of fern spores by subtle alterations in the type and concentration of phosphate salts added to the medium. In view of the later knowledge of the buffering capacity of these salts, their favorable effects on germination can be attributed to pH changes. It is to be expected that germination of spores, as a process concerned with living cells, is limited in its ability to occur within a narrow range of pH on either side of neutrality. On the other hand, survival and growth of various ferns in unusual soil types and in ecological niches subjected to extremes of climate make it likely that pH changes of the substratum have at best only a minor influence on spore germination. A survey of published accounts of pH ranges for optimum germination shows that spores of most ferns germinate at slightly acidic or neutral pH range, as seen, for example, in *Pteridium aquilinum* (Conway, 1949). On the other hand, spores of *Gymnogramme* (=*Pityrogramma*) *sulphurea* (Adiantaceae) and *Dryopteris marginalis* germinate well in nutrient solutions ranging in pH from 4 to 7 and 3.2 to 8.2, respectively (Guervin and Laroche, 1967; Otto, Crow and Kirby, 1984). In some cases, the percentage of spore germination is reduced at low pH and the elongation of rhizoid and protonema initials is very much stymied (Courbet, 1955). The tolerance of spores of some cheilanthoid ferns such as *Notholaena cochisensis* and *Pellaea* (=*Notholaena*) *limitanea* to pH in the range of 9 to 10 has been found to correlate with the high alkalinity of the soil in which they grow (Hevly, 1963). According to Weinberg and Voeller (1969*b*), the inhibitory effects of high pH on the germination of spores of *Anemia phyllitidis* incubated in suboptimal concentration of GA are overcome by increasing the concentration of the hormone in the medium; interestingly enough, high pH of the medium does not inhibit light-induced spore germination in this species.

Gases

The composition of the ambient atmospheric gas is also an important feature of the environmental control of germination of fern spores, although few published accounts of the effects of particular gaseous phases on germination exist. Respiratory processes are probably stimulated soon after spores are hydrated and so it is reasonable to assume that they should have access to oxygen during this period. This has been found to be the case; for example, if spores of *Onoclea sensibilis* are allowed to imbibe in an atmosphere of N_2, they gradually lose their sensitivity to red light. Recovery from anaerobiosis is quick however, and spores attain half-maximal sensitivity to red light after about one hour in the air. Administration of N_2 to photoinduced spores of *O. sensibilis* also inhibits their germination, maximum inhibition being obtained when N_2 is applied six to eight hours after red light. Compared to the rapidity of recovery of spores from anaerobiosis during the preinduction phase, photoinduced spores subjected to anaerobiosis recover relatively slowly suggesting that the characteristics of oxidative processes during the two phases of germination are probably quite different (Towill and Ikuma, 1975*a*).

The presence of ethylene in the ambient atmosphere also inhibits both dark and light-induced germination of spores of *O. sensibilis* (Edwards and Miller, 1972*b*; Fisher and Miller, 1975, 1978; Rubin and Paolillo, 1984). Exposure of spores to the gas within the first six to eight hours of irradiation is most effective in blocking germination in a population. Although this inhibition is accompanied by such catastrophies as failure of DNA synthesis, nuclear movement, and cell division, spores recover if ethylene is removed from the milieu. The gas also interacts with other environmental factors in its effects on germination. If ethylene-treated spores are exposed to nonlethal doses of CO_2 (Edwards, 1977) or to red or white light (Fisher and Miller, 1975; Fisher and Shropshire, 1979), the effect of the hydrocarbon is partially cancelled and spores germinate normally (Fig. 3.8). Since experiments using CO_2 and ethylene involve different pretreatments to spores, the precise stages in the germination episode when aerobic processes are sensitized by the gases, cannot be determined. In view of the extreme sensitivity of spores of *O. sensibilis* to the antagonistic effects of ethylene and CO_2, conflicting reports on the extent of dark germination of spores by different researchers may be ascribed to the presence of varying amounts of these gases in the ambient atmosphere. The effect of ethylene on fern spore germination is the opposite of its effect on seed germination. The mechanism of ethylene

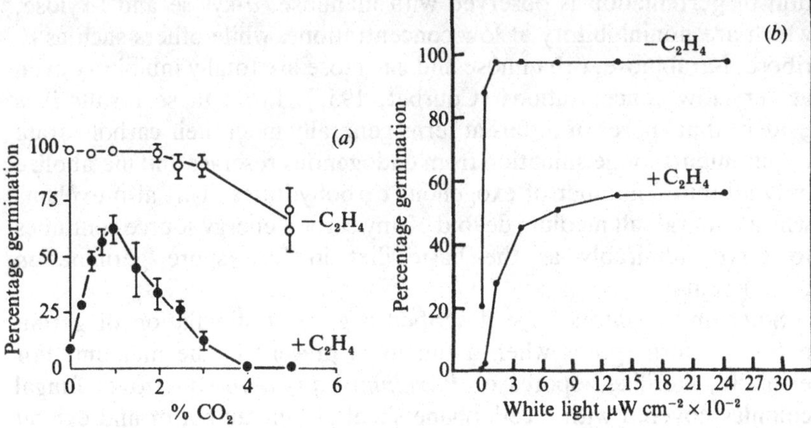

Figure 3.8. Reversal of ethylene-induced inhibition of germination of *Onoclea sensibilis* spores by CO_2 and white light. (*a*) Reversal by CO_2; spores germinated in 1 μl/l of ethylene. In the control ($-C_2H_4$), ambient ethylene was absorbed by mercuric perchlorate solution. (From Edwards, 1977.) (*b*) Reversal by white light. (From Fisher and Miller, 1975.)

action in the germination of seeds and spores is in no way clear enough to account for these differential effects. It should be kept in mind that ethylene acts as a hormone in controlling a variety of growth processes in higher plants. It thus seems likely that the primary target of ethylene action on spore germination concerns a fundamental reaction common to several processes mediated by this hormone.

Defined and undefined chemicals

A number of unrelated chemicals have been tested for their effects on the germination of spores of diverse ferns. In these trials, the test substance is generally incorporated in a range of concentrations into a basal medium in which spores are sown under conditions propitious for germination. No evidence was however sought in these studies to determine whether the tested compound is essential to trigger germination or can substitute for the requirement for a critical factor such as light. These facts should be kept in mind while evaluating the following data.

To begin with, we will consider the effects of carbohydrates. Hurel-Py (1955*b*) found that while supplementation of the medium with 2% sucrose or 1% glucose is without any appreciable effect on the germination of spores of *Alsophila australis* (Cyatheaceae), addition of 1% fructose is inhibitory. In contrast, glucose at a range of concentrations between 0.5 and 1.5% and fructose between 0.5 and 4.0% promote germination of spores of *Athyrium filix-femina*. No comparable promotion of germination is observed with mannose, *d*-xylose and *l*-xylose, which are noninhibitory at low concentrations, while others such as *d*-ribose, *l*-arabinose, *l*-rhamnose and galactose are totally inhibitory even at very low concentrations (Courbet, 1957). From these results it is evident that spores of different ferns generally meet their carbohydrate requirements for germination from endogenous reserves and metabolize only a limited number of exogenous carbohydrates. This also explains why a mineral salt medium devoid of any carbon energy source continues to serve admirably as the basic diet in fern spore germination experiments.

Some investigators have described a general stimulation of germination of fern spores when a fungus is present in the medium. For example, sterilized spores of *Pteridium aquilinum* sown over fungal colonies covered with a cellophane sheet germinate better and exhibit more robust protonemal growth than spores sown in a pure culture. The presence of the fungus also seems to halt, but not prevent completely, the loss of viability that accompanies storage of spores (Hutchinson and Fahim, 1958). According to Bell (1958), a chance contamination of cultures of *Thelypteris palustris* by an unidentified fungus increases their germination. From these results it appears that fungal association imparts to the spore some vital substance necessary to promote germination. The nature of this substance is not known, but a compound isolated from cultures of *Fusarium oxysporum* causing change in the morphology of gametophytes of *Polypodium vulgare* has been identified as ethanol (Smith and Robinson, 1969). It will indeed be surprising if the germination-promoting substance released by fungi turns out to be ethanol.

Most compounds tested are without effect or, more frequently, inhibit germination of spores. An example of this group of compounds is ABA; despite the fact that the hormone is present in spores and protonemata of *Anemia phyllitidis*, red light-induced germination of spores is not affected by exogenous ABA (Cheng and Schraudolf, 1974). This observation, as well as reports of the lack of sensitivity of spores of *Matteuccia struthiopteris* (Jarvis and Wilkins, 1973) and *Mohria caffrorum* (Chia and Raghavan, 1982) to ABA indicate that the hormone probably does not function as an endogenous regulator of spore germination. Maleic hydrazide (Sossountzov, 1953; Khare and Roy, 1977), coumarin (Sossountzov, 1961; Schraudolf, 1967*c*), colchicine (Rosendahl, 1940; Mehra, 1952; Yamasaki, 1954; Nakazawa, 1959; Singh and Roy, 1984*a*), morphactin (Sharma and Singh, 1983), lycorine, an antagonist of ascorbic acid extracted from plants belonging to Amaryllidaceae (Kumar and Roy, 1985), methylisocyanate (Singh and Roy, 1987) and various alcohols and other solvents (Miller, 1987) have also been reported to delay or inhibit germination of spores of various ferns, but the relevance of these compounds to the endogenous control of germination leaves much to be desired.

Pringle (1970) found that certain simple saturated fatty acids starting with valeric acid and ascending the homologous series up to caproic acid and the unsaturated oleic and linoleic acids inhibit germination of spores of *O. sensibilis*. The addition of culture filtrate of the gametophytes of *P. aquilinum* however overcomes this inhibition and allows full germination of spores. Since these same fatty acids which inhibit spore germination also enhance the antheridium-inducing potency of the culture filtrate (Pringle, 1961), one can envisage an interaction between substances in the latter and fatty acids in the regulation of spore germination and antheridia formation on gametophytes under natural conditions. According to Yamane *et al.* (1980), linoleic acid and (+)-8-hydroxyhexadecanoic acid which are endogenously present in *Lygodium japonicum* spores inhibit their germination in culture.

Finally, we come to the role of undefined chemical compounds in fern spore germination. Substances which inhibit spore germination are reported to be present in the centrifuged sediment of gametophytes of *Dryopteris filix-mas*, but a water extract of the gametophyte has the opposite effect on germination (Bell, 1958). Spores of some ferns are sensitive to water extracts of their leaves or of leaves, spores and sporangia of other ferns (Froeschel, 1953; Munther and Fairbrothers, 1980). Based on the drastic inhibitory effect of water extracts of *Lantana camara* on the germination of spores of *Cyclosorus dentatus* (=*Christella dentata*) (Thelypteridaceae), it has been suggested that an allelopatheic reaction probably contributes to the decline in density of the fern population in areas where it grows side by side with *Lantana* (Wadhwani and Bhardwaja, 1981).

A possible effect of population density on the germination of spores of *Polypodium vulgare* was described by Smith and Robinson (1971), who found that the greater the density of spores at sowing, the greater is the inhibition of their germination. This has been ascribed to the secretion by spores of certain volatile and nonvolatile substances into the medium. Inhibitory compounds separated from gametophytes of *P. vulgare* grown in red light also prevent spore germination (Smith, Robinson and Govier, 1973). Identification of the inhibitors could provide the missing information that will explain clearly the population effect on spore germination and light effect on gametophyte morphogenesis in this species.

Requirement for an after-ripening treatment for germination, involving storage under dry conditions at ordinary room temperatures or storage at low temperatures, is common among seeds, but has rarely been mentioned in ferns. Freshly collected spores of a strain of *Ceratopteris richardii* apparently require an after-ripening treatment under dry storage conditions for several months before they germinate fully, but germination is accelerated by treatment of immature spores with the ethylene-releasing chemical, (2-chloroethyl)phosphonic acid (CEPA) (Warne and Hickok, 1987*b*).

Germination of spores of heterosporous pteridophytes

There have been few investigations on the physiology of germination of spores of heterosporous ferns. It has long been known that sporocarps of *Marsilea* require an after-ripening treatment, since freshly harvested sporocarps do not yield viable gametophytes. After-ripening treatment is best secured by treating sporocarps for short periods at 65 °C (Bloom, 1955; Bhardwaja and Sen, 1966). Although germination of megaspores of *Marsilea* can occur under a wide range of environmental conditions, they generally exhibit a preference for low intensity white light, a temperature of about 25 °C and pH near neutrality for optimum germination (Mahlberg and Yarus, 1977). Among other heterosporous pteridophytes, spores of certain species of *Isoetes* exhibit differences in the temperature pretreatments required for germination (Kott and Britton, 1982). For example, both megaspores and microspores of *I. macrospora* and *I. tuckermanii* germinate at 23 °C without a cold pretreatment while megaspores of *I. acadiensis, I. echinospora* and *I. riparia* germinate best at 23 °C after treatment at 12 °C for 12 weeks. Microspores of *I. riparia* require a cold pretreatment to germinate, while those of *I. echinospora* which do not exhibit a clear-cut requirement for cold pretreatment, germinate in large numbers after such a treatment. Although the physiological basis for the differences in temperature sensitivity between megaspores and microspores of *Isoetes* has not been established, it appears to have some ecological value in ensuring that viable microspores are available for fertilization throughout the growing season for megaspores with receptive eggs.

General comments

The subject of physiology of fern spore germination is relatively vast, with several contributions having been made prior to the exciting period when photomorphogenetic studies in angiosperms blossomed and general principles applicable to spore germination emerged. To do proper justice to the more recent studies, coverage of earlier investigations has been less than complete. As pointed out in this chapter, the overall response of spores to light quality and chemicals has many striking similarities to the behavior of seeds. At the same time, to emphasize the distinctiveness of the regulatory processes in fern spore germination, there is also evidence to indicate that additional control mechanisms not hitherto established during seed germination, may be operational in the fern spore. Because of the diversity of observations and the small number of species investigated, the general conclusions about the physiological control of fern spore germination that we can draw at this stage are limited.

The ability of fern spores to germinate in response to multiple signals provides an assay that is crucial in defining at the cellular level the mechanism of initiation of development in a quiescent or a dormant system. That the photoreceptor involved in germination of spores of some ferns has authentic properties of phytochrome shows that the significance of the pigment in promoting germination is not so much its presence as in the orderly conformational changes which it can seemingly undergo in response to red and far-red irradiation. When the full story of phytochrome changes during fern spore germination is told, it will not be surprising if it closely corresponds to what is currently known about the molecular transformations of the pigment in other systems.

4

Cell determination and morphogenesis during germination

It is now reasonably clear that the dormant fern spore is a thick-walled, tetrahedral or bilateral cell enclosing a partially dehydrated cytoplasm with a centrally placed nucleus surrounded by other organelles and storage granules. The relationship of this basic structure to the first asymmetric division during germination is the subject of this chapter in which we shall also define those inherent and experimentally detectable aspects of spore polarity that may serve as a basis for the morphological differentiation that follows. The cytology of spore germination leaves little doubt that identical genetic information is transmitted to the two cells which are born out of a simple mitotic division of the spore nucleus. Yet, these cells follow dissimilar pathways of differentiation – one gives rise to the rhizoid, the other to the protonema initial – suggesting that each nucleus is exposed to a different milieu. Thus, it seems likely that cell differentiation during spore germination may be understood in terms of visible structural or cryptic physiological or biochemical differences in the cytoplasm.

The areas of discussion delineated for this chapter have their roots in classical morphological investigations on the germination of spores of diverse ferns made during the past hundred years or so. However, only recently has it been possible to pose the relevant questions in physiological terms. For this reason, and for a different perspective on cytoplasmic asymmetry, we will also delve into the morphology of the germinated spore toward the end of the chapter.

Polarity of the spore

This section is concerned with two principal questions: Does differentiation of the rhizoid and protonema initials result from a segregation of specific subcellular structures at germination, or are there biochemical or physiological markers suggestive of a predetermined polarity in the spore? If polarity is already resident in the spore, does it respond to extracellular signals and undergo subtle changes? These questions underlie some of the fundamental issues in early differentiation of a single cell, be it a spore, zygote, pollen grain or root hair.

Is polarity inherent in the spore?

The structural homogeneity of the dormant spores of homosporous ferns as discussed in Chapter 2 implies the absence of an inherent polarity. Nonetheless, there is some compelling physiological evidence for the existence of polarity in the spore which becomes more pronounced with the progress of germination. An intermediate stage between the dormant and fully germinated spore is the appearance of the rhizoid initial which also provides a dramatic illustration of the establishment of polarity in the spore. Some of the most extensive experiments on the development of polarity determining the site of rhizoid initiation have been conducted with spores of *Equisetum arvense* (Equisetaceae; see Nakazawa, 1952, 1961 for review). In seeking a morphological marker to explain the site of origin of the rhizoid initial, Nakazawa (1958) identified a peculiar thickening of the spore membrane adjacent to the cytoplasm at the site of the presumptive rhizoid. This part selectively stains with basophilic dyes such as methylene blue and neutral red. With the progress of germination, the rhizoid pole becomes more deeply stained. The part of the spore membrane designated as the 'rhizoid point' is also identified upon treatment of fresh spores of *E. arvense* and *Dryopteris erythrosora* with concentrated KOH (Kato, 1957*d*). The rhizoid point thus appears to have special properties not displayed by the rest of the spore. Moreover, the rhizoid point as well as the cytoplasm in the immediate vicinity has a high affinity for various metallic ions (for example, Mn, Fe, Co, Ni, Cu, Zn, Sn and Pb) (Nakazawa, 1960*b*). The cytoplasm adsorbing metal ions has been designated as 'metallophilic cytoplasm', which is present uniformly in a newly formed rhizoid initial, but persists only in the basal or sub-basal part of an elongating rhizoid.

In a normally germinating *Equisetum* spore, chloroplasts are gathered at the side opposite to where the rhizoid initial appears. If the stratification of cytoplasmic organelles is important in rhizoid determination, it should be possible to displace organelles and cause the rhizoid to arise at a new position. When spores of *E. arvense* are centrifuged at × 50 000 *g* for 5 min, a marked stratification of the spore contents into an oily layer, a chloroplast and nuclear layer, and a clear cytoplasmic zone containing mitochondria, results. However, when the rhizoid initial duly develops on the centrifuged spore, there is no relationship between the site of its origin and the point of stratification of the spore contents (Nakazawa, 1952, 1957). Similarly, when spores are allowed to germinate in an environment devoid of any gradients, such as that obtained on a rotating clinostat, there is no change in the site of rhizoid initiation (Mosebach, 1943; Nakazawa, 1956). These experiments are generally interpreted to mean that in the dormant spore of *E. arvenese* there is a feeble predetermined site of rhizoid origin, but determination becomes stronger as germination proceeds. The chemical and ultrastructural elaboration during germination of this site obviously holds the key to so much that is characteristic of polarity of spores of ferns and other pteridophytes, that all interested in this problem would do well to ponder.

Since polarity is not evident in dormant fern spores, it must arise epigenetically during the early phase of germination. The first sign of the establishment of polarity during spore germination is the migration of the nucleus from its central location to the proximal pole. This occurs several hours after the spore is potentiated to germinate. In spores of *Onoclea sensibilis*, the migrating nucleus surrounded by mitochondria and ribosomes ends up near the equatorial pole where the first division wall carving out the rhizoid initial is laid down (Vogelmann and Miller, 1980). Thus, a regional cytoplasmic differentiation is established on one side of the spore early during the germination episode. Several lines of evidence indicate that spores of *O. sensibilis*, unlike spores of other ferns, possess an inherent stable polarity which makes one end of the spore different from the other. Evidence in favor of this view is that treatments which prevent nuclear migration also block rhizoid initiation (Vogelmann, Bassel and Miller, 1981). Moreover, the inherent axis of polarity is capable of withstanding centrifugation which greatly disrupts the spore contents. Upon centrifugation, spores of *O. sensibilis* settle randomly in the centrifugal field and thus become stratified in all possible relations to the axis of polarity. However, when the centrifuged spore germinates, irrespective of the direction of stratification of its contents, the nucleus faithfully migrates to the proximal face before ending up near the equatorial pole (Bassel and Miller, 1982). This could not have happened if the integrity of preexisting polarity was lost in the midst of centrifugal traffic. A recent work (Robinson, Miller, Helfrich and Downing, 1984) has shown that the proximal face of *Onoclea* spore has a greater affinity for nickel ions than the distal face; at specific times during germination, the inner part of the exine at the proximal face also reacts avidly to a sulfide-silver stain which localizes heavy metals. When spores so stained are examined ultrastructurally, an arresting feature emerges: pore-like channels extend from the inner layer to the outer surface of the exine at the proximal face. Because Ca^{2+} might react to the sulfide–silver stain, it seems probable that localized changes in the concentration of Ca^{2+} ions through the pores might be involved in establishing spore polarity and in directing nuclear migration. As we noted in Chapter 3, other studies have implicated Ca^{2+} in the phytochrome-controlled germination of *Onoclea* spores.

Experimental control of polarity

Several workers have shown that when a light gradient is established across spores of *Equisetum*, the rhizoid initial always originates farther from the lighted side (Nienburg, 1924; Mosebach, 1943; Nakazawa, 1952; Haupt, 1957, 1958). In partially illuminated spores, the rhizoid initial predictably appears on the less illuminated pole (Etzold and Jaffe, 1963; Meyer zu Bentrup, 1964). The division wall delimiting the protonema initial at the strongly illuminated side is laid down at right angles to the gradient of light absorption. According to Nienburg (1924), during rhizoid initiation, all cytoplasmic organelles in the spore probably evince some kind of redistribution prior to alignment of the mitotic spindle parallel to the direction of incident light, but none show a more precise and programmed movement than the chloroplasts (Fig. 4.1). The sensitivity of this organelle to light is apparent as early as four hours after sowing, even before the appearance of any morphological symptoms of germination (Mosebach, 1943; Haupt, 1957).

A study of polarity induced by high intensities of light has added to our knowledge of the properties of the photoreactive system in the spores of some species of *Equisetum*. When spores are exposed to unilateral light, increasing light fluences do not lead to a saturation of the response, but account for a decrease in the percentage of responding spores. This result has been interpreted to suggest that as absorption of radiation on the lighted side becomes saturated there is a concomitant increase in the amount of light absorbed on the dark side, leading to a weakening of the gradient (Haupt, 1957, 1958). A second view of these results is that high energy light applied for a short period inactivates the photoreactive system rendering it incapable of triggering the reaction for fixing polarity (Meyer zu Bentrup, 1964). A suggestion that saturating light doses might result in a mixture of inactive and active photoreceptors is entirely compatible with this view.

A complementary approach has made use of polarized light to determine the orientation of the rhizoid and protonema initials during germination. The method offers the important advantage that spores respond not only to the direction of illumination but also to the vibration of the electric vector, thus enabling deductions to be made about the disposition of the photoreceptors. When *Equisetum* spores are irradiated with plane polarized light, the rhizoid initial appears parallel to the axis of vibration of the electric vector. If the preferred origin of the rhizoid initial corresponds to the least light absorbing region, the results apparently indicate the presence of surface-parallel dichroic photoreceptors in the spore (Haupt and Meyer zu Bentrup, 1961; Meyer zu Bentrup, 1963; Etzold and Jaffe, 1963). Action spectrum studies of polarity induction with plane polarized light are consistent with the view that two different

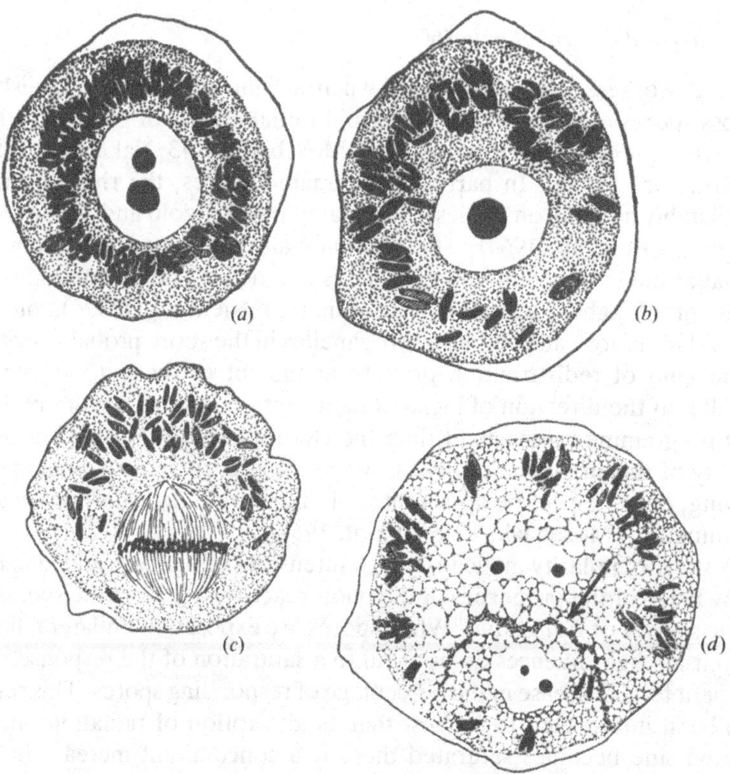

Figure 4.1. Chloroplast movement and origin of polarity in germinating spores of *Equisetum*. (*a*) Unpolarized spore showing uniform distribution of chloroplasts around the nucleus. (*b*) Beginning of polarized movement of chloroplasts away from the site of the presumptive rhizoid initial. (*c*) Spore nucleus in mitosis to form the rhizoid initial. (*d*) Formation of the rhizoid initial (arrow). (From Nienburg, 1924.)

pigments function as photoreceptors. Apart from the difference in action spectra, the main line of evidence for this dualistic concept of photoreceptors lies in the order of magnitude of light fluence required for polarity induction. The chemical nature of the photoreceptors has been difficult to ascertain in part because they are not readily separated from the other components of the spore; however, as far as can be determined from action spectra, the likely candidate pigments for the high intensity and low intensity responses are, respectively, riboflavin and carotene (Meyer zu Bentrup, 1963).

Observations on the origin of the rhizoid and protonema initials in relation to the direction of incident light in germinating spores of *Dryopteris filix-mas* (Mohr, 1956b), *Athyrium filix-femina*, and *Matteuccia struthiopteris* (Pietrykowska, 1963) have affirmed the view that the dark and lighted sides favor the rhizoid and protonema initials, respectively. The work of Pietrykowska (1963) has shown that exposure of spores of *A. filix-femina* to unilateral light for four days is necessary to fix the polarity of the rhizoid and protonema initials, even though the direction of incident light is changed during subsequent periods. If the direction of light impinging on the spore is changed during the first four days of germination, the rhizoid initial experiences random orientation on the spore. The general view suggested by this study is that there is a requirement for a minimum duration of exposure to light before polarity is irreversibly stabilized in the spore. Claims have been made that unilateral illumination sets up an auxin gradient in the spore which would, of course, offer the prospect of identifying the internal factor fixing polarity (Mohr, 1956*a*; Pietrykowska, 1963); in the absence of any data, these claims are fraught with difficulties.

The action of polarized white, red, and blue light on the polarity of rhizoid initiation in fern spores is similar to that described in *Equisetum* spores, and irrespective of the quality of light, the rhizoid initial always appears parallel to the plane of polarization (Bünning and Etzold, 1958; Jaffe and Etzold, 1962; Pietrykowska, 1963). Jaffe and Etzold (1962) have presented compelling evidence to show that the response of spores of *Osmunda cinnamomea* to blue light is due to an ordered arrangement of dichroic photoreceptors parallel to the surface of the spore. In this situation, it is conceivable that when spores are irradiated with polarized light, the rhizoid initial will appear on the side which absorbs the least light. On the basis of the limited extent to which it has been possible to orient the rhizoid initial in spores of *Onoclea sensibilis* in the direction of unilateral white light or polarized red light, it has been argued that determination of the rhizoid angle does not depend upon a directional stimulus (Miller and Greany, 1974).

In summary, it seems that polarity is generated in spores by asymmetries in their cytoplasmic organization. No generalizations can be made about the factors controlling polarity in spores of ferns and other pteridophytes. In some, polarity may be determined by a chemical gradient, while in others it may be due to a gradient set up by differential light absorption. If spores possess an inherent polarity, it is probably conferred from the outside by its position in the tetrad.

Preparations for germination

A convenient starting point for a general outline of the cytology of germination is the fully imbibed spore ready to respond to a germination inducing stimulus. This spore also provides a standard of reference for evaluating the changes that occur as a prelude to the formation of the rhizoid and protonema initials as opposed to the changes due to hydration of the cell contents. Due to the presence of the thick spore coats, cytological events of fern spore germination such as organelle elaboration, their movement and the division process itself have proved difficult to analyze. The paucity of work in this area is symbolized by the fact that thus far detailed data are available for only a single species, *Onoclea sensibilis*, which has proved to be an exceptional experimental system.

Spores of *O. sensibilis* are chlorophyllous, and germinate under different experimental conditions and in various media including distilled water. The spore is bounded by a relatively colorless, thick exine and a close, brown perine, both of which could easily be removed by swirling spores in commercial sodium hypochlorite solution. This treatment, if timed carefully, does not affect the integrity of the cell contents, but renders the nucleus visible in the living state during the time course of spore germination (Vogelmann and Miller, 1980). A major problem in the analysis of ultrastructural cytology of the spore is the presence of the intine which leads to shrinkage, deformation and other fixation artifacts, and thus makes the spore recalcitrant to electron microscopy. Recently, these problems have been overcome by treatment of glutaraldehyde-fixed spores with a sodium hypochlorite–mannitol solution which removes the exine–perine complex and causes a break in the intine allowing dehydration without shrinkage (Bassel *et al.*, 1981). As an additional aid in the cytological study, spores are sensitive to chemicals which, in appropriate dilutions, block specific steps in the germination process. By fixation in aldehyde fixatives and embedding in epoxy resins, sections suitable for light microscopic analysis of germination can be obtained from spores of ferns with thick spore coats without subjecting them to any pretreatments (for example, Endress, 1974; Raghavan, 1976).

Nuclear migration

In a fully imbibed spore, the nucleus is a dynamic organelle which responds rapidly to a germination inducing stimulus. The nucleus becomes large and metabolically active, acquires increased granularity and begins to move. By studying living spores in the light microscope and by electron microscopy, Miller and colleagues (Vogelmann and Miller, 1980; Bassel *et al.*, 1981) have provided a wealth of details about nuclear migration during germination of spores of *O. sensibilis*. Substantial light, however, has not been shed on the question as to why the nucleus migrates the way it does and what propels this organelle to its ultimate destination. After exposure of spores to about 16 hours of light, the nucleus is seen in the center as a clear area. At the ultrastructural level, the distinguishing features of the spore at this stage are the increasing proximity of starch-containing chloroplasts to the lipid layer and the appearance of mitochondria as a ring around the nucleus. Soon the nucleus begins to move in the cytoplasm displacing the chloroplasts and other organelles. Assumption of an elongate form by the nucleus in association with the presence of lobes or extensions around its periphery that bear little or no resemblance to the organelle at its stationary phase is graphic proof of nuclear movement. Coincident with nuclear migration, a new cell wall is reconstituted beneath the intine of the spore.

The characteristic pathway of nuclear movement is to the proximal pole of the spore as one would expect. But upon reaching this point, the nucleus continues to move to the equatorial end of the spore (Fig. 4.2). An association of mitochondria and ribosomes with the migrating nucleus as seen in the electron microscope can perhaps be singled out to illustrate the energy relations of the nucleus in motion. The end of nuclear migration signals the beginning of the preparation phase for the asymmetric division of the spore.

The cytoplasmic microtubule system is generally concerned with the motility and shape of cells and it seems likely that there is a similar role for this network of tubules in the germinating fern spore. Although microtubules are not aligned parallel to the migrating nucleus in the spore of *O. sensibilis*, they are found in close association with the nuclear envelope and surrounding mitochondria (Bassel *et al.*, 1981). Supporting the idea that microtubules are the driving force in nuclear migration is the observation that compounds which block microtubule assembly or organization, interfere with nuclear migration (Vogelmann *et al.*, 1981). The most striking of these compounds is colchicine which disrupts microtubule polymerization by combining with their tubulin subunits. As shown in Fig. 4.3, colchicine treatment is inimical to spores that develop into enlarged, mostly uninucleate cells. Indeed, the picture is not different from what would be expected of spores in which polar nuclear movement and cell division are prevented. When colchicine-treated spores are examined internally, the failure of nuclear migration is found to correlate with the complete failure of microtubule assembly. Comparable results are obtained with isopropyl-N-3-chlorophenylcarbamate (CIPC) and griseofulvin. Failure of nuclear migration in the spore and its germination into two more or less equal cells as a result of treatment with these inhibitors follow the prediction that microtubule assembly is inhibited. These results do not prove that microtubules alone control nuclear movement during germination of the spore. The fact that the path of migration of the nucleus in the germinating spore is inverse of that followed by the same organelle during spore maturation (p. 21) has led to the suggestion (Vogelmann and Miller, 1980) that a persistent cytoskeletal track laid down at an earlier stage in the ontogeny of the spore might serve as a discrete pathfinder for the nucleus during germination. If this corresponds in any way to reality, we have a nice

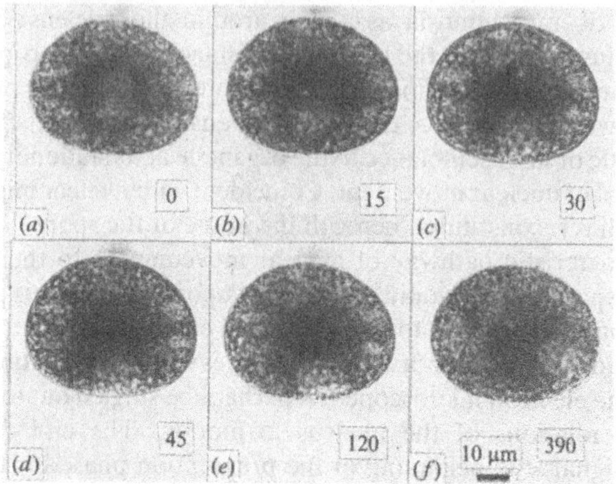

Figure 4.2 (*a–f*) Time-lapse photographs of a germinating spore of *Onoclea sensibilis* showing migration of the nucleus (visible as a clear area among the darker chloroplasts). Proximal pole of the spore is towards the bottom of the page. Time (minutes) after onset of nuclear migration is given in the lower right hand corner of each photograph. (From Vogelmann and Miller, 1980; photographs supplied by Dr J. H. Miller.)

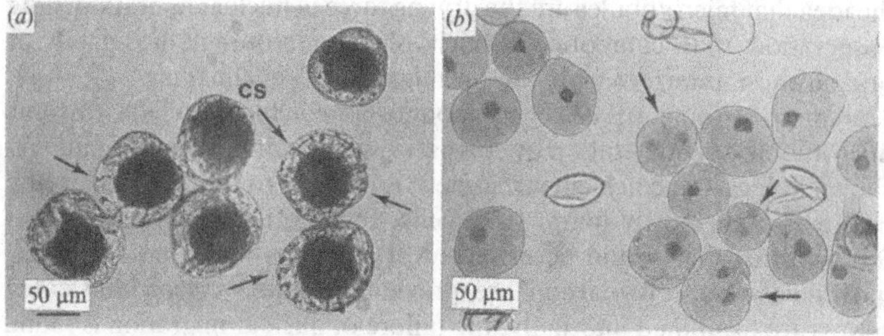

Figure 4.3. Inhibition of germination of spores of *Onoclea sensibilis* by colchicine. (*a*) The centrally placed nucleus is surrounded by chloroplasts. Arrows point to the cytoplasmic strands (cs). (*b*) Colchicine-treated spores stained with aceto-carmine-chloral hydrate to show the position of the nucleus. Some spores (arrows) have divided once symmetrically, but neither cell differentiates into a rhizoid. (From Vogelmann *et al.*, 1981; photographs supplied by Dr J. H. Miller.)

illustration of how the cytoplasmic support system coordinates the oriented movement of an organelle in both directions at different phases in the life of a cell.

Obviously, the directional migration of the nucleus and its arrival at the new address are turning points in the germination of the spore. These

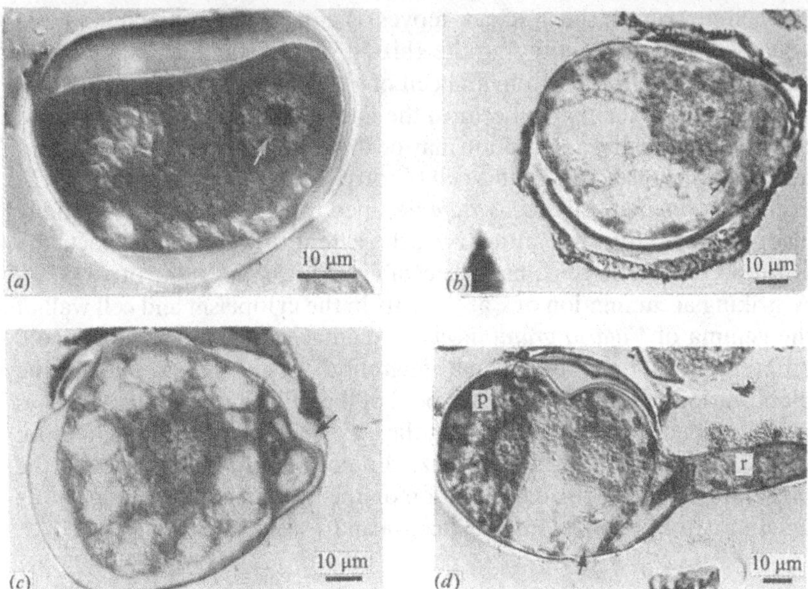

Figure 4.4. Sections showing different stages in the germination of spores of *Onoclea sensibilis* giving rise to the rhizoid and protonema initials. Proximal pole of the spore is towards the top of the page. (*a*) Nucleus (arrow) poised at one end of the spore. (*b*) Formation of a cell wall (arrow) delimiting the rhizoid initial. (*c*) Rhizoid initial (arrow) breaking out of the exine. (*d*) A germinated spore with an elongated rhizoid (r) and a protonema initial (p) formed by the division of the spore cell (arrow). Sometimes, following rhizoid initiation, the spore cell may directly function as the protonema initial. Photographed with Nomarski optics. (From Raghavan, 1987.)

events create an atmosphere in which polarity is stabilized and the nucleus is poised to undergo mitosis. As shown in Fig. 4.4, the mitotic apparatus tends to be positioned toward one end of the spore and the axis of the spindle parallels the long axis of the cytoplasmic mass. In this position of the spindle, when cytokinesis occurs, there is not much of a choice for the spore other than to generate two disproportionate cells. As to the fate of these cells, the small cell becomes the rhizoid initial and the large cell becomes the protonema initial.

A digression must be made here to point out much that is common between the origin of the rhizoid initial during spore germination and the origin of secondary rhizoids from different parts of the gametophyte. Paralleling rhizoid initiation during spore germination, it is remarkable that preparatory to secondary rhizoid formation from the spore, protonemal cell, mature prothallial cell, and vegetative propagule (gemma), the nucleus migrates to one side of the target cell (Crotty, 1967; Dyer and Cran, 1976; Kotenko, Miller and Robinson, 1987). However, the spot in the cell to which the nucleus moves is not predetermined and the asymmetric division initiating the rhizoid occurs wherever the nucleus happens to be. In the prothallial cell of *Pteris vittata*, various structural and cytochemical changes occur in the nucleus before it migrates; given this pattern, the first assumption may be that a mRNA involved in rhizoid initiation is transcribed by this cell (Crotty, 1967). Although the nucleus of the protonemal cell of *P. vittata* destined to produce a rhizoid readily incorporates ^{3}H-uridine into RNA, the extent to which this reflects the synthesis of mRNA remains to be established (Cohen and Crotty, 1979). A striking accumulation of Ca^{2+} occurs in the cytoplasm and cell wall of the gemma of *Vittaria graminifolia* (Adiantaceae) at the site of rhizoid differentiation prior to nuclear migration. Since this is due to the mobilization of the internal Ca^{2+} pool of the cells, nuclear migration has been ascribed to a decrease in cytosolic Ca^{2+} (Kotenko *et al.*, 1987; Miller and Kotenko, 1987). Like the rhizoid point of germinating spores, the presumptive site of rhizoid initiation on the protonemal cells also shows an affinity for metallic ions (Nakazawa and Tsusaka, 1959*a*, *b*).

Mitosis and cytokinesis

Not many detailed observations have been made on mitosis and cytokinesis during fern spore germination. Although these two processes appear to be closely linked, dissociation of cytokinesis from mitosis has been achieved in spores of *O. sensibilis* by treatment with caffeine, a drug which thwarts cell plate formation during division. In the presence of this drug, most of the spores complete mitosis in anticipation of cytokinesis, but the latter event is blocked, resulting in multinucleate spores (Miller and Bassel, 1980). The idea that geometrical asymmetry and the orientation of the plane of cell division are pivotal factors in cell differentiation has a long history, and a good measure of support from assorted examples in the plant kingdom, as Bünning (1958) has pointed out in an important review. Yet, there has been little by way of direct verification of this view based on the pattern of differentiation in cells induced to divide symmetrically. According to Miller and Greany (1976), a method that allows symmetrical division of spores of *O. sensibilis* is treatment with 2.5% methanol. When spores pretreated with the alcohol for three days are subsequently reared in the basal medium, the developmental consequence of cytokinetic symmetry is revealed by the formation of twin protonema initials instead of a rhizoid initial and a protonema initial (Fig. 4.5). Judging from a later work of Vogelmann and Miller (1981) it appears that nuclear migration to the equatorial end of the spore is arrested by methanol; when division occurs at the site where the nucleus comes to lie – usually at the proximal pole – the resulting cells are

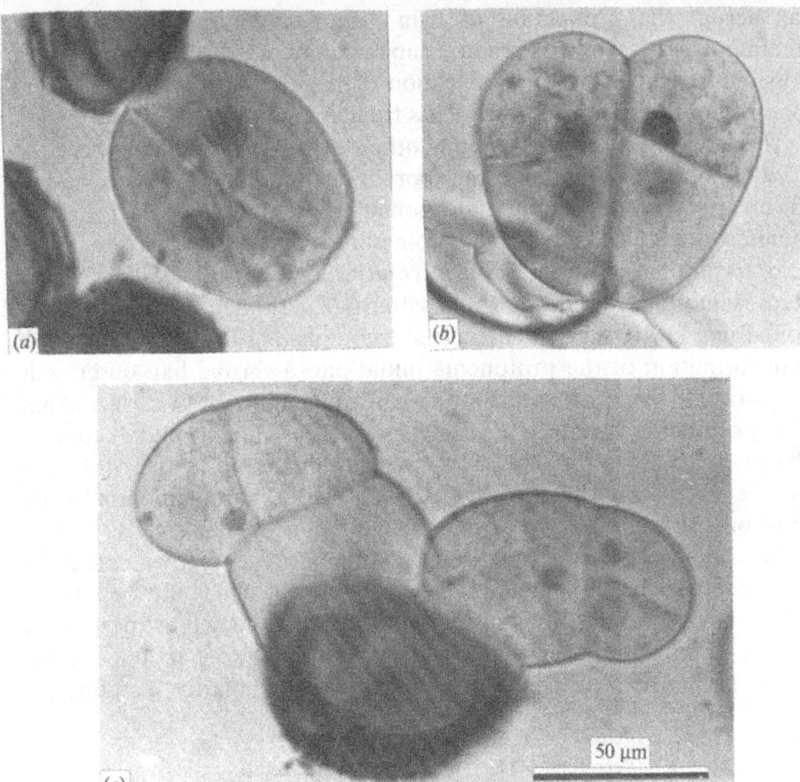

Figure 4.5. Formation of twin protonemata in spores of *Onoclea sensibilis* germinated in a medium containing methanol. (*a*) An early stage of germination resulting in two almost identical cells. (*b*) Division in both cells to form filaments. (*c*) Initiation of planar growth in both filaments. Scale bar in (*c*) applies to all. (From Miller and Greany, 1976; photographs supplied by Dr J. H. Miller.)

symmetrical. As shown in Fig. 4.6, the most sensitive period for inhibition of nuclear migration and rhizoid initiation in the germinating spores is just prior to the onset of the actual nuclear movement; methanol must also be present until completion of cell division in order to prevent rhizoid differentiation. Centrifugation of spores if timed to coincide with mitosis also induces the production of two identical cells neither of which differentiates into a rhizoid (Bassel and Miller, 1982). The results of experiments on methanol effect and centrifugation have a consistency which strengthens their validity. It has been suggested that methanol acts by disrupting a special region of the cell membrane that is probably involved in guiding the nucleus to the equatorial pole of the spore. We can accept the production of twin protonema initials by methanol treatment as a fact and its relationship to failure of nuclear migration as a possibility. However, the association of the chemical with membrane integrity of the spore is a hypothesis fraught with pitfalls.

There are also examples among other ferns and fern allies where the normal polar germination of the spore is disrupted by chemicals leading to a variety of developmental anomalies. Perhaps the most striking manifestations of abnormal syndrome are those induced by the amino acid tryptophan in spores of *Dryopteris erythrosora* (Kato, 1957*a*). Representative types of abnormal patterns of germination noted in media containing low to moderate levels of tryptophan (1.0–50.0 mg/l) include transformation of the protonema initial into a rhizoid-like outgrowth, formation of twin protonema initials and appearance of twin rhizoid and twin protonema initials in the same spore. These abnormalities contrast sharply with those observed in spores germinated in media containing high levels of tryptophan (300–500 mg/l). Here, germination is for the most part restricted to the development of giant spherical cells. In these cells there is a permanent obliteration of polarity so that when they are transferred to the basal medium they regenerate rhizoids and protonemal cells from all around their surface. In other treated spores, despite the loss of polarity, a propensity for division seems to persist but the wall formed is in a plane opposite to the normal axis of polarity. Treatment of

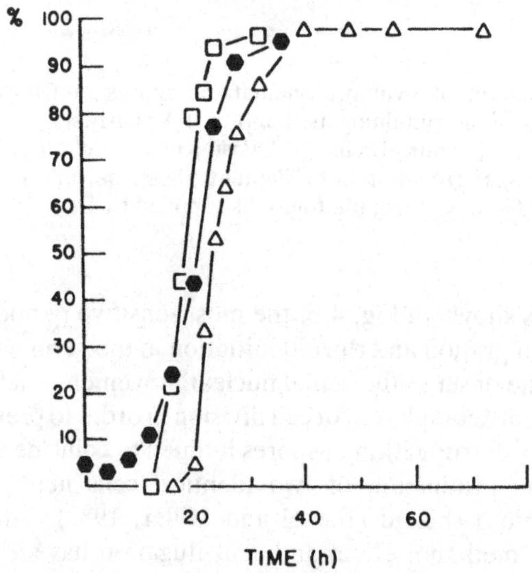

Figure 4.6. Time course of nuclear migration (□) and cell division (△) in spores of *Onoclea sensibilis* sown in the basal medium are plotted along with the percentage of spores that form rhizoid initials (●) at different times after transfer to a medium containing methanol. (From Vogelmann and Miller, 1981.)

spores of a variety of other ferns with colchicine (Rosendahl, 1940; Mehra, 1952; Mehra and Loyal, 1956; Yamasaki, 1954; Kato, 1957*c*; Nakazawa, 1959; Miller and Stephani, 1971) has also provided a picture of formation of giant cells associated with a loss of polarity. There is little, if any, significant difference in the type of colchicine-induced hypertrophy between spores of different species. Generally, rhizoid initiation is inhibited in the enlarged colchicine-treated spores, but the latter do not swell themselves to death. Even after spores enlarge abnormally in the presence of colchicine, they divide and revert to a semblance of normal gametophytes if they are taken out of the drug (Nakazawa, 1959). Rarely, as seen for example in *D. erythrosora*, colchicine treatment not only allows normal elongation of the rhizoid, but also provokes its ramification (Kato, 1957*b*, *c*). Whether these changes in the germination patterns are based on nuclear perturbations in the spore, or are due to a compensatory effect on the existing polarity gradient in the cytoplasm, has not been determined. Offhand, these data seem to suggest that rhizoid and protonema initials are extreme morphological variants of one basic kind of outgrowth.

A network of other chemicals also modifies the polarity of germinating spores, but in view of the assorted nature of the compounds tested and the fortuitous occurrence of the abnormalities, the overall probability for their role in the regulation of cell morphogenesis would seem to be marginal. Included in the list are naphthaleneacetic acid (NAA) which causes inhibition of rhizoid initiation and promotion of growth of secondary rhizoids from the protonemal cell of *D. erythrosora* spores (Kato, 1957*d*) and ethanol and acetaldehyde which induce twinning of protonema unaccompanied by rhizoid growth on spores of *Polypodium vulgare* (Smith and Robinson, 1969). Multiple rhizoids are formed on spores of *Pteris vittata* germinated under a variety of conditions, but as seen in Table 4.1, addition of glucose, sucrose or starch to the medium appears to promote the outgrowth of maximum number of rhizoids in a large number of spores (Kato, 1973*a*). In this species, outgrowth of the rhizoid and protonema initials appears to be related to the osmotic milieu of the spore, since spores germinated in relatively high concentrations of sucrose (15% or 20%) lack rhizoids but form twin protonema initials. While these abnormalities are noted infrequently in populations of spores which do not germinate or which germinate normally in the presence of the chemical, germination of spores of *Anemia phyllitidis* in allo-gibberic acid evokes twin protonema initials in as high as 80% of the population (Schraudolf, 1966*a*).

Among extracellular factors, radiations of varying character constitute another broad class of agents known to modify the morphogenesis of germinating spores. Administration of low intensity light to spores during germination (Pietrykowska, 1962*a*) or exposure of spores to UV-light

Table 4.1 *Effects of addition of different supplements to distilled water on rhizoid formation during germination of spores of* Pteris vittata[a]

Additives to distilled water	Number of rhizoids[b]										Total	M^c
	1	2	3	4	5	6	7	8	9	10		
None (floating condition)	5	81	80	4	1	1	0	0	0	0	172	2.5
None (submerged condition)	110	120	10	0	0	0	0	0	0	0	240	1.6
Agar (0.8%)	19	36	103	30	18	10	9	3	6	2	229	3.5
Sucrose (1%)	3	98	84	77	78	56	9	8	7	10	429	4.0
Glucose (1%)	14	14	56	70	14	16	22	6	8	6	226	4.3
Soluble starch (1%)	5	60	60	24	31	20	22	9	6	5	242	4.1
Agar (0.8%) +sucrose (1%)	7	34	58	150	72	37	4	2	1	5	360	4.3
Mannitol (1%)	98	37	11	0	0	0	0	0	0	0	146	1.4

[a] From Kato (1973*a*).
[b] Number of germinating spores with the respective number of rhizoids.
[c] Average number of rhizoids per spore.

before germination (Kato, 1964*a*) has been shown to generate two or more protonema initials. According to Palta and Mehra (1973), irradiation of spores of *Pteris vittata* with a high dose of X-rays (75 600 r and above) leads to the loss of polarity during germination and formation of giant apolar cells lacking rhizoids, regeneration of twin protonemal cells and interconversion of the rhizoid and protonema initials.

The multiplicity of agents that interfere with germination of spores indicates that the character of the reacting system is more important than the particular chemical or physical agent that affects the process. Although cell differentiation during fern spore germination has been studied in painstaking detail, only spores of *Pleurosoriopsis makinoi* (Adiantaceae) germinating in an undisturbed normal environment appear to have lost their sense of polarity and form two protonemal initials (Masuyama, 1975*c*). This is reminiscent of a similar chemically-induced change observed in spores of other ferns.

Summarizing, we see that the first division of the spore does more than produce two cells; it also determines the fate of cells formed. Despite the fact that partitioning of genetic information is apparently normal at spore mitosis, cells with different phenotypic traits appear. In the present climate of opinion, this means that differences in the cytoplasmic package around each nucleus must be responsible for the different modes of nuclear expression and cell differentiation. The nature of the two sets of cytoplasmic signals leading to rhizoid and protonema functions is not clear; we can push all this back one more level and say that the differences in function demanded of them imply that changes in the expression of genetic information are involved.

The germinated spore

As the visible manifestation of germination is the production of the rhizoid and protonema initials, our concern here is to describe what these cells have in common and how they differ, structurally and physiologically, to justify their divergence in development.

The rhizoid initial

One of the most interesting cells in the early gametophytic phase of ferns and other pteridophytes and one that has received very little attention is the rhizoid initial. Two features characterize the growth of this cell. First, it is a cell which typically elongates by apical extension like pollen tubes, root hairs, fungal hyphae and other tip-growing cells. Second, the rhizoid initial is programmed for terminal differentiation and does not normally divide after it is cut off. As was mentioned previously, until midway through the germination time course, there are no signs of polarization of

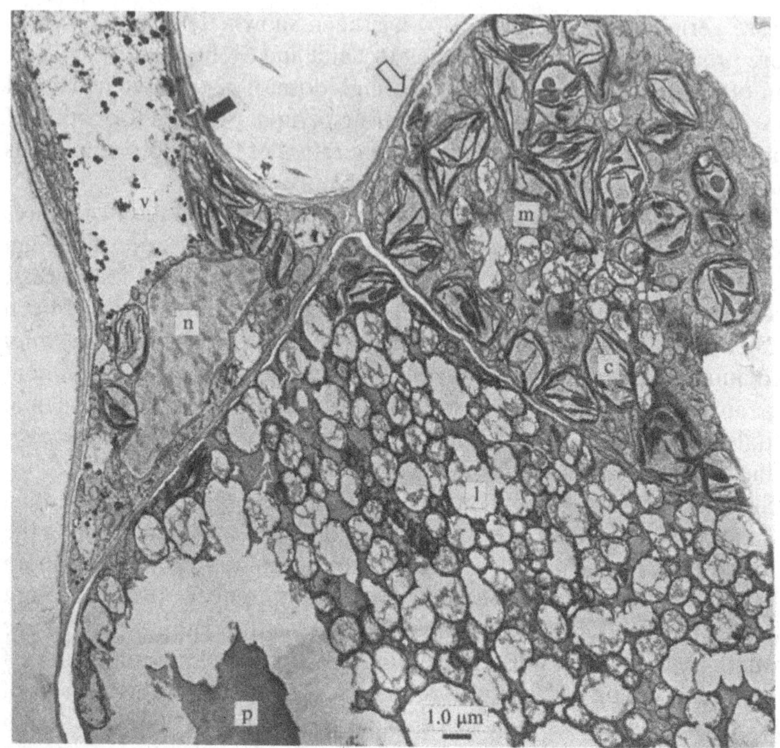

Figure 4.7. Electron micrograph of a germinated spore of *Blechnum spicant* showing the relationship between the rhizoid (solid arrow) and protonema (hollow arrow) initials. c, chloroplast; l, lipid body; m, mitochondria; n, nucleus; p, protein storage body; v, vacuole. (From Beisvåg, 1970; photograph supplied by Dr T. Beisvåg.)

the spore contents suggestive of the formation of a cell fated to elongate but not divide. According to Bassel *et al.* (1981), in the germinating spore of *Onoclea sensibilis*, ribosomes and mitochondria are parceled out to the rhizoid initial in greater numbers than to the protonema initial, whose cytoplasm is filled with much of lipids and protein granules besides chloroplasts.

Rhizoids show variable structural features and are not fundamentally different from normal elongating cells. Figs. 4.7 and 4.8 typically illustrate the relationship between the spore cell and the rhizoid and protonema initials in two representative species. The rhizoid and protonema initials may not be in direct contact with each other as in *Blechnum spicant* (Beisvåg, 1970) and *Dryopteris borreri* (=*D. affinis*; Dyer and Cran, 1976) or the protonema initial and the spore cell may share a common wall of the rhizoid initial as in *Polypodium vulgare* (Fraser and Smith, 1974).

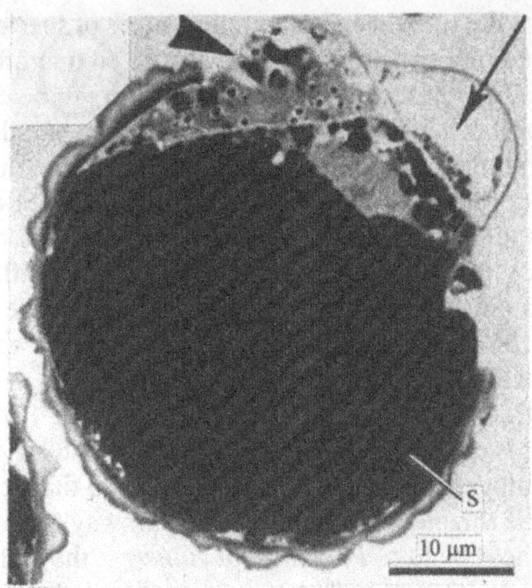

Figure 4.8. Section of a germinated spore of *Polypodium vulgare*. The basal wall of the rhizoid (arrow) is in contact with the protonema initial (arrow head) and the spore cell (s). There is a large mass of lipid bodies (seen in black) in the spore cell. (From Fraser and Smith, 1974.)

The nucleus is confined to the base of a newly cut-off rhizoid initial. It is surrounded by a blend of organelles including chloroplasts, mitochondria, ribosomes, golgi, ER, vesicles, and spherosome-like bodies. Structural change associated with the elongation of the rhizoid is mainly a matter of extensive vacuolation and a concurrent accumulation of cytoplasm in the apical region. As the rhizoid begins to elongate, chloroplasts show signs of degeneration such as loss of integrity of the thylakoids and escape of starch grains (Gantt and Arnott, 1965; Beisvåg, 1970; Fraser and Smith, 1974; Dyer and Cran, 1976). In the apical region of the rhizoid of *D. borreri* (=*D. affinis*) there are numerous small vacuoles, vesicles of small dimensions, golgi bodies and ER oriented parallel to the cell wall. The vesicles of this region are notable for some degree of complexity. They are smooth-surfaced, but some have an electron-transparent core, while others have a density similar to the golgi cisternae. The preponderance of vesicles, presumably derived from the golgi cisternae in the elongating part of the rhizoid is consistent with the need to synthesize cell wall materials for growth (Dyer and Cran, 1976). By two-dimensional electrophoretic analysis of the soluble proteins of germinating spores of *O. sensibilis*, Huckaby and Miller (1984) have identified close to nine rhizoid-specific polypeptides, in the sense that these polypeptides are substantially more prominent in the rhizoid protoplasts than in whole spores.

The cell wall of the rhizoid exhibits some degree of specialization since it is not composed of a system of regularly oriented microfibrils as in the protonema initial. Rather, the rhizoid wall appears to be equipped with two layers of fibrillar matrix distinguished mainly by the dense material deposited within the fibrils. In its chemical composition, the rhizoid wall differs from that of the protonema initial in having higher concentration of pectic substances and a lower cellulose content. As the tip of the rhizoid elongates, the remainder of the cytoplasm of the cell passes through a programmed disintegration, characterized by the accumulation of deposits and of an undefined tubular system in the vacuoles and a concentration of degenerating chloroplasts at the tip (Smith, 1972*b*; Fraser and Smith, 1974; Cran and Dyer, 1975; Cran, 1979; Miller and Bassel, 1980). It is conceivable that in a rhizoid which has completed its growth, the disintegration proceeds acropetally and that at this stage the metabolism of the rhizoid is switched on to a pathway of self-destruction.

A feature of the rhizoid of *Polypodium vulgare* is the high activity for a number of phosphatases such as acid and alkaline phosphatase, adenosine triphosphatase (ATPase), 5-nucleotidase and glucose-6-phosphatase. It has been suggested that these enzymes are involved in the active uptake of materials into the rhizoid (Smith, 1972*a*). Based on staining reactions, it has been proposed that the rhizoid wall of *P. vulgare* contains free carboxyl groups which bestow on it cation exchange properties. This property, not detected in the cells of the protonemal filament, is probably related to the absorptive function of the rhizoid which becomes more refined, focused and adaptive with time (Smith, 1972*b*). Two other points of general significance in the function of the rhizoid are its greater wall pore size than the cells of the protonemal filament (Miller, 1980*a*) and the absence of a multilayered lipid coat detected over the latter (Wada and Staehelin, 1981).

During growth in length of the rhizoid, the nucleus migrates from its basal location to a mid-position or laterally. Although nuclear migration in the rhizoid is an uneventual act, occasionally the nucleus becomes spindle-shaped or coiled like a tube (Schraudolf, 1983*a*). As shown in Fig. 4.9, with the progressive elongation of the rhizoid, the nucleus keeps on migrating towards the tip so that the distance from the tip of the rhizoid to the nucleus is always constant (Kato, 1957*b*). An apparent lack of shift of charcoal particle markers placed on the rhizoid of *Pteridium aquilinum* with a micromanipulator has led to the view that elongation is confined to the very extreme tip of the cell – perhaps to within 1 μm (Takahashi, 1961).

What factors govern the continued growth of the rhizoid initial on the germinated spore? This question is important in determining whether specific factors are involved in promoting growth of the rhizoid initial as opposed to its differentiation during germination. Obviously, the rhizoid

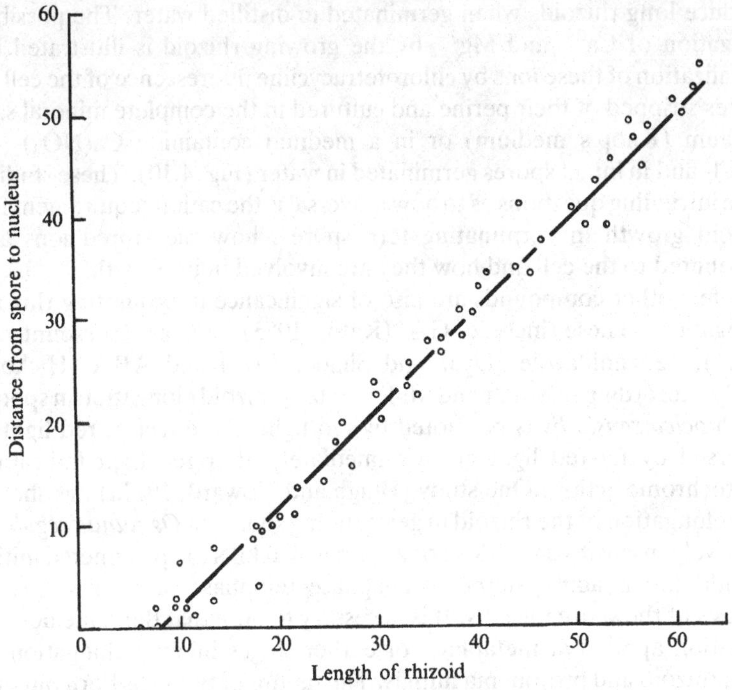

Figure 4.9. Relationship between the length of the rhizoid and the position of the nucleus in germinating spores of *Dryopteris erythrosora.* For all rhizoid lengths, the distance from the tip to the nucleus is 9–10 μm. (From Kato, 1957*b*.)

initial draws upon the stored reserves of the spore, but this is not an inexhaustible supply. Some experiments have implicated single salt ions supplied in the nutrient medium in rhizoid growth. When spores of *Pteris vittata* are germinated in distilled water, the rhizoid elongates poorly, but $CaCl_2$, $CaNO_3$, $MgSO_4$, KNO_3, and H_3BO_3 added singly promote its elongation (Kato, 1970*b*). The role of mineral ions in promoting growth of the rhizoid in germinating spores has been given a sharp focus in a recent study by Miller, Vogelmann and Bassel (1983). These investigators have proposed that in spores of *Onoclea sensibilis*, the perine serves not only in the traditional role as a protective covering but also as a source of ions which are released to sustain rhizoid elongation. The experimental basis for this hypothesis is the observation that spores divested of their perine germinate normally in glass-redistilled water but form short rhizoids. Rhizoids could be 'rescued' in such spores if they are germinated in the complete mineral salt medium or in a medium containing single salt ions of Ca^{2+}, Mg^{2+} or Mn^{2+}. Consistent with the view that the spore coat is an indispensible part of the picture as a reservoir of mineral ions is the observation that intact spores which retain their perine produce long rhizoids when germinated in distilled water. The possible utilization of Ca^{2+} and Mg^{2+} by the growing rhizoid is illustrated by visualization of these ions by chlorotetracycline fluorescence of the cell in spores stripped of their perine and cultured in the complete mineral salt medium (Knop's medium) or in a medium containing $Ca(NO_3)_2$ or $MgCl_2$ and in intact spores germinated in water (Fig. 4.10). These studies raise intriguing questions as to how universal is the cation requirement for rhizoid growth in germinating fern spores, how the stored ions are distributed to the cell and how they are involved in its growth.

A few other compounds are also of significance in promoting rhizoid elongation. These include GA (Kato, 1955), auxins (Sossountzov, 1957*c*), benzimidazole (Dyar and Shade, 1974) and ABA (Hickok, 1983). According to Miller and Miller (1963), rhizoid elongation in spores of *Onoclea sensibilis* is promoted by red light; the effect of red light is reversed by far-red light given immediately after red light indicating phytochrome action. One study (Haigh and Howard, 1973*a*) has shown that elongation of the rhizoid in germinating spores of *Osmunda regalis* is relatively insensitive to a dose of X-rays which kills the protonema initial if radiation is administered at anaphase–telophase or later stages of mitosis of the spore nucleus. It is necessary to note that the same dose of radiation applied at metaphase or earlier stages inhibits elongation of both rhizoid and protonema initials. The failure of two cells born out of a simple mitotic division to share radiation damage if irradiated at anaphase–telophase or later stages defies any simple explanation of X-ray effects on this system. A recent report (Huckaby, Bassel and Miller, 1982) of successful isolation of protoplasts from the rhizoid of *O. sensibilis* and cell wall regeneration in the isolated protoplasts provides an excellent opportunity to study the controlling factors in rhizoid elongation without interference from the protonema initial or the spore, and to demonstrate totipotency of the rhizoid.

Certain data on the growth of rhizoids on the female gametophyte of *Marsilea* have been difficult to fit into the picture described in the preceding paragraphs. In several species of *Marsilea*, rhizoids originate from a small group of cells at the base of the multicellular cushion of the female gametophyte. The location of rhizoids tends to be very regular at the lower side of horizontally placed stationary gametophytes while in

---

Figure 4.10. Chlorotetracycline fluorescence in the rhizoid of spores of *Onoclea sensibilis* germinated in different media. Fluorescent micrograph (*a–c*) and light micrograph (*a′–c′*) of the same fields are presented for comparison. (*a–a′*) Spores germinated in distilled water (no fluorescence seen). (*b–b′*) Spores germinated in Knop's solution. (*c–c′*) Spores germinated in a medium containing 100 mM Ca^{2+}. Scale bar in (*c′*) applies to all. (From Miller *et al.*, 1983; photographs supplied by Dr J. H. Miller.)

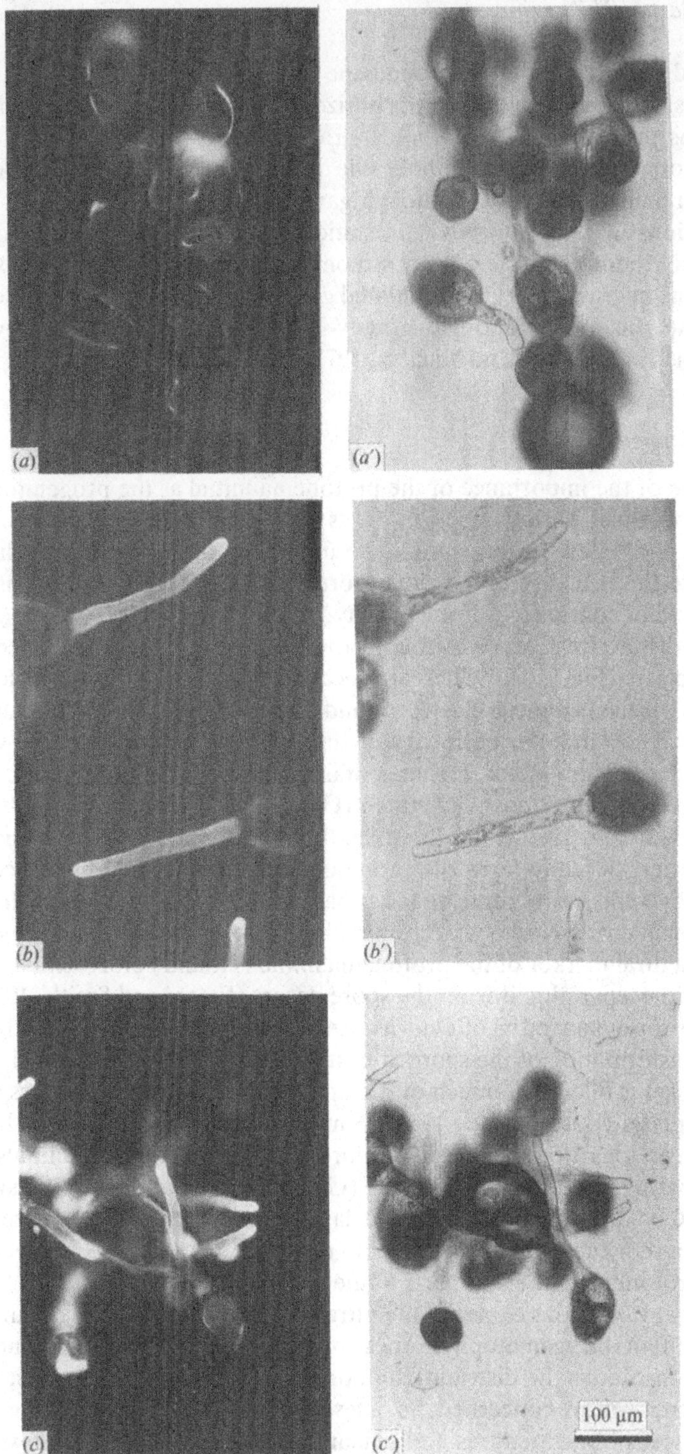
(a)
(a′)
(b)
(b′)
(c)
(c′)
100 µm

agitated cultures, cells all around the base of the gametophyte form rhizoids. Rhizoid initiation on the horizontally placed megagametophyte has thus the hallmarks of a typical gravitropic response, involving the migration of auxin from the upper side of the cushion to the lower side as an interesting possibility. Although gametophytes exposed to low concentrations of IAA and NAA in stationary cultures generate rhizoids from all around the base of the cushion, in the absence of data on auxin transport in gravitropically stimulated gametophytes, it is not possible to attribute rhizoid initiation to gravity-induced changes in auxin levels (Bloom, 1962; Bloom and Nichols, 1972*a, b*).

The protonema initial

Because of the importance of the protonema initial as the progenitor of the prothallus, various aspects of its growth and differentiation are treated separately in this book. The discussion that is developed here refers to the structure of the newly born protonema initial poised on the threshold of continued mitotic activity.

The distinctive feature of the protonema initial is the abundance of chloroplasts (Fig. 4.11). They are recognizable in an elliptical or lens-shaped profile in a loose cluster around the nucleus, becoming gradually dispersed towards the periphery of the cell. The chloroplasts are well differentiated into grana regions containing between 2 and 25 thylakoids and have limited deposits of starch (Gantt and Arnott, 1965; Gullvåg, 1968*b*; Beisvåg, 1970; Fraser and Smith, 1974; Cran, 1979). As pointed out earlier, in many ferns the protonema initial is formed by a second division of the spore after the latter has cut off the rhizoid initial. In an electron microscopic study of germinating spores of *Polypodium vulgare*, no structural marker of the protonema initial is found before actual wall formation separating it from the spore occurs (Fraser and Smith, 1974). In the protonema initial of *Onoclea sensibilis* which is formed directly by the transformation of the spore after it has cut off the rhizoid initial, the cytoplasm is filled with much of the lipid droplets, protein granules and chloroplasts (Bassel *et al.*, 1981). Similar ultrastructural features have also been described in cells purported to be protonema initials in germinating spores of other ferns (Gantt and Arnott, 1965; Beisvåg, 1970; Gullvåg, 1968*b*, 1971*b*), but lack of information on the precise division sequence makes it difficult to determine whether we are dealing with protonema initials or parts of the spore.

In many essential respects, the ultrastructure of a cell of the filamentous or plate-like gametophyte does not differ from that of the protonema initial, insofar as the distribution of organelles, types of organelles, and their structure are concerned. So, a few fine structural observations made on the cells of filamentous and planar gametophytes are of interest. In

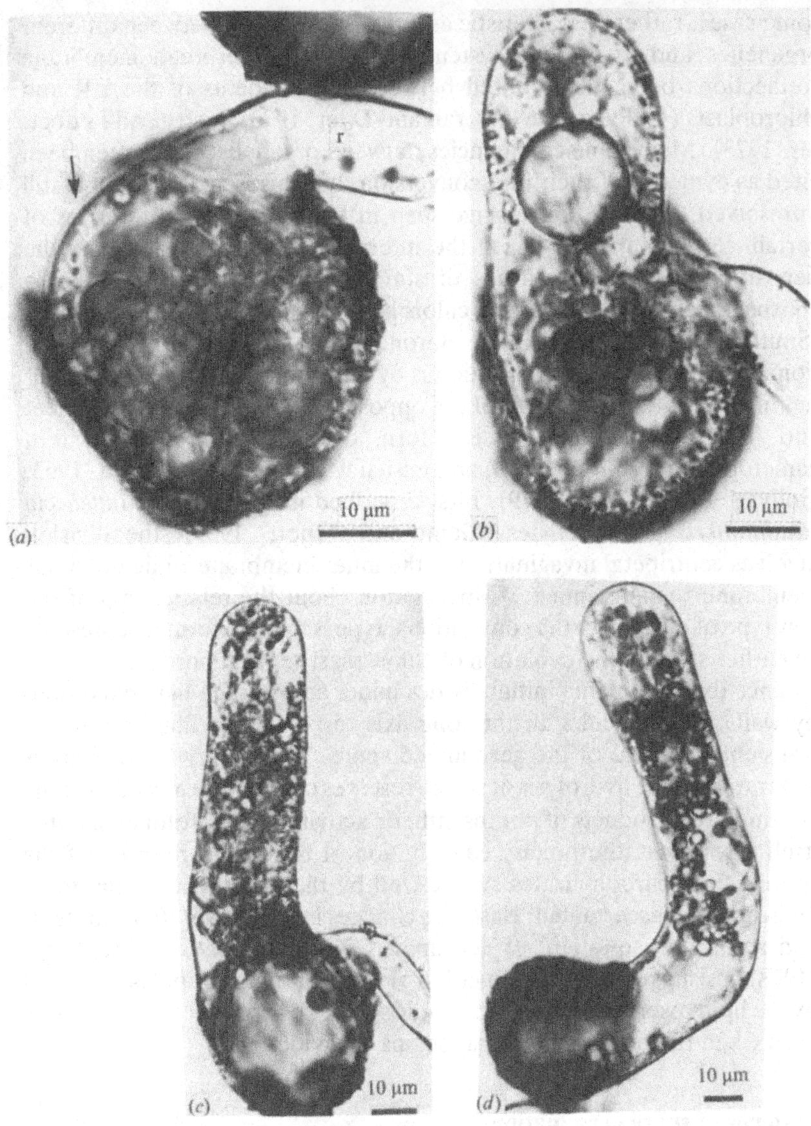

Figure 4.11. Formation of the protonema initial during germination of spores of *Pteridium aquilinum* in red light. (*a*) After 36 hours in red light. Arrow indicates the protonema initial emerging from the spore coats; r, rhizoid. (*b*) After 48 hours in red light. (*c*) After 72 hours in red light. (*d*) After 96 hours in red light. At this time, the protonema initial is poised to undergo the first transverse division.

some cases, rather characteristic associations are seen between different organelles and membrane systems, the most widespread membrane connections being those noted between the elements of the ER and chloroplasts (Gullvåg, 1968*b*; Cran and Dyer, 1973; Crotty and Ledbetter, 1973). Membrane confluencies between organelles have often been cited as evidence of their interconversion, but this is an issue that is still unresolved. A structural change seen in light-grown gametophytes of certain ferns is a reduction in the internal membrane system and the deposition of large quantities of starch. These changes represent a conspicuous remodeling of the chloroplast to give rise to an amyloplast (Smith and Smith, 1969; Faivre-Baron, 1977*a*). According to the traditional view, chloroplast replication is by the formation of a circumferential furrow which deepens until the opposing sides of the envelope meet and fuse. Besides this classical form of division, plastids of fern gametophytes also replicate in an unusual way (Gantt and Arnott, 1963; Gullvåg, 1968*b*; Cran, 1979). First described in the cells of *Matteuccia struthiopteris* gametophytes (Gantt and Arnott, 1963), the division involves centripetal invagination of the inner membrane while the outer membrane remains intact. A speculation about the relationship of the new type of division to the conventional type is that the former represents an earlier stage in the evolution of chloroplast replication.

Since the protonema initial divides under appropriate light conditions by walls perpendicular to the long axis to produce a filament, it is a transient structure of the germinated spore. The energy for division is generated by the hydrolysis of stored reserves of the spore as well as of the accumulated products of photosynthetic activity in the protonema initial itself. However, the precise contribution of the stored reserves of the spore *versus* carbohydrates synthesized by the protonema initial to its division is not ascertained. Based on changes in dry weight, lipid content, and respiratory quotient of germinated spores of *Pteris vittata*, Sugai (1968) has suggested that elongation of the protonema initial is sustained by the lipid reserves of the spore; photosynthetic byproducts are probably used when the protonema initial begins to divide.

Patterns of spore germination

Patterns of germination yielding the rhizoid and protonema initials have been described in spores of a number of homosporous ferns. A complete account of this work is beyond the scope of this chapter, but a brief orientation is provided as a basis for discussion. The first cell division occurs when the protoplasm is still enclosed within the spore coats and before any overt symptoms of germination appear. As described in members of Polypodiaceae, this division is asymmetric and is perpendicular to the polar axis of the spore. The small cell that is formed differentiates into the rhizoid initial, while the protonema initial is formed by another division of the spore perpendicular to the first. In spores of other ferns, there are variations in the sequence of wall formation as well as in the fate of the daughter cells formed (Nayar and Kaur, 1968). A generous portion of the past research on the cell division sequence during germination of spores of homosporous ferns has been based on observations made from whole mount preparations. In recent years, a new set of data on the patterns of germination of spores of members Parkeriaceae (Endress, 1974), Schizaeaceae (Raghavan, 1976; Rutter and Raghavan, 1978; Raghavan and Huckaby, 1980; Schraudolf, 1980*b*; Montardy-Pausader, 1982; von Aderkas and Raghavan, 1985; Nester, 1985), Cyatheaceae (Huckaby and Raghavan, 1981*b*), Thelypteridaceae (Huckaby and Raghavan, 1981*c*), Adiantaceae (Huckaby, Nagmani and Raghavan (1981), and Polypodiaceae (Nagmani and Raghavan, 1983) based on cleared materials and on sections of materials embedded in epoxy resins and glycol methacrylate has questioned the conclusions derived from whole mounts.

Naturally, for a given genus or for a group of allied genera, the relationship of one division to another is a carefully determined genetic trait. One consideration of this statement has been to classify the various known patterns of germination of spores of homosporous ferns. On the basis of the planes of cell division in relation to the polarity of the spore, and the direction of growth of the rhizoid and protonema initials, Nayar and Kaur (1968) have classified the germination patterns into 'Osmunda', 'Vittaria', 'Anemia', 'Gleichenia', 'Cyathea', 'Hymenophyllum', 'Trichomanes' and 'Mecodium' types. Briefly stated, while the first division giving rise to the rhizoid initial is perpendicular to the polar axis of the spore in both 'Osmunda' and 'Vittaria' types, it is the second division yielding the protonema initial that distinguishes them. In the 'Osmunda' type, this division is parallel to the first, while in the 'Vittaria' type it is perpendicular to the first. In the 'Anemia' type, a small proximal cell is born out of the first division of the spore perpendicular to its polar axis; the newly formed cell cuts off rhizoid and protonema initials by a wall perpendicular to the first. The rhizoid initial is formed by a wall parallel to the polar axis of the spore in both 'Gleichenia' and 'Cyathea' types, but the protonema initial originates by a wall perpendicular to the first wall in the 'Cyathea' type and parallel to the first wall in the 'Gleichenia' type. Two cell divisions by walls parallel to the polar axis of the spore giving rise to three cells characterize the 'Hymenophyllum' type of germination; in each of the three cells a lens-shaped cell which may grow into a rhizoid or a protonemal cell, is cut off. The feature that distinguishes the 'Trichomanes' type is the apparent absence of the first two divisions found in the 'Hymenophyllum' type so that the spore directly gives rise to three lens-shaped cells. In the 'Mecodium' type, not only do the first two divisions of the 'Hymenophyllum' type persist, but additional divisions occur in each of the three cells so that a plate of 9 to 12 cells is formed.

The germination patterns schematically depicted in Fig. 4.12 include 'Osmunda', 'Vittaria', and 'Gleichenia' types described by Nayar and Kaur (1968). These types have been confirmed in sections along with 'Ceratopteris', 'Lygodium', and 'Dicksonia' types which are also included here. In the 'Ceratopteris' type, the division sequences are identical to the 'Vittaria' type, but the fate of the daughter cells formed is reversed. The 'Lygodium' type exhibits a pattern in which the first division of the spore is perpendicular to its polar axis, followed by a division of the proximal cell by a wall parallel to the polar axis. This results in the formation of the protonema initial and an intermediate cell; the rhizoid initial is born out of a division of the intermediate cell which also forms a wedge-shaped cell lying between the rhizoid and protonema initials. The 'Anemia' type of Nayar and Kaur (1968) has been omitted since it has not been identified in sections and the 'Cyathea' type of these authors has been changed to the 'Dicksonia' type. 'Hymenophyllum', 'Trichomanes' and 'Mecodium' types described by Nayar and Kaur (1968) are included here for completeness, although they have not been confirmed in sections.

General comments

We have attempted here to set the stage for much that is to follow in the next few chapters. We have described the structural and physiological basis for spore polarity, have dissected the germinated spore into its component cells and have focused on their salient structural features. When a spore is primed to germinate, it passes from a stage where its properties are determined exclusively by its cellular inclusions to one where it is subject to influences of the external milieu. The environmental influences act perhaps not so much to control the pattern of germination as to enable the spore to express its innate genetic make-up. This becomes reasonably clear when we consider the morphogenetic processes that accompany spore germination. Germination initiates structural and physiological changes in the spore that carry it toward the first mitosis. This division seems to converge upon the primary objective of giving rise to cells vastly different from the spore from which they are formed.

The region of the spore from which a rhizoid initial is cut off during germination does not have any characteristic structural features, but it acquires some soon after the first division. The question is how a particular location of the spore is chosen for the outgrowth of the rhizoid initial, and what role do the acquired structural features play in the

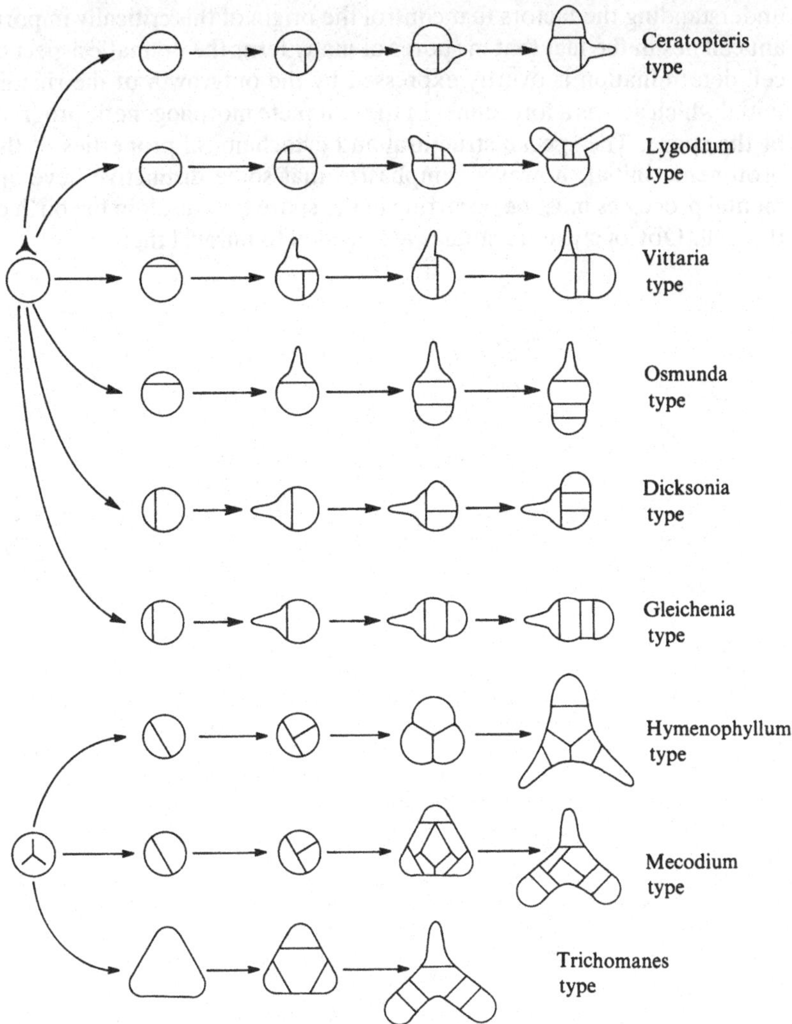

Figure 4.12. Diagram illustrating the different patterns of spore germination in homosporous ferns as seen in whole mount preparations and in sections. The triradiate mark on the spores at left indicates the position of the proximal pole in all spores in that group. Arrows show progress of germination. (Based on Nayar and Kaur, 1971.)

metamorphosis of the newly formed cell into a rhizoid with a limited growth potential. Now that we are beginning to get a handle on this problem, it is reasonable to expect advances in our basic knowledge of this simple form of cellular differentiation.

The origin of the protonema initial has not been a target of as much experimental inquiry as that of the rhizoid initial. The main difficulty in understanding the factors that control the origin of this critically important cell lies in the fact that in spores of many ferns the dramatic aspect of cell determination is overtly expressed by the outgrowth of the rhizoid initial which acts as a forerunner of the complete morphogenetic program of the spore. The special structural and cytochemical properties of the protonema initial, however, emphasize that some distinctive developmental processes must be occurring in the spore that result in the birth of this cell. Obviously more studies are needed to unravel them.

5

Biochemical cytology and biochemistry of germination

The potential of appropriately induced spores to germinate depends upon the sequential unfolding of a predetermined program of metabolic processes and regulatory mechanisms that control cell differentiation. Since germination of fern spores is accomplished in a closed system – that is, without the intake of any external nutrients, the synthetic processes must occur at the expense of the stored reserves. Thus, the first order of business of a germinating spore is hydrolysis of its storage reserves to simple compounds. Equally pressing is the need to replenish the structural proteins of the cytoplasm and the complement of enzymes required for general cellular metabolism. At the same time, a sustained biogenesis of membrane systems and organelles also occurs in the spore. Not long after these events, cell division ensues with all the complex biochemical processes it entails, including deoxyribonucleic acid (DNA) and protein synthesis and assembly of the mitotic spindle. Unlike the angiosperm seed, where major catabolic and anabolic activities during germination are segregated to morphologically different tissues, there is no conceivable division of labor between cells in the fern spore. Here, both degradative and synthetic processes preparatory to germination occur in the same cell.

This chapter is devoted to an analysis of the metabolic and synthetic activities during germination of spores of a selected number of ferns. This restriction in the number of species to be examined is of necessity due to the fact that our knowledge of the biochemical aspects of fern spore germination has been in general rather superficial and has lagged behind advances made in the study of seed germination. Hopefully, a few common principles applicable to spores of a large number of ferns will emerge from this account. The initial period of germination before the onset of the first cell division is of particular biochemical significance, because it is then that molecular events that determine the fate and functions of daughter cells presumably occur.

Histochemical changes

Light and electron microscopic histochemistry has contributed importantly to the study of metabolic changes during germination of fern spores and has thrown some light on the manner in which storage granules and their hydrolytic products are appropriated by daughter cells formed from a mitotic division of the spore nucleus. In the first application of histochemical methods to study cell metabolism during spore germination, it was found that although starch granules are completely absent in the dormant spores of *Matteuccia struthiopteris*, they appear in small numbers in spores imbibed in the dark and in great abundance after exposure to a dormancy-breaking light stimulus (Gantt and Arnott, 1965). Starch accumulation is later followed by extensive degradation of storage protein granules and by the time of the first mitosis, massive accumulation of starch and complete hydrolysis of protein granules are characteristically observed. The inverse relationship between starch synthesis and protein breakdown might in theory suggest a connection between the two, but this remains to be demonstrated. Nonetheless, the abundance of protein granules in the dormant spore and their hydrolysis during germination support the contention that they are the primary storage products which provide the raw materials for germination.

Another histochemical change associated with the germination of spores of *M. struthiopteris* is an increase in the concentration of cytoplasmic ribonucleic acid (RNA) surrounding the nucleus at the site of the presumptive rhizoid. RNA concentration remains high in the rhizoid initial soon after it is born, but as it begins to elongate, there is a decrease in RNA concentration with a concomitant increase in cytoplasmic proteins. Moreover, in contrast to the protonema initial with its rich endowment of starch, proteins and lipid granules, the rhizoid initial is almost impoverished of food reserves. These observations lead to the inference that the differentiation of rhizoid and protonema initials in the germinating spore is associated with gradients in the distribution of food reserves, RNA and proteins, but the steps that intervene between the appearance of macromolecules and the first observable sign of differentiation are not fully defined.

Although spores of *Anemia phyllitidis* contain substantial amounts of stored proteins and lipids, it is the histochemical transformation of the former that serves as a marker for GA-induced germination. In fully imbibed spores, the first sign of germination seen as early as 12 hours after addition of the hormone is hydrolysis of protein granules. This is initiated in the immediate vicinity of the nucleus so that with the progress of germination there is a gradient in the distribution of protein granules with respect to the nucleus (Raghavan, 1976). A sustained degradation of protein reserves is also seen in spores of *Pteris vittata* exposed to a saturating dose of red light (Raghavan, 1977*b*). Proteolysis is relatively rapid, so that 24 to 36 hours after red light exposure, most of the granules disappear leaving spores as empty shells criss-crossed by numerous cytoplasmic strands.

Electron microscopic examination of chlorophyllous spores of *Equisetum fluviatile* (Gullvag, 1968*a*, 1971*b*) has confirmed the general picture of breakdown of storage protein granules during the early phase of germination. A particularly striking ultrastructural aspect of disintegrating protein bodies is their close association with layers of membranous ER which generally surround them like a ring. A possible role for ER in the translocation of breakdown products of protein bodies to regions of high metabolic activity in the spore is suggested by this finding. According to Beisvåg (1970), in spores of *Blechnum spicant*, membrane-bound lipid bodies which form the main storage reserves with proteins break up into small granules during imbibition preparatory to germination. In spores of *Polypodium vulgare* (Fraser and Smith, 1974), *Dryopteris filix-mas* (DeMaggio, Greene, Unal and Stetler, 1979) and *A. phyllitidis* (Gemmrich, 1981), microbodies are found in close contact with the disintegrating lipid granules. Since similar associations of microbodies with lipids have been well characterized in germinating seeds, it has been suggested that microbodies of fern spores are akin to glyoxysomes that contain enzymes for the conversion of fatty acids to succinate. In spores of *E. arvense* and *E. fluviatile* microbodies are found close to chloroplasts suggesting that they are peroxisomes that harbor enzymes for the metabolism of glycolate (Gullvåg, 1971*a*). Microbodies found in the proximity of chloroplasts in the protonemal initial of *A. phyllitidis* also apparently function as peroxisomes (Gemmrich, 1981).

It is reasonable to assume that histochemical changes observed in germinating spores should be reflected in the presence of an array of enzymes, although the task of characterizing them has hardly begun. The only published works on enzyme histochemistry of germinating spores relate to acid phosphatase and catalase activities. Acid phosphatase activity in *M. struthiopteris* spore resides primarily in the protein bodies and in the ER which maintain direct contact with each other. Since high acid phosphatase titer coincides with the period of intense metabolic activity in the spore, it has been suggested that liberation of phosphate which is bound to the protein granules is one of the functions of the enzyme (Olsen and Gullvag, 1971). In spores of *E. arvense,* catalase activity is detected in the microbodies as well as in certain peculiar membrane-bound outgrowths of the chloroplasts (Olsen and Gullvåg, 1973).

It is notable that histochemical changes characterized during spore germination involve mostly breakdown of storage reserves. Does this reflect an inherent limitation of this kind of approach in unraveling the tangle of biochemical events? Perhaps so: as we shall see in the next section, there is more to the biochemistry of spore germination than metabolic adjustments focused on the hydrolysis of storage granules.

General metabolism of germinating spores

When the spore finds itself in a dry state after maturation, it shuts down a wide range of its metabolic activities. We can view this as a mechanism for surviving hard times, since the spore husbands its resources by engaging in a minimum of metabolic functions until germination is initiated. The corollary to this is that hydration of the spore followed by the administration of a germination stimulus results in a surge of biochemical and metabolic activities to ensure a smooth germination. These activities mostly include respiration, metabolism of storage products, and synthesis of nucleic acids and proteins.

Respiratory changes

Since spore germination involves energy-requiring reactions, it is to be expected that O_2 will be indispensible for the process. However, no data are available on the O_2 tension of the medium necessary to sustain germination and on the limits of O_2 availability within which spores of various species germinate. As mentioned briefly in Chapter 3, failure of spores of *Onoclea sensibilis* imbibed in an atmosphere enriched with N_2 to become photosensitive suggests the involvement of an oxidative phase during the pre-induction period. Germination of photoinduced spores also comes to a standstill in a N_2 atmosphere. The pre-induction and post-induction oxidative processes appear to be different in their recovery kinetics following residency in the N_2 milieu and by implication, in their ability to derive energy from anaerobic reactions for metabolic purposes (Towill and Ikuma, 1975*a*).

Exposure of spores of *O. sensibilis* to an inductive light stimulus increases the rate of respiration to 30% to 50% over that observed prior to irradiation. However, the accelerated activity is found to precede an enhanced carbohydrate metabolism. Surprisingly, adenosine triphosphate (ATP) content of spores shows a steady increase during dark imbibition, while following irradiation, it declines after a transient increase. These observations are consistent with the view that in germinating spores of *O. sensibilis* there is provision for ATP utilization in carbohydrate metabolism, particularly in the synthesis of starch. Clearly, the succession of events during spore germination is intimately controlled by a chain of oxidative reactions, but it is difficult to attribute the effect of light to any particular segment of the reaction chain. It is possible that the same activated pigment independently induces multiple reactions such as respiratory surge and carbohydrate breakdown and resynthesis (Towill and Ikuma, 1975*c*).

As shown by Reynolds (1982), there is an apparent shift in respiratory pathways during the early phase of germination of spores of *Sphaeropteris cooperi* (Cyatheaceae). Evidence for this comes from the observation that O_2 uptake during the first 120 hours of germination is insensitive to KCN, but is inhibited by salicyl-hydroxamic acid (SHAM); after this time, germination becomes KCN-sensitive and SHAM-insensitive. Exposure of spores to an elevated temperature during the cyanide-resistant phase appears to impart sensitivity to the drug, but inhibits germination. From these observations a cytochrome pathway does not seem to be operating in the germination process; direct proof for the existence of an alternate respiratory pathway that is integrated in some fashion with the overall metabolism of the germinating spore, is however lacking.

It is evident from the foregoing that tremendous advances made in the field of the biochemistry of respiration have not caught up with fern spores. For example, we do not have any information regarding the electron transport systems operating during respiration of spores. Equally large is the void relating to mitochondrial properties and function during spore germination. These investigations remain as important tasks in the years ahead and their impact on our understanding of the metabolism of germinating spores might be decisive.

Fate of storage reserves

Histochemical studies reviewed earlier have shown that the most conspicuous feature of spore germination is the disappearance of storage reserves. However, because of differences in the type of storage materials, the timing and pattern of their disappearance and the associated metabolic events are not the same in spores of different species. For instance, in dark-germinating spores of *Pteridium aquilinum*, starting about 12 hours after sowing, there is a decrease in protein content for the next 36 hours. During this period, appearance of the rhizoid initial is the primary morphogenetic event. From a period of 72 to 96 hours after sowing when the protonema initial is formed, a small increase in protein content is registered (Raghavan, 1970). Protein content of photoinduced spores of *Onoclea sensibilis* also decreases with the progress of germination, while in dark-imbibed spores, it remains surprisingly constant. Concomitant with the disappearance of proteins, the concentration of free amino acids increases. A similar relationship is seen between hydrolysis of endogenous sucrose and starch accumulation, with the difference that low levels of sucrose hydrolysis and starch synthesis occur in spores during dark imbibition (Towill and Ikuma, 1975*c*; Wiebe, Towill and Campbell-Domsky, 1987).

Investigations into the compositional changes in photoinduced spores of *Adiantum capillus-veneris* have reinforced the idea that hydrolysis of proteins and sugars and synthesis of starch are common biochemical landmarks of germination. However, changes in the activity of hydrolytic enzymes concerned with the breakdown of proteins and sugars do not always parallel the observed decrease in the amounts of the latter compounds (Minamikawa, Koshiba and Wada, 1984; Koshiba, Minamikawa and Wada, 1984). According to Cohen and DeMaggio (1986), temporal changes during germination of spores of *Matteuccia struthiopteris* correlate well with the breakdown and mobilization of storage protein reserves, activity of proteases and increase in the levels of free amino acids.

In spores of *Gymnogramme* (=*Pityrogramma*) *sulphurea* (Guervin, 1972), *Polypodium vulgare* (Robinson, Smith, Safford and Nichols, 1973), and *Anemia phyllitidis* (Gemmrich, 1977*a*) which contain abundant lipophilic materials, germination is accompanied by the breakdown of endogenous tri-glycerides and a resynthesis of several classes of di- and tri-glycerides. *De novo* synthesis of new tri-glycerides following the breakdown of reserve lipids has been established in the spores of *P. vulgare* by incorporation of labeled acetate predominantly into neutral lipid fractions. These newly synthesized lipids are the same ones found in mature sporophyte fronds, suggesting the presence of enzyme systems for the fabrication of sporophytic type of lipids even in the very young gametophyte (Robinson *et al.*, 1973).

In general, products of lipid breakdown end up as constituents of the carbohydrate pool of the cell. In fern spores, the mechanism that oversees lipid degradation and conversion of the byproducts to carbohydrates has proved to be remarkably similar to that described in seeds. There are several experiments which show that in seeds the initial steps in lipid degradation are mediated by lipases and that the terminal phase of these reactions leads to the production of glycerol and fatty acids. The latter are in turn oxidized to acetyl-coenzyme A which serves as a primer for the glyoxylate cycle resulting in sucrose as one of the end products. It is well-known that in seeds, enzymes for lipid degradation are compartmentalized in specialized microbodies known as glyoxysomes. As photosynthetic systems are activated, there is a gradual loss of glyoxysomes, a decrease in glyoxylate enzyme activity and the concurrent appearance of structurally similar microbodies called peroxisomes harboring the enzymes of glycolate metabolism. Since isocitrate lyase (ICL) and malate synthase (MS) are key enzymes of the glyoxylate cycle, it may be useful to examine the evidence that bears on their activity as an indication of lipid degradation during fern spore germination. Although previous histochemical and electron microscopic studies have assumed the existence of lipid-degrading enzymes in fern spores, Gemmrich

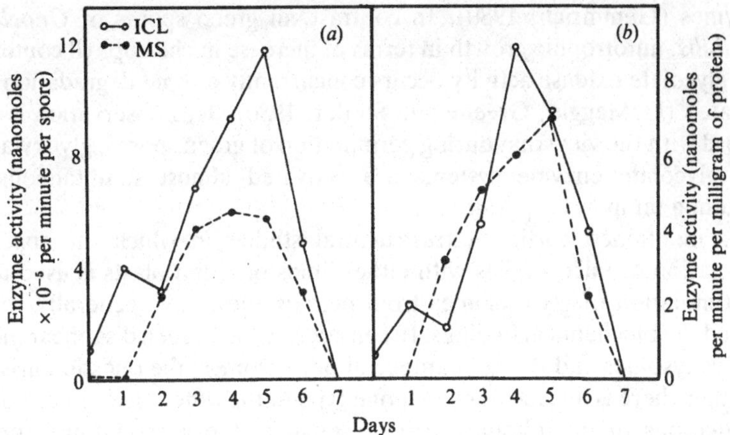

Figure 5.1. Time course of changes in the activities of isocitrate lyase (ICL) and malate synthase (MS) during germination of spores of *Dryopteris filix-mas*. Results are expressed as (*a*) activity per spore and (*b*) activity per mg protein. (From DeMaggio *et al.*, 1979.)

(1979*b*) and DeMaggio *et al.* (1979) provided conclusive evidence for the involvement of glyoxylate cycle enzymes in the germination of nongreen spores of *A. phyllitidis* and *Dryopteris filix-mas,* respectively. In dry spores of both species, enzyme activity is barely detectable; however, during the first few days of germination, storage lipids are degraded rapidly to suggest as though the spore concentrates on catabolic activity to the neglect of synthetic processes. The period of lipid degradation is found to neatly coincide with the development of ICL activity and by the time enzyme activity disappears, spores are completely depleted of their lipid reserves. ICL activity appears on schedule in *A. phyllitidis* spores, irrespective of whether they are potentiated to germinate by red light or by GA, but a more active lipid breakdown occurs in red-potentiated spores (Gemmrich, 1982). As seen in Fig. 5.1, the activity of MS during germination of spores of *D. filix-mas* is strikingly similar to that of ICL (DeMaggio *et al.*, 1979).

In the germination of nongreen spores there are two fundamental plans of enzyme activity which coincide roughly with the heterotrophic and autotrophic phases of germination. During the heterotrophic phase, the spore utilizes energy from the catabolism of storage reserves while autotrophic activity beginning with the biogenesis of the first few plastids in the protonema initial marks a sharp turning point in enzyme activities. The main outline of enzyme changes during germination of spores of *Pteris vittata* indicates that the beginning of photosynthetic activity is signaled by an increase in activity of hydroxypyruvate reductase, a typical peroxisomal enzyme and by a gradual disappearance of glyoxylate cycle enzymes (Gemmrich, 1980). In contrast, in green spores of *Onoclea sensibilis*, autotrophic growth in terms of increase in chlorophyll content and glycolate oxidase activity occurs concurrently as lipid degradation is initiated (DeMaggio, Greene and Stetler, 1980). This observation is in accord with the view that during germination of green spores, glyoxylate and glycolate enzyme systems are activated almost simultaneously without a temporal separation.

As mentioned earlier, ultrastructural studies in which the spatial association of microbodies with either lipids or chloroplasts is used for discrimination of glyoxysomes from peroxisomes, have generally confirmed the biochemical findings. In considering the sudden disappearance of glyoxysomes and the appearance of peroxisomes, the question arises whether there is a transition from one type to the other or whether both populations of microbodies arise *de novo* in the germinating spore. Ultrastructural observations of glyoxysomes and peroxisomes along with the general features of spore germination have been woven into a discussion by Gemmrich (1981) to suggest that both groups of microbodies arise *de novo*, but this view cannot be considered as anything but tentative.

An implicit assumption in the investigations described above is that catabolic and anabolic functions of germination are initiated in the spore only after administration of a germination stimulus. Later studies have shown that this is not so. For example, in spores of *A. phyllitidis*, lipid synthesis as monitored by ^{14}C-acetate incorporation is set in motion soon after imbibition, even in the absence of a germination stimulus (Gemmrich, 1979*a*). The picture is just what one would expect if enzymes for lipid synthesis pre-exist in the dry spore. Moreover, several hydrolytic enzymes as well as enzymes involved in the interconversion of amino acids and conversion of amino acids and hydroxy acids into oxo acids have also been shown to be present in dry spores of certain other ferns (Lever, 1971; Gemmrich, 1975; Koshiba *et al.*, 1984; Cohen and DeMaggio, 1986). In fully imbibed spores of *O. sensibilis*, in which starch synthesis occurs even in the absence of photoinduction, the presence, before imbibition, of functional starch-synthesizing and sucrose-degrading enzymes appears likely (Towill, 1980; Wiebe *et al.*, 1987). If enzymes mediating in macromolecule synthesis and degradation are in fact stored in the dry spore, it will be a remarkable demonstration of their conservation in a single cell. But we are left with a key question: what role (if any) do these biosynthetic activities play in the well-being of the dormant spore if they do not lead to germination?

Nucleic acid and protein metabolism during germination

It is a widely held notion that cell differentiation is modulated by the transcription of specific information carrying mRNA molecules and their translation into proteins. For this reason, an analysis of nucleic acid and protein metabolism during spore germination is germane for our discussion insofar as one can determine from this information the extent to which cell morphogenesis is inextricably coupled to changing patterns of macromolecule synthesis and degradation. As DNA is the well-known genetic material of the cell capable of replicating itself and being expressed as protein, this account will logically begin with a consideration of DNA synthesis during germination.

DNA synthesis

When fully imbibed spores are potentiated to germinate, in addition to respiratory activity and hydrolysis of storage reserves, events that culminate in the division of the spore nucleus are also started. The nucleus of the dry spore consists of a mass of chromatin that does not show any granularity. Only after imbibition has commenced does the nucleus become granular and come to resemble a normal nucleus in its morphology. As studied by autoradiography of ^{3}H-thymidine incorporation (Howard and Haigh, 1970; Raghavan, 1976, 1977*b*; Fisher and Miller, 1978; Rutter and Raghavan, 1978), the essential condition for the division of the spore nucleus is DNA synthesis appropriate for the haploid genome. Generally, soaking nondormant spores or administration of a germination stimulus to hydrated dormant spores commits them to DNA replication and division.

One striking feature of fern spore germination is the occurrence of extranuclear DNA synthesis. This was first demonstrated in *Anemia phyllitidis* in which DNA is synthesized both in the nucleus and the immediately surrounding cytoplasm of the undivided spore. Cytoplasmic DNA synthesis continues for several hours in the daughter cell born out of the first mitosis, but as this cell elongates, a strictly nuclear incorporation of ^{3}H-thymidine is observed (Raghavan, 1976, 1977*a*). In spores of *Pteris vittata*, cytoplasmic DNA synthesis is initiated after the S phase of nuclear DNA synthesis is under way and remains quite vigorous even after the latter ceases following the formation of the rhizoid initial. With the emergence of the protonema initial, there is no further ^{3}H-thymidine incorporation in the spore cytoplasm (Raghavan, 1977*b*). Spores of *Lygodium japonicum* (Rutter and Raghavan, 1978) and *Onoclea sensibilis* (Raghavan, unpublished) also display cytoplasmic DNA synthesis during germination. In the latter, the pattern of DNA synthesis is especially intriguing. Within four hours after exposure of spores to red light, a modest level of ^{3}H-thymidine incorporation found exclusively in the cytoplasm attains a peak at about 16 hours; subsequently, when the nucleus begins to migrate to the proximal end of the spore, a strictly nuclear labeling of ^{3}H-thymidine occurs (Fig. 5.2). Considering the biochemical aspect of the problem, we must assume that each species

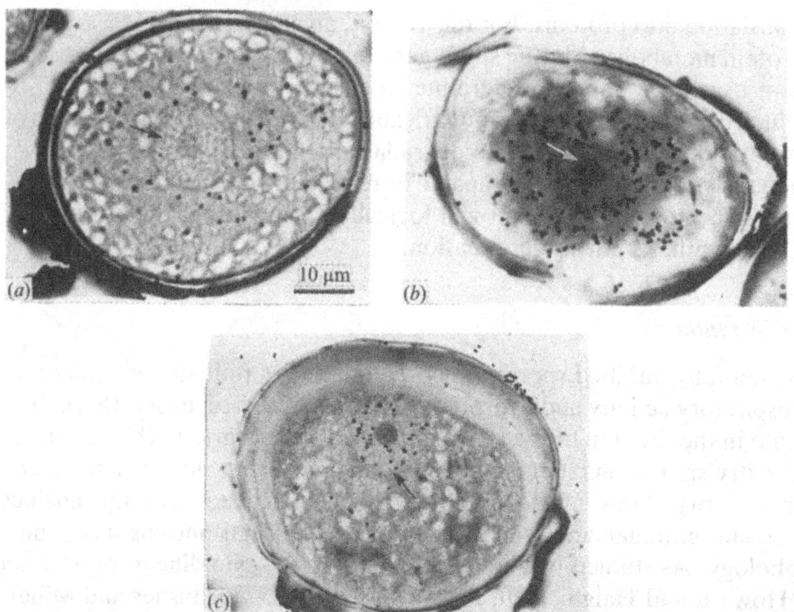

Figure 5.2. Autoradiography of ^{3}H-thymidine incorporation in germinating spores of *Onoclea sensibilis* showing cytoplasmic and nuclear DNA synthesis. (*a*) Section of a spore showing the beginning of cytoplasmic DNA synthesis, 8 hours after exposure to red light. (*b*) Section of a spore showing intense cytoplasmic DNA synthesis 18 hours after irradiation. (*c*) Section of a spore showing nuclear DNA synthesis, 36 hours after irradiation. Arrows point to the nucleus. Scale bar in (*a*) applies to all.

utilizes local differences in cell metabolism to activate cytoplasmic DNA synthesis during germination. More work is needed before it can be determined whether cytoplasmic DNA synthesis occurs in the plastids or in the mitochondria.

There are indirect indications, based on the use of inhibitors, that DNA synthesis is not a prerequisite for germination of fern spores. Hydroxyurea is an inhibitor of DNA synthesis which blocks the action of ribonucleotide reductase, an enzyme that converts ribonucleotides to deoxyribonucleotides. A remarkable feature of GA-induced germination of spores of *A. phyllitidis* is that hydrolysis of storage reserves, breakage of the exine, and enlargement of the spore simulating germination occur even when nuclear and cytoplasmic DNA synthesis is inhibited by hydroxyurea (Raghavan, 1977*a*). Aphidicolin, a reversible inhibitor of DNA polymerase action, is even less effective than hydroxyurea in inhibiting germination of *A. phyllitidis* spores, since the appearance of a protonema initial is the norm in spores in which nuclear DNA replication is inhibited by the drug (Fechner and Schraudolf, 1986). From these results one gains the impression that a role for GA at the level of DNA synthesis during germination of spores of *A. phyllitidis* is unlikely.

RNA synthesis

RNA synthesis not only accompanies spore germination, it is necessary for completion of this event. There are, to be sure, many aspects of this problem such as timing and regulation of RNA synthesis, role of mRNA synthesis and so on, that are of concern here.

RNA synthesis is initiated in dormant spores several hours after the administration of a germination stimulus. In autoradiographs of spores of *A. phyllitidis* fed with ^{3}H-uridine, a precursor of RNA synthesis, the first silver grains are seen in about 36 hours after the addition of GA and thereafter the magnitude of incorporation increases with the progress of germination (Fig. 5.3). RNA synthesis continues at a reduced rate in the original spore cell, the rhizoid and the basal cells of the protonemal filament even after they cease to divide (Raghavan, 1976). The ribosomes thus generated permit protein synthesis in these cells necessary for housekeeping functions. Although RNA synthetic system is totally inactive during imbibition of dormant spores of most species examined, in spores of *Pteris vittata,* ^{3}H-adenosine incorporation into low molecular weight RNA occurs as early as two hours after soaking (Zilberstein, Gressel, Arzee and Edelman, 1984).

We do not find a common pattern of RNA synthesis during spore germination, and only superficially can we correlate the cytological events of germination with the synthesis of certain species of RNA. When spores of *Pteridium aquilinum* are pulse-labeled with ^{3}H-uridine at different times after sowing, precursor incorporation into RNA is found to increase slowly beginning about 36 hours until the end of the time course (84 hours). Sucrose-gradient centrifugation analysis of labeled RNA has revealed that while the period of rhizoid initiation (36 to 48 hours after sowing) in the spore is characterized by the transcription of 4S and lighter species of RNA, during rhizoid elongation (48 to 60 hours after sowing) there is an accelerated transcription of all types of RNA including 23S and 16S species (Raghavan, 1970). In spores of *Pteris vittata*, there is a burst of ^{3}H-adenosine incorporation into low molecular weight RNA soon after photoinduction, followed later by accelerated ribosomal RNA (rRNA) synthesis, but these activities are registered long before any overt signs of germination occur (Zilberstein *et al.*, 1984).

One approach to identify the species of nuclear RNA involved in spore germination is to use actinomycin D; if a stage of germination proceeds in the presence of this inhibitor, the assumption is made that the event studied is probably independent of mRNA biosynthesis. When spores of

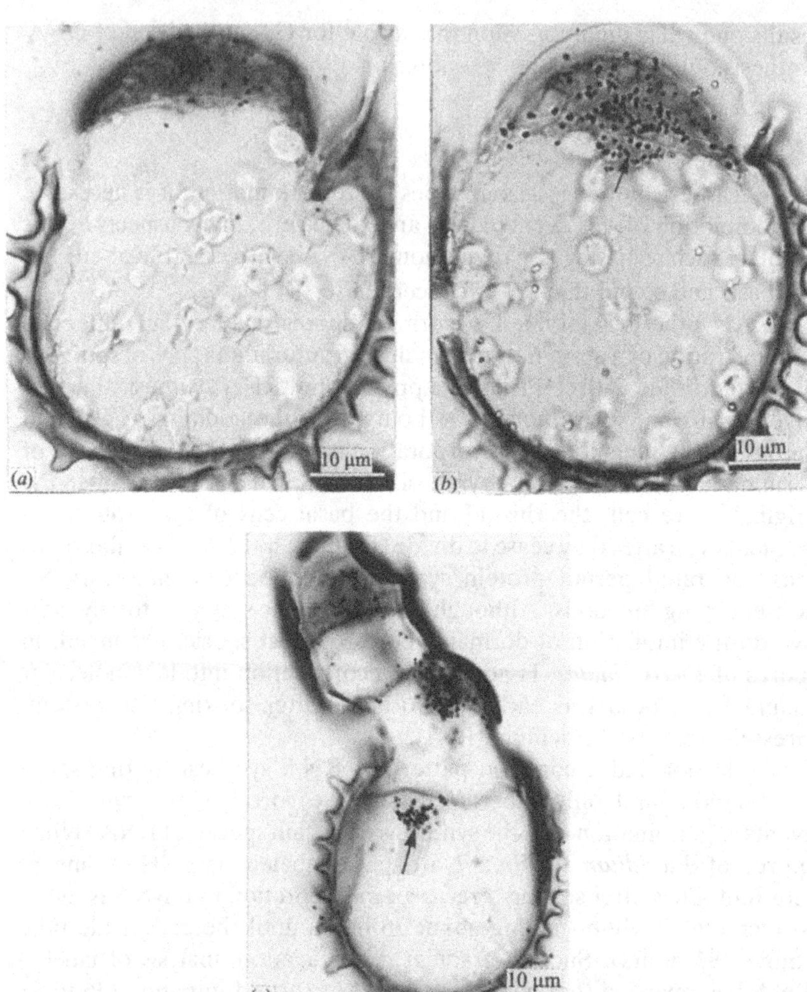

Figure 5.3. Changing patterns of RNA synthesis during germination of spores of *Anemia phyllitidis* as studied by autoradiography of ^{3}H-uridine incorporation. (*a*) Section of an undivided spore showing the label in the nucleus. (*b*) Section of a germinated spore showing nuclear and cytoplasmic incorporation of the label in the daughter cell. Arrow points to the nucleus of the original spore. (*c*) Section of a germinated spore with a filamentous protonema showing incorporation of the label in the nucleus of the original spore (arrow) and in both nucleus and cytoplasm of the protonemal cells.

P. aquilinum are incubated in high concentrations of the drug, elongation of the rhizoid is completely blocked, but the exine is found to rupture at the site of the presumptive rhizoid indicating that some events prior to rhizoid elongation are unaffected by the drug (Raghavan, 1970). Effects of actinomycin D on the initiation and growth of the protonema initial during germination of spores closely parallel those on rhizoid initiation and growth (Raghavan, 1971*b*). In a similar way, normal germination accompanied by a general inhibition of growth of the gametophyte is reported to occur when spores of *Equisetum arvense* and *Osmunda japonica* are treated with high concentrations of actinomycin D (Nakazawa and Tanno, 1967). These observations reinforce the thought that concomitant synthesis of mRNA is not necessary for the induction of germination which occurs even when mRNA synthesis is inhibited.

Based on the work reviewed above, it is generally assumed that the early events of germination, however limited they are structurally, cytologically, and biochemically, are programmed, operated and controlled by informational macromolecules present in the spore at the time of sowing. Biochemical demonstrations of the existence of mRNA in the dry spore and its translation in a cell-free system have been important recent developments that lend credence to this view. Since most mRNAs of plant, animal and viral origin possess a sequence of polyadenylic acid [poly(A)] tracts at the 3′ end, poly(A)-containing RNA [poly(A) + RNA] isolated from polysomal preparations is generally considered as mRNA. Because adenine is naturally complementary to uracil and thymine, a convenient technique to isolate poly(A) + RNA from biochemical preparations is by affinity chromatography on polyuridylate or oligo-deoxythymidine immobilized on a solid support material like Sepharose or cellulose. Included in the list of spores from which mRNA has been isolated by this method are those of *Anemia phyllitidis* (Fechner and Schraudolf, 1982), *Pteris vittata* (Zilberstein *et al.*, 1984), *Cyathea spinulosa* (Miura, Koshiba and Minamikawa, 1986), and *Onoclea sensibilis* (Raghavan, 1987). Poly (A) + RNA has also been localized in the nucleus and cytoplasm of the dry spore of *O. sensibilis* by *in situ* hybridization using ^{3}H-polyuridylic acid as a probe.

Evidence has been provided for the coding function of mRNAs isolated from dry spores of *A. phyllitidis* (Fechner and Schraudolf, 1984) and *C. spinulosa* (Miura *et al.*, 1986) by cell-free (*in vitro*) translation using rabbit reticulocyte lysate and wheat germ extract, respectively. Messengers isolated from dry spores of *A. phyllitidis* apparently direct the synthesis of a spectrum of proteins which qualitatively resemble those synthesized *in vitro* by mRNA from spores primed to germinate, and *in vivo* by noninduced, imbibed spores. Differences in gene expression between dry and imbibed or induced spores, if they exist, are thus too subtle to be detected by cell-free translation, but the point is that, like fertilized animal eggs and seeds of angiosperms and gymnosperms, mature fern spores also come equipped with stored, template-active mRNA. As we will see later in this section, the extent to which the spore relies on stored mRNA and newly synthesized mRNA to trigger germination is a moot question.

This leads us to a consideration of the changes in mRNA population

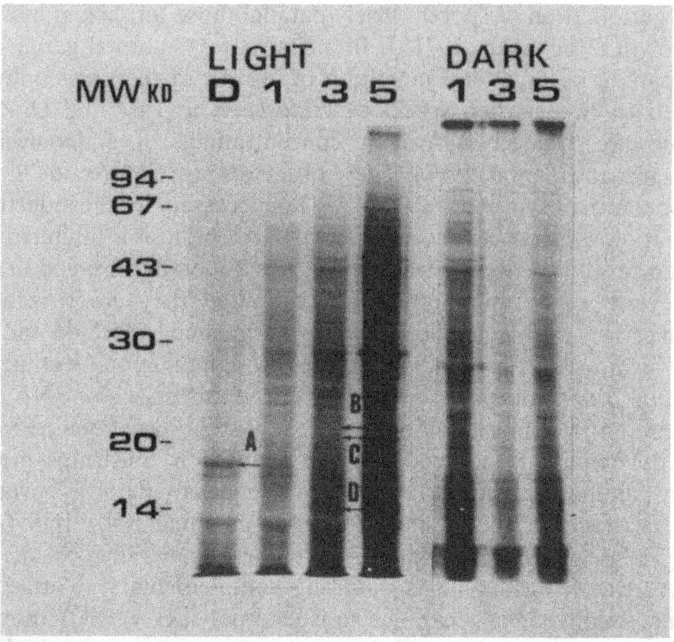

Figure 5.4. Fluorographs showing the electrophoretic separation of cell-free translation products of RNA preparations from spores of *Cyathea spinulosa*. The translation products were labeled with ^{35}S-methionine. Lanes (from left to right) represent dry spores, spores cultured in continuous light for 1, 3 and 5 days, and spores cultured in the dark for 1, 3 and 5 days. A, indicates a band present in dry spores and day 1 spores cultured in light. B, C, and D are bands which appear in day 3 and day 5 spores cultured in light. (From Miura *et al.*, 1986; photograph supplied by Dr M. Miura.)

during the germination of spores. As alluded to earlier, a measurable level of RNA synthesis occurs in imbibed spores of *P. vittata*; despite this, mRNA with *in vitro* template activity does not appear in spores until 16 hours after photoinduction (Zilberstein *et al.*, 1984). Poly(A) + RNA synthesis is initiated in spores of *C. spinulosa* after germination is under way; its template activity increases with the progress of germination of photoinduced spores, while in noninduced spores, following an initial surge of template activity, no further activity appears (Fig. 5.4). There are few qualitative changes in the profile of *in vitro* translation products; among the changes noted are the disappearance of a band corresponding to stored mRNA of the dry spore and appearance of three new bands characteristic of mRNA of differentiating cells of the germinating spore (Miura *et al.*, 1986). Biochemical analysis and *in situ* hybridization mapping have shown that germination of spores of *O. sensibilis* is accompanied by a decline in poly(A) + RNA content (Raghavan, 1987).

Further advancement in our understanding of RNA metabolism during spore germination requires novel approaches such as the use of cloned sequences to probe specific messenger transcripts. Obviously, there should be quantitative and qualitative changes in specific mRNA populations during germination. Isolation and characterization of these messages, especially those unique to spores, is necessary to understand their developmental regulation and function during spore germination.

Protein synthesis

As germination is initiated, a pattern of protein synthesis takes shape in the spore. The changes in protein synthetic activity have been monitored by autoradiographic, biochemical, and electrophoretic analyses of spores at various stages of germination. Autoradiography indicates that ^{3}H-leucine, a precursor of protein synthesis, is incorporated into both nucleus and cytoplasm of undivided spores of *Anemia phyllitidis* (Raghavan, 1977*a*) and *Pteris vittata* (Raghavan, 1977*b*). Later, as the nucleus migrates to the proximal pole of the spore establishing the future site of the first cell, protein synthesis is particularly heavy in the cytoplasm in this region (Fig. 5.5). It thus seems that there is a division of labor between the presumptive site of the first cell and the rest of the spore with regard to biochemical activities.

When spores of *Pteridium aquilinum* are pulsed at different times after sowing with ^{14}C-leucine, there is no detectable incorporation of the label into proteins until about 36 hours when breakage of the exine and protrusion of the rhizoid ensue; thereafter there is a steady increase in the amount of precursor incorporated into proteins (Raghavan, 1970). Essentially the same pattern of protein synthesis prevails during germination of spores of *Pteris vittata* although the morphological events are not circumscribed so clearly (Zilberstein *et al.*, 1984). The necessity for protein synthesis for germination of *P. aquilinum* spores has been confirmed by treatment with cycloheximide, which interferes with the initiation and formation of nascent polypeptides on ribosomes. Exposure of spores to cycloheximide at different times after sowing has indicated that proteins for rhizoid initiation are synthesized during a sensitive period beginning about 26 hours after sowing. In this experiment it was found that introduction of the drug any time up to 23 hours inhibits rhizoid initiation in the entire population of spores. If the drug is applied between 26 to 41 hours after sowing, 15 to 40% of spores form visible rhizoid initials, while addition of the drug at 44 hours or later after sowing has little effect on the percentage of spores with visible rhizoid initials (Table 5.1). Elongation of the rhizoid, however, depends upon the continued synthesis of proteins, as exposure of spores with a rhizoid initial to cycloheximide inhibits its subsequent elongation (Raghavan, 1970). Treatment with cycloheximide of spores poised to form a protonema initial prevents this morphogenetic event also (Raghavan, 1971*b*).

Table 5.1 *Effects of addition of cycloheximide at different times on the initiation and growth of the rhizoid during germination of spores of* Pteridium aquilinum[a]

Time after sowing[b] (hours)	Percentage of spores with rhizoid or with ruptured exine	Length of the rhizoid, µm ± standard error
16	0	—
17	0	—
19	0	—
23	10.4	11.0 ± 1.1
26	15.3	11.0 ± 1.3
36	20.5	17.8 ± 1.8
41	41.8	24.7 ± 2.1
44	66.0	34.2 ± 2.6
48	86.4	41.4 ± 2.6
60	84.5	45.2 ± 3.3
Control	86.4	65.8 ± 3.3

[a] From Raghavan (1970).
[b] Red light was administered for 3 h at 12 h after sowing. Measurements were made 72 h after sowing.

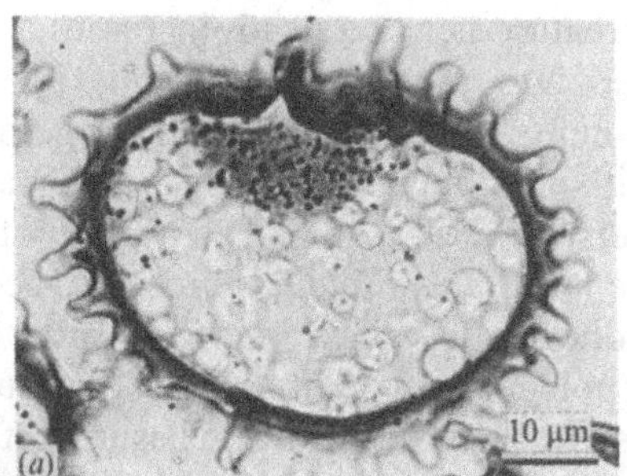

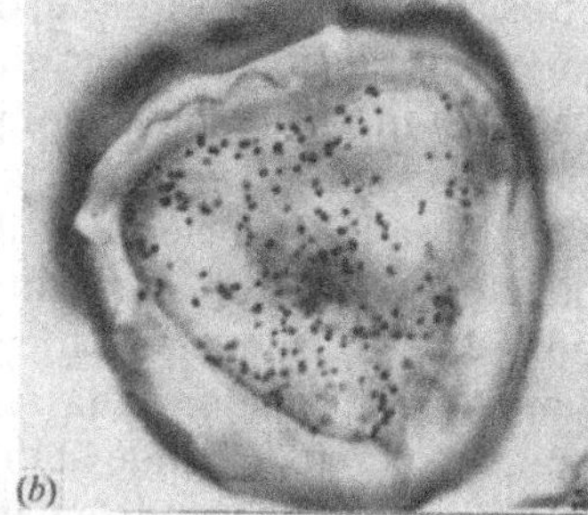

Figure 5.5. Autoradiography of incorporation of ^{3}H-leucine in germinating spores. (*a*) Section of an undivided spore of *Anemia phyllitidis* showing incorporation of the label in the nucleus and the surrounding cytoplasm. (*b*) Section of an undivided spore of *Pteris vittata* showing nuclear and cytoplasmic incorporation of the label. Scale bar in (*a*) applies to both.

The problem of protein synthesis during photoinduced germination of spores of *Onoclea sensibilis* is complicated by the requirement for continued enzyme activity for the maintenance of sensitivity to irradiation (Towill and Ikuma, 1975*b*). This is seen from the observation that exposure of spores to cycloheximide for 30 to 45 min during the pre-induction phase followed by photoinduction, or during the early post-induction phase, is sufficient to cause the same extent of inhibition of germination. The similarity in the kinetics of recovery of spores from pre-induction and post-induction cycloheximide treatments has led to the suggestion that the same enzymes with half-lives of 30 to 45 min are required for both phases of germination. When spores are treated with cycloheximide at different times after irradiation, escape from inhibition yields a biphasic curve with a sharp deflection at 10 hours after irradiation. Taken together, these interesting results are in accord with a model in which a short-lived enzyme initiates germination processes in the spore as early as 10 hours after photoinduction. One recurrent feature of germination of fern spores is a lag period between initiation of enzyme activity and detectable protein synthesis.

As expected, electrophoretic analyses have revealed some shifts in the profile of extant and newly synthesized proteins during spore germination. This is illustrated in Fig. 5.6 which is a diagrammatic rendition of

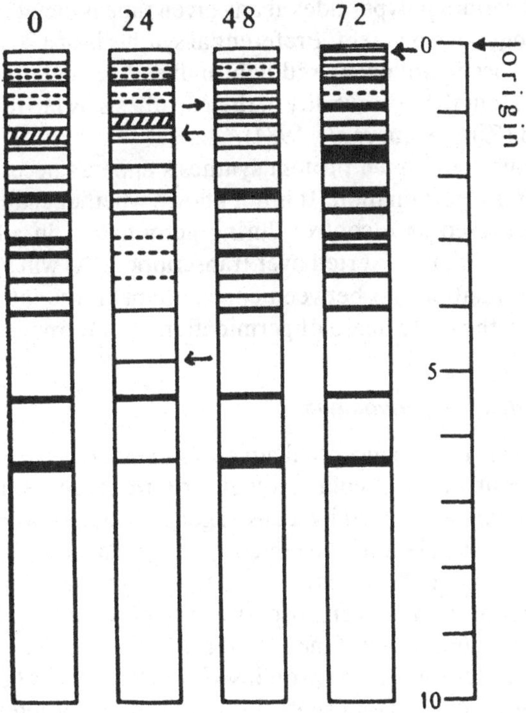

Figure 5.6. Electrophoretic separation of proteins from spores of *Pteridium aquilinum* during germination. Lanes from left to right represent dry spores, and spores 24, 48 and 72 hours after sowing. New polypeptides that appear during germination are indicated by arrows. Intensity of bands; ----, light; ▨, diffuse, and ■, dark. (From DeMaggio and Raghavan, 1972.)

densitometric tracings of stained gels of soluble proteins of *Pteridium aquilinum* spores. In addition to the protein bands common to the dry spore and germinated spore, note the bands indicated by arrows that appear at some point during germination (DeMaggio and Raghavan, 1972). By two-dimensional gel electrophoresis, Huckaby and Miller (1984) have identified more than 500 spots of soluble proteins from spores of *O. sensibilis*; nearly 25% of the spots change during germination (Fig. 5.7). The characteristics of these changes are consistent with the idea that germination represents a concerted activity involving many proteins, possibly orchestrated by differential organelle distribution as well as by differential protein synthesis. In keeping with phytochrome control of germination, the synthesis of a 42 kd and a 36 kd polypeptide and the activity of a pyridine-linked dehydrogenase, isocitrate dehydrogenase, are stimulated by exposure of spores of *Dryopteris filix-mas* to red light and inhibited by far-red light (Nagy, Paless and Vida, 1978; Nagy, Bocsi, Paless and Vida, 1984; Paless, Vida and Nagy, 1984). This illustrates the dynamic nature of cellular activity in the germinating spore in which the presence of certain polypeptides at any given time is dictated by conformational changes of a pigment. Preferential synthesis of a 20 kd and a 42 kd protein also occurs following red-light-induced germination of spores of *Pteris vittata*, but the reversibility of their synthesis by far-red light has not been tested (Zilberstein *et al.*, 1984).

In summary, continued protein synthesis appears necessary for completion of spore germination. It is not known whether changes in protein synthesis are used as a control during germination in addition to the primary control that is exerted over transcription. As will be seen below, the possible relationship between concomitant transcription and translation during the early phase of germination is an intriguing question.

The first proteins of germination

In the wake of recent conceptual advances in biochemistry and genetics, thoughts about the molecular biology of fern spore germination have been profoundly influenced by ideas of gene action, information transfer and metabolic regulation. Sustained investigations on the post-fertilization development of various animal embryos and on the germination of seeds of gymnosperms and angiosperms and spores of fungi have convincingly demonstrated the existence of preformed, template-active mRNAs which carry codes for the first proteins of development as these reproductive units are aroused from their dormant state. As mentioned earlier, there is some evidence for the existence of stored mRNA in dormant fern spores, but the extent of utilization of this information pool by the spore during the first hours of germination is not fully understood.

That stored mRNAs are utilized during germination is indirectly

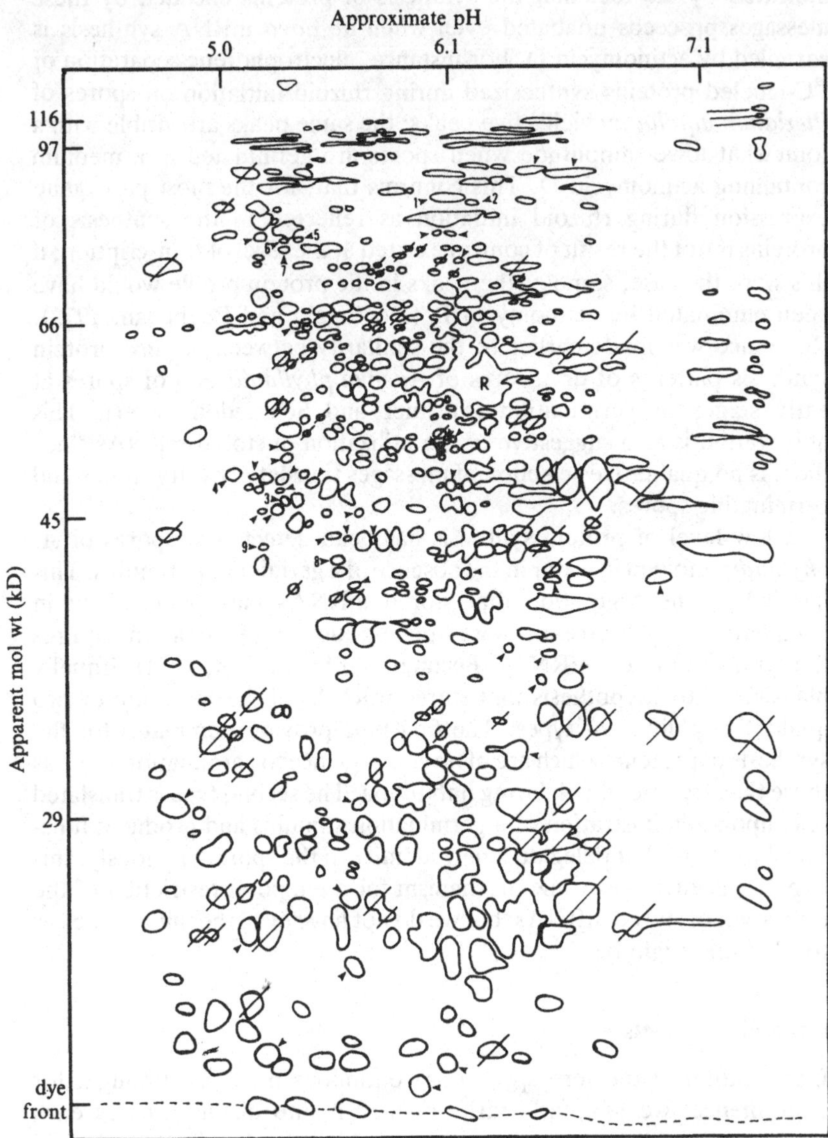

Figure 5.7. Diagrammatic representation of the changes in the soluble protein pattern of germinating spores of *Onoclea sensibilis*. Spots that are attenuated or that disappear during germination are marked by diagonal lines. Arrowheads point to spots that appear or increase in intensity during germination. Putative rhizoid-specific proteins are indicated by numbers. R, large subunit of Rubisco. (From Huckaby and Miller, 1984.)

indicated by the fact that the synthesis of proteins encoded by these messages proceeds unabated even when *de novo* mRNA synthesis is curtailed by actinomycin D. For instance, electrophoretic separation of ^{14}C-labeled proteins synthesized during rhizoid initiation on spores of *Pteridium aquilinum* yields five peaks; the same peaks are visible with a somewhat lower amplitude when spores are germinated in a medium containing actinomycin D. This confirms that, for the most part, gene expression during rhizoid initiation as reflected in the synthesis of proteins is not the result of control exerted at the level of transcription; if this were the case, some of the peaks in the protein profile would have been eliminated by actinomycin D (DeMaggio and Raghavan, 1972). Reference was made earlier to the similarity between *in vitro* protein synthesis patterns of dry spores of *Anemia phyllitidis* and of spores at early stages of germination (Fechner and Schraudolf, 1984). This observation is also suggestive of the utilization of stored mRNAs since there is no qualitative selection of messages translated by dry spores and germinating spores.

A low level of protein synthesis has been detected in spores of *A. phyllitidis* imbibed in water in the absence of a germination stimulus. This has led to the suggestion that stored mRNAs may be involved in completing the processes of spore ripening and that germination requires the synthesis of new mRNAs (Fechner and Schraudolf, 1984). Equally plausible is the hypothesis that stored mRNAs of the spore are of two qualitatively different types. The first type provides templates for the synthesis of proteins which are of no consequence to germination, such as those that are assembled during imbibition. The second type is translated only upon administration of a germination stimulus and produces functional proteins that delegate a new identity to the spore. Obviously, this hypothesis envisages some mechanism for a temporal restriction of the utilization of stored mRNAs, but we do not have even the faintest idea as to what this might be.

General comments

Germination of the fern spore is a continuous process, although, for convenience, we have arbitrarily divided it into distinct phases each expressing a morphogenetic event. Viewed in a biochemical, cytological and molecular sense, germination is triggered soon after the spore is hydrated or after it has been subjected to a dormancy-breaking treatment. Clearly, the metabolic regulation of spore germination is not a simple phenomenon since multiple pathways of enzyme synthesis and substrate degradation are involved.

Based on the available evidence, a key feature of gene activation during germination of the spore appears to be the utilization of mRNAs transcribed during sporogenesis. Evidently, it is assumed that activation of stored mRNAs in the spore sets off reactions leading to protein synthesis and a general surge in metabolic activity required for cell division and morphogenesis. If the validity of this concept is established by more definitive experiments, utilization of stored mRNAs can be looked upon as a survival strategy of the spore before its synthetic machinery becomes functional for the production of complex metabolites. The implementation of this strategy also depends upon the existence in the spore of a mechanism that triggers transcriptional control of germination with a minimum of overlap with translational control.

Part II

Growth and maturation

6

Developmental physiology of gametophytes – the germ filament

In all homosporous ferns, except those included in the Marattiaceae and Osmundaceae, the protonema initial generates a uniseriate, elongate filament composed of a varying number of chlorophyllous cells. All division planes leading up to the formation of the filament are oriented perpendicular to the long axis of the cell and thus lie parallel to one another. At each division, new walls are laid down between cells to hold them together as an elongating filament. When the filament has produced a certain number of cells, as determined by the conditions of growth and other factors, its terminal or subterminal cell divides by a wall oblique to the long axis to give rise to a bidirectionally dividing plate of cells. This marks the beginning of planar or prothallial growth. By the continued meristematic activity of this cell, aided later by the establishment of a pluricellular meristem, a flat, often cordate or exceptionally, ribbon-shaped, structure is formed. Rarely, as in *Schizaea* and *Trichomanes* (Hymenophyllaceae), no change ever occurs in the plane of cell divisions of the germ filament, which thus retains the basic filamentous morphology throughout the gametophytic phase. In this book, the terms 'germ filament', 'protonemal filament' or 'protonema' (plural, protonemata) are applied to the filamentous gametophyte and the term 'prothallus' is used to denote the structure with the planar morphology.

Our discussion of the developmental physiology of fern gametophytes is divided into three chapters. In this chapter we shall be concerned with the physiology of growth of the germ filament, that is, with the intracellular and extracellular control of filamentous growth. In the next two chapters we shall consider planar growth, encompassing the transduction of light signals to initiate morphogenesis.

Role of cell division and cell elongation in germ filament growth

As a system to study the control of cell division and cell elongation, the germ filament is elegant in its simplicity and sweeping in its heuristic value. Growth of the germ filament begins with the appearance of the protonema initial. A dominant feature of this cell is the presence of abundant chloroplasts which makes it autotrophic. Two recent studies on the germ filament of *Adiantum capillus-veneris* (Yatsuhashi, Kadota and Wada, 1985; Yatsuhashi, Hashimoto and Wada, 1987) have demonstrated a light-oriented movement of chloroplasts with both red and blue parts of the spectrum being active in modulating the movements. The effect of red light is partially reversed by far-red light, indicating phytochrome control of the reaction. The persistence of the blue light effect shows that now for the first time we can realistically accept the involvement of a photoreceptor other than phytochrome in chloroplast movement in a cell. This is not surprising, as we are beginning to realize that there is hardly any photomorphogenetic reaction system in the fern gametophyte that does not involve blue light.

The protonema initial generally undergoes a limited round of divisions and a limited extent of elongation to form the germ filament. As will be seen below, the number of cells formed in the filament before it initiates planar growth as well as the extent of elongation attained by cells are strictly dependent upon the quality and intensity of light in the ambient environment. This point needs emphasis because there has been an unfortunate tendency to describe the morphology of the germ filament in homosporous ferns in terms of their cell number and filament length without reference to the light regimens in which they are grown. However, there are some species, especially in the Grammitidaceae in which filamentous growth is rather prolonged, and irrespective of the growth conditions, it results in the formation of 15 to 20 cells (Stokey and Atkinson, 1958). A minor morphological feature of the germ filament of certain ferns is the crowning of its terminal cell by a unicellular papillate hair. The papilla is a club-shaped cell containing a nucleus, chloroplasts, mitochondria, and a large vacuole. One reason for the restricted occurrence of the papilla is that it is possibly an evolutionary remnant. Formation of the papilla precociously terminates the propensity of the terminal cell for further transverse divisions. According to Nayar and Kaur (1971), papilla formation is characteristic of a type of prothallial development designated as the 'Aspidium' type.

The involvement of cell division and cell elongation in the formation of the germ filament ensures that localized differences in mitotic rate and elongation rate might account for its growth. During the life of the germ filament, mitosis occurs only in the terminal cell which divides at a constant position from the tip, regardless of its length, except for very long or very short cells. In a growing filament, this is accomplished by the migration of the nucleus to the tip of the cell at the same rate as cell growth, so that the distance between the nucleus and the apical dome of the cell is more or less constant (Ito, 1969; Wada, Mineyuki, Kadota and Furuya, 1980). As shown in Fig. 6.1, for the terminal cell of the germ

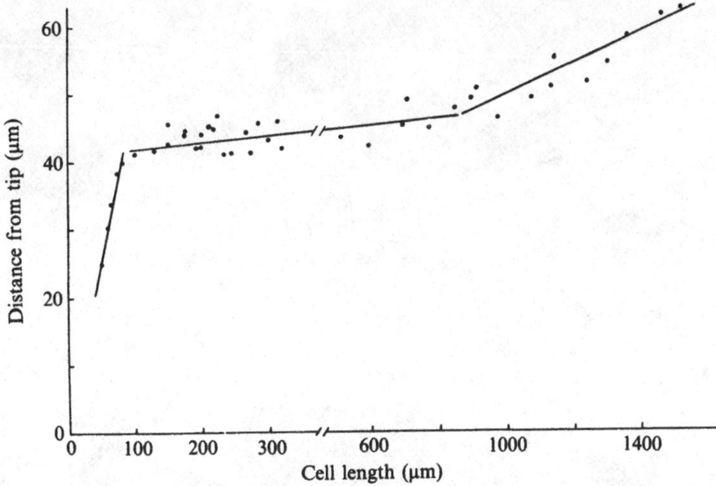

Figure 6.1. Relationship between the length of the growing terminal cell of the germ filament of *Pteris vittata* and the position of the newly formed cross wall. (From Ito, 1969.)

filament of *Pteris vittata* this critical region is 40 to 50 µm from the tip. Since the apical dome of the cell harbors the nucleus and much of the cytoplasm, it is tempting to explain the occurrence of the division wall in this framework. Thus, when the nucleus is moved from the tip of the terminal cell of the germ filament by a dark treatment, position of the cell plate during a subsequent division in white light is altered accordingly (Ito, 1969).

The germ filament morphology is considered suggestive of exhibiting typical 'tip' growth, that is, cell wall synthesis and growth are limited to the hemispherical dome of the terminal cell. In the germ filament of *Dryopteris filix-mas* the growing zone is restricted to within 20 µm of its tip (Mohr, 1956*a*). When measurements are made of the shift of carbon particles on the germ filament of *Pteridium aquilinum*, the dome of the terminal cell is found to elongate maximally. Growth in length decreases further from the tip and completely ceases at about 30 µm from this point (Takahashi, 1961; Davis, Chen and Philpott, 1974). Similar results have also been obtained in the growth of the germ filament of *Pteris vittata* (Furuya, Ito and Sugai, 1967). Freeze-fracture electron microscopy has revealed that on the inanimate outer surface of the terminal cell of the germ filament of *A. capillus-veneris* where maximum growth occurs, there is a concentration of particle rosettes arrayed against the plasma membrane; these are believed to be the key to the biosynthesis of cellulose microfibrils in certain eukaryotes (Wada and Staehelin, 1981).

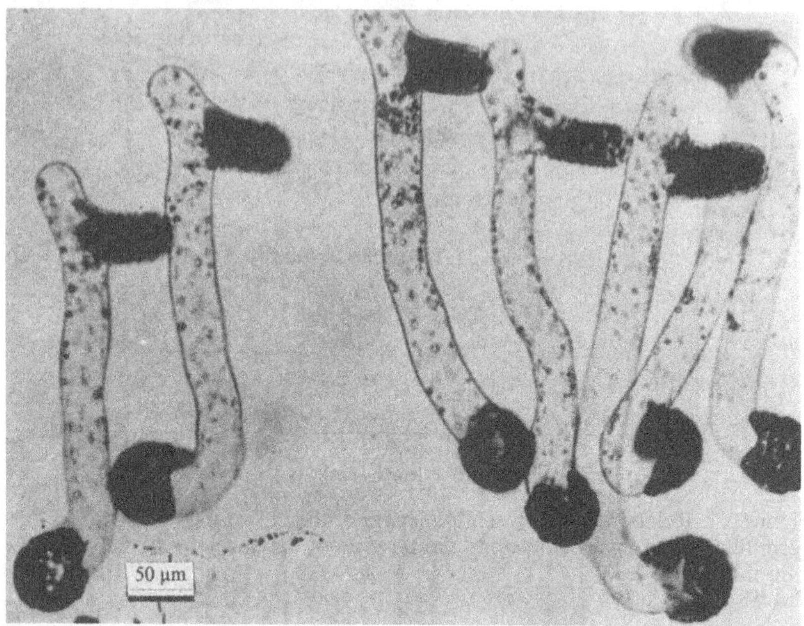

Figure 6.2. Induction of lateral elongation in the terminal cell of the germ filament of *Pteris vittata* by a sequential treatment involving darkness and exposures to white and red light. (From Ito, 1969.)

On both morphological and cytological grounds, therefore, it seems certain that the tip of the germ filament elongates because that is where cell wall biosynthesis occurs.

It is perhaps more than a coincidence that the nucleus resides at the tip of the terminal cell of the filament where maximum extension growth occurs. Reference was made earlier to an experiment in which the nucleus was pulled down from the apical dome of the terminal cell of the germ filament of *P. vittata* and was induced to divide by a dark treatment followed by exposure to white light (Ito, 1969). In an extension of this work it was found that if the newly divided protonemal filament is exposed to red light, the terminal cell of the filament elongates laterally at the site of the nucleus just above the cross wall, instead of terminally (Fig. 6.2). Similarly, when one- to three-celled germ filaments of *P. vittata* and *A. capillus-veneris* are centrifuged at × 500 to 50 000 *g*, the protoplasm of the terminal cell is stratified at its basal end. Tip growth is inhibited in such cells which take off laterally from where the protoplasm is sedimented (Ootaki, 1963; Mineyuki and Furuya, 1980). In general, the polarized elongation of the terminal cell appears to reflect an intrinsic property of the entire germ filament, since cells farther from the tip of the filament also exhibit, albeit at a decreasing rate, a tendency to elongate (Takahashi, 1961). What is of interest here is the inference that cells of the germ filament are not behaving as isolated units, but function as truly integrated parts of an organized structure.

An additional type of germ filament described in *Gymnogramme* (= *Pityrogramma*) *sulphurea* deserves note and this is one where the main contribution is from the subterminal cell. The activity of the terminal cell stops at the three- to five-celled stage when the subterminal cell begins to divide in the transverse plane. This, as well as the elongation of the basal cell lead to growth in length of the germ filament. Continued divisions of the terminal cell have been observed only under unfavorable culture conditions (Guervin, 1968, 1971).

Light and growth of the germ filament

A great deal of effort has been expended in attempts to determine the mechanism governing the growth of the germ filament. A basic discovery made in these studies is that if germinated spores are exposed to red light or maintained in complete darkness, continued elongation of the protonemata accompanied by a low rate of cell division ensues. This unique situation provides investigators with a system that is well-suited to study cell elongation without complications due to concomitant divisions as well as one in which cell divisions can be induced at will. Indeed, a few laboratories, especially in Japan, are currently deep in the study of the mechanism of cell division and cell elongation in fern protonemata using sophisticated optical methods. The experiments relevant to our present discussion fall into four general categories: (1) regulation of growth of the germ filament by red and far-red irradiances; (2) blue light effects on growth; (3) interaction between favorable light quality and growth hormones and (4) phototropism and polarotropism.

Red and far-red light effects

Different wavelengths of light from green to far-red are known to promote growth in length of the germ filament appreciably over dark controls. In the first important utilization of relatively pure wavelengths of light to study photomorphogenesis of fern protonemata, Mohr (1956*a*) showed that yellow and far-red light are superior to red light in inducing elongation of the germ filament of *Dryopteris filix-mas*, while blue light is inhibitory. In later experiments, it was found that when red and far-red irradiances are applied sequentially, the effect of red light is negated by a subsequent irradiation with far-red light (Schnarrenberger and Mohr, 1967). These results fit the bill for involvement of phytochrome as the primary photoreceptor in the process.

Miller and associates (Miller and Miller, 1961, 1963, 1964*a*, 1965,

1967*a*, *b*; P. M. Miller and J. H. Miller, 1966; Miller and Wright, 1961) have studied in detail the effects of light quality on the elongation of the germ filament of *Onoclea sensibilis* under different experimental conditions. An action spectrum obtained by irradiating filaments for different times at constant fluence is found to exhibit peaks at 400 to 420 nm, 580 to 599 nm and 640 to 660 nm. However, at saturating light doses, maximum effect is achieved in the yellow (560 to 580 nm) and far-red (720 nm) regions of the spectrum (Miller and Miller, 1967*a*). The implication of these results in the identity of the photoreceptor will be discussed below.

When germinated spores of *O. sensibilis* are irradiated for four days with a single period per day of red light of 5 min duration, filaments normally exhibit a certain amount of elongation over dark controls. Far-red light of the same duration is more effective than red light in promoting filament elongation. When filaments are exposed to red and far-red irradiances in sequence, their growth is reduced to a level normally produced by the last light treatment in the sequence and not to the level obtained in darkness. Thus, if far-red is followed by red light, the filaments behave as if they were irradiated with red light alone. Similarly, in the reverse sequence, far-red does not merely negate the red light effect, but causes some growth promotion as well (Miller and Miller, 1963). These results suggest that the effect of red and far-red irradiances on the elongation of the germ filaments of *O. sensibilis* cannot be explained by the general logic of phytochrome action. A vexing problem is to reconcile the observation that far-red light, which converts active P_{fr} to the inactive P_r form, produces the greatest growth promotion, and red light which initiates the formation of P_{fr} yields only a small bit of elongation.

The possible existence of two different photoreceptors in the germ filament of *O. sensibilis* has been evoked to explain the paradox posed by these results (Miller and Miller, 1967*a*, *b*; Greany and Miller, 1976). One of these is phytochrome and the other is an unusual pigment absorbing in the yellow region of the spectrum (580 to 600 nm), but having a long wavelength absorption tail extending to at least 729 nm. This pigment, designated as P_{580} presumably interacts with phytochrome in such a manner that whenever the latter is present as P_{fr}, the action of P_{580} is blocked. This consideration seems to imply that elongation of the germ filament in red light (phytochrome present as P_{fr}) is controlled by the usual phytochrome system, while elongation in far-red and yellow irradiances (phytochrome present as P_r) is due to the activation of P_{580}. Based on this view, the reversal of red light-induced elongation of the filament ordinarily expected following administration of far-red light is masked by the simultaneous activation of P_{580}.

That some caveat is needed in postulating the existence of specific pigments based on ostensibly reliable dose-response curves is emphasized by the photobiological re-evaluation of germ filament elongation in *O. sensibilis* by Cooke and Paolillo (1979*a, b*). By designation the cross-sectional area at the base of the apical dome of the terminal cell as the growth parameter under direct photocontrol, these investigators have shown that dose-response curves for yellow light parallel but lag behind those for red light. Moreover, there is a clear reversal by far-red light of the effects of both red and yellow light on the cross-sectional area of the filament. It is not inconceivable from this study, which is as important for the concept as for the results, that filament elongation in the yellow, red, and far-red parts of the spectrum is mediated by phytochrome without interference from other pigments. The possibility of actinic radiation influencing cross-sectional area to cause elongation of the germ filament does not seem to have been previously appreciated.

The immediate cause of filament elongation in red and far-red light is a narrowing of the diameter of the terminal cell which occurs in a framework of profound cytological changes. It has been reasoned that the low level of P_{fr} present in the filament exposed to far-red light restricts its growth to the extreme apical end of the terminal cell to produce an unusually slender cell. On the other hand, the existence of a high level of P_{fr} in red light is conducive for growth at a broader zone at the apex to form a short but broad filament (Schnarrenberger and Mohr, 1967). In the absence of data on the gradient of pigment distribution in the germ filament, the effects of red and far-red light cannot be understood in terms of qualitatively different actions of the photoreceptor. A recent *in vivo* spectrophotometric demonstration of phytochrome in chlorophyll-deficient germ filaments of *Anemia phyllitidis* takes us a step closer to providing such data (Grill and Schraudolf, 1981).

Miller and Stephani (1971) have proposed a model in which light-induced changes in microfibril orientation in the cell wall might possibly lead to inhibition of lateral expansion and promotion of elongation growth of the germ filament. Since the relationship between cell wall architecture and form of plant cells is well-known, it will not be reviewed here except in a general way. Random microfibril orientation in the cell wall generally results in spherical cells while cylindrical cells with a potential for diffuse growth have a predominantly transverse orientation of microfibrils. The proposed model assumes that in the dark-grown germ filament, arrangement of microfibrils is not strictly transverse, while irradiation with red or far-red light imposes their rigidly ordered transverse configuration permitting the formation of a long, narrow filament. The role of light quality in determining the structural properties of the cell wall is attributed to its effect in maintaining the orientation of cytoplasmic microtubules which are presumed to be the prime movers in the alignment of cellulose microfibrils. The basis for this view is the observation that spores of *O. sensibilis* remain as giant, undivided spherical cells when they are allowed to germinate and grow in white light in a medium containing colchicine. Irradiation of colchicine-poisoned spores with red or far-red light induces a modest polar elongation of cells. Since colchicine is known to disrupt the apico-basal microtubule alignment, it has been suggested that long wavelength light restores microtubules to their proper configuration to cause ordered cell wall microfibril deposition.

One piece of critical information that we need about this model concerns specific ultrastructural differences in microtubule arrangement between the cells of elongating and non-elongating germ filaments. This has been provided by Stetler and DeMaggio (1972) who found that in red light-grown filament of *D. filix-mas* microtubules line up axially in the basal half of the terminal cell, while in blue light which permits lateral expansion, their orientation is more or less random. In another investigation relevant to this issue, Murata, Kadota, Hogetsu and Wada (1987) have shown by immunofluorescence microscopy the presence of a circular array of cortical microtubules in the subapical part of the tip cell of the *Adiantum capillus-veneris* filament. As this is radically different from the parallel arrangement of microtubules seen in the rhizoid, the mechanism of tip-growth in the protonema is considered to be different from that in the rhizoid. These observations will add a new dimension to our concept of light-induced changes in cell shape in the fern protonema if it can be demonstrated that orientation of mirofibrils in the cell wall parallels the arrangement of microtubules.

Quite a different role for phytochrome is seen in the growth of germ filaments of *P. vittata* (Furuya *et al.*, 1967) and *A. capillus-veneris* (Kadota and Furuya, 1977). As the most studied and the most impressive responses to red and far-red irradiances are those of *Adiantum*, this work deserves extended attention. For a background and some details, see the reviews by Furuya (1978, 1983). When single-celled germ filaments are transferred from red light to dark, their growth in length decreases in about 24 hours and comes to a complete halt in 72 hours. By monitoring the growth of individual filaments before, during and after light treatments, a clear phytochrome action on filament elongation during the first 24 hours in the dark, uncomplicated by the occurrence of simultaneous cell division, has been established. Thus, cell elongation in the dark is strongly inhibited if far-red light is given immediately before dark incubation of the filaments and the effect of far-red is reversed by a subsequent red light. A later work (Kadota and Furuya, 1981) has also shown a pronounced effect of red light on the resumption of elongation of the terminal cell of non-growing two-celled filaments and its reversal by far-red light. On the basis of these results, it is not difficult to visualize localized competition for phytochrome molecules at the tip of the germ filament which causes changes in the rate of cell elongation in the appropriate light regimens. Although exposure of germ filaments to near UV and blue light also inhibits their elongation in the dark, this inhibition is not reversed by a subsequent red light (Kadota, Wada and Furuya, 1979).

Now, let us turn to the problem of cell division in the germ filament of *A. capillus-veneris* and its control by phytochrome. Single-celled filaments reared in red light divide very slowly due to a blockage in the early G_1 phase of the cell cycle. However, when the filaments are transferred from red light to the dark condition, the cell cycle progresses and synchronous divisions occur. In such filaments a cell cycle of 36 hours is typical of the first division in the dark, with 20 hours devoted to G_1, approximately 5 hours to S, 7 hours to G_2 and 4 hours to M. This division is delayed by a brief exposure to far-red light at the beginning of the dark treatment. Red light following far-red light negates the effect of the latter by allowing cells to resume the cycle and in a repeated irradiation sequence, the imprint of the final light quality prevails in determining the percentage of dividing cells. The use of ^{3}H-thymidine has provided impressive confirmation of the fact that in this case phytochrome controls the timing of cell division by regulating the duration of the G_2 phase of the cell cycle; apparently, far-red exposure at the beginning of the dark period extends the G_2 phase and a following red light shortens it (Wada and Furuya, 1972; Miyata, Wada and Furuya, 1979).

The precision of cell cycle arrest in germ filaments grown in continuous red light has made it possible to identify the nature of the photoreceptor involved. Here again, as in the progression of the cell cycle in the dark, we find an apparent antagonism between red and far-red irradiances, implicating phytochrome. Using the number of new cells formed in a population of one-celled filaments as an index, it has been shown that once the cell cycle takes off in the dark, it is returned to the beginning of G_1 by short exposure to red light; far-red light given immediately after red light veers the phase of the cell cycle in the opposite direction (Wada, Hayami and Kadota, 1984). A two-dimensional gel electrophoretic profile of *in vivo* synthesized proteins has brought us encouragingly close to associating the G_1 phase of the cell cycle with the synthesis of a new protein (Nagatani, Suzuki and Furuya, 1983).

Further insight into the photocontrol of division in the single-celled filament of *A. capillus-veneris* has been gained by the use of ingenious optical techniques combined with electron microscopy, leading to a minute-by-minute account of the fate of the nucleus and other organelles during the cell cycle (Fig. 6.3). When the growth of the filament is monitored with the help of a time-lapse video system, the nucleus is found to keep up with cell elongation during the many hours of the G_1 phase by sliding just so much toward the tip, thus keeping a constant distance to the

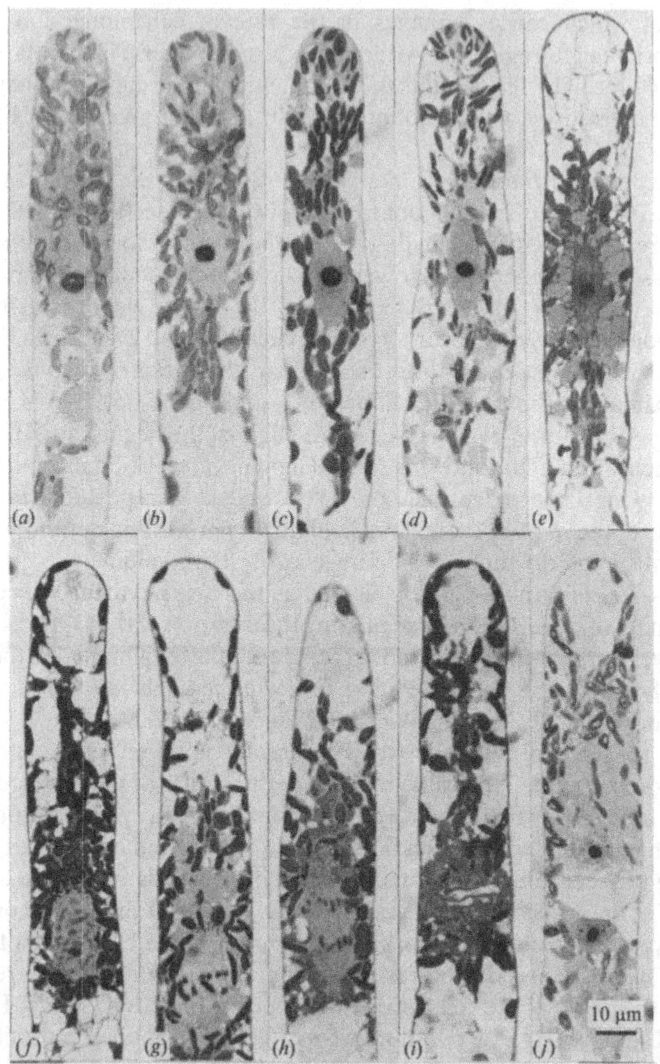

Figure 6.3. Longitudinal sections of the terminal cell of the germ filament of *Adiantum capillus-veneris* showing the changes in intracellular structure during cell cycle. (*a*) At the end of 6 days in red light. (*b, c*) Cultured in the dark for 12 hours and 20 hours, respectively, after 6 days in red light. (*d, e*) Cultured in the dark for 28 hours and 32 hours, respectively, after 6 days in red light. (*f, g, h, i*) Nuclei in prophase, metaphase, anaphase and telophase, respectively; collections made after 36 hours and 42 hours in the dark following 6 days in red light. (*j*) Terminal cell after cytokinesis. Scale bar in (*j*) applies to all. (From Wada *et al.*, 1980; photographs supplied by Dr M. Wada.)

dome of the filament. Forward movement of the nucleus during one phase of the cell cycle encourages backward movement during another phase and this occurs at the S or G_2 phase. Two pieces of evidence have indicated that microtubules may be somehow involved in giving directionality to nuclear movement in the filament. First, by painstaking counts made from longitudinal sections of filaments in the electron microscope, Wada *et al.* (1980) showed that the cortical microtubules aligned along the circumference are positioned mainly in the apical dome of the cell during the G_1 phase of the cell cycle and that a change in the direction of nuclear migration at G_2 coincides with the appearance of microtubules between the apical dome and the nucleus. As shown in Fig. 6.4, there are indeed differences in number that one would predict if microtubules were sliding past one another to take up new positions in the

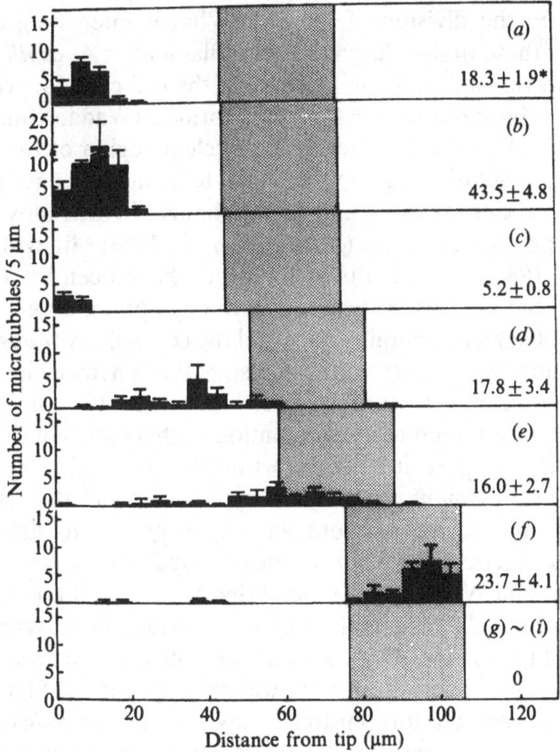

Figure 6.4. Distribution of cortical, circumferentially aligned microtubules in the terminal cell of the protonema of *Adiantum capillus-veneris* during the progression of cell cycle. Cells in (*a–i*) correspond to those shown in Fig. 6.3. Stippled regions indicate the position of the nucleus in the cell. The total number of microtubules in each cell (± standard error) are also given. (From Wada *et al.*, 1980; print supplied by Dr M. Wada.)

cell during division. Second, treatment of the filament with colchicine stops nuclear migration so that the distance between the nucleus and the tip of the filament becomes longer with time (Mineyuki and Furuya, 1985). Centrifugation combined with colchicine treatment has uncovered yet another role for microtubules during cell cycle in the germ filament of *A. capillus-veneris* and this is in the positioning of the premitotic nucleus. Centrifugation of the filament at the G_2 or M phase of the cell cycle does not displace the nucleus suggesting that this organelle is anchored by some cytoplasmic elements. The identity of the latter with microtubules is indicated by the observation that the nucleus moves around in the cytoplasm when the filament is treated with colchicine (Mineyuki and Furuya, 1986).

Other cell organelles such as oil droplets and chloroplasts also move about in the terminal cell of the germ filament during the progress of division. It was therefore natural to believe that this is related in some remote way to the division of the cell. What is interesting about the movement of these organelles in the germ filament of *A. capillus-veneris* is that it slows down during the M phase of the cell cycle and comes to a complete stop by the time a cell plate is formed (Wada, Mineyuki and Furuya, 1982). When organelles in the nuclear region of the cell were analyzed by digital image processing techniques combined with microscopy, striking temporal and spatial changes in their movement are observed during the cell cycle (Mineyuki *et al.*, 1983; Mineyuki, Takagi and Furuya, 1984). The results of this study have been interpreted to indicate that the nucleus is anchored in the cytoplasm by microtubules positioned between its lateral surface and the cell wall. Although there is as yet no sound evidence from these results that phytochrome controls microtubule assembly to bestow directionality on the nucleus, these observations have served to draw attention to this possibility.

To conclude this part, it will be worthwhile to consider other studies relating to germ filament growth and say a little about the relationship between the age of the filament and its responses to light quality. Experiments described above have mostly used one or two celled filaments. Miller and Wright (1961) found that the effect of red light on the growth in length of the filaments of *Onoclea sensibilis* varies with their age at the time of irradiation. Dark-grown filaments form no more than one to three cells over a period of three weeks. Exposure of filaments to a short period of red light for four to five days at the beginning of the dark period promotes cell elongation, while irradiation after seven days in the dark causes little growth promotion. Irradiation with red light after 12 days in the dark is even inhibitory to filament elongation. However, age is not a barrier to promotion of cell elongation by far-red light which always produces greater elongation than red light (Miller and Miller, 1963). A case for inherent age-dependent changes in photosensitivity of filaments, implicit in Miller and Wright's (1961) work has been weakened by the completely conflicting data of Cooke and Paolillo (1980*a*). The main evidence adduced by these latter investigators is that the cross-sectional area of the growing tip of the filament declines with age over a period of time in the dark at the same time as the filament elongates, indicating that the dark-mediated tapering of the cross-sectional area of the cell might account for its seeming age-dependent change in photosensitivity.

Blue light effects

Much evidence indicates that growth of the germ filament is modified by short wavelength blue light. Most importantly, as we will see in the next chapter, this is manifest in an altered pattern of cell division leading to planar growth. Under certain experimental conditions, continuous or intermittent irradiation of the germ filament with blue light also results in the promotion of filament elongation unaccompanied by planar growth (Mohr, 1956*b*; Sobota and Partanen, 1966; Miller and Miller, 1964*a*). As seen in *O. sensibilis*, continuous exposure of photoinduced spores to low fluences of blue light enhances germ filament elongation slightly over dark controls. Optimum growth promotion is observed following exposure to 30 min/day of blue light for four days, while longer periods of exposure cause progressive inhibition of filament elongation coupled with induction of planar growth. There is also a peculiar interaction between blue light and far-red light in the elongation of the germ filament. When the filament is exposed to blue light either before or after far-red light, the former is found to interfere with far-red light induced promotion of elongation (Miller and Miller, 1964*a*). One might deduce from these results the existence of two effects of blue light in the germ filament of *Onoclea,* one leading to elongation growth and the other to planar growth.

Another blue light effect concerns the timing of cell division in the germ filament of *Adiantum capillus-veneris* (Wada and Furuya, 1972, 1974; Miyata *et al.*, 1979). Phytochrome control of cell division in this system was described earlier, but the picture of blue light action that emerges is most complex. If red light grown filaments are exposed to blue light, the G_1 phase is skewed in a subsequent period in the dark and cell divisions occur at significantly earlier times than in those exposed to far-red light. Not surprisingly, a reversible far-red/red light reaction governs this blue light effect. The data in Table 6.1 show that far-red light lowers the frequency of cell divisions induced by blue light and a subsequent red light reverses the effect of far-red light and returns the progressing cell cycle back to G_1. As we begin to understand the interaction of phytochrome with blue light in inducing cell divisions in the filament, the ubiquitous pigment $P_{b\text{-}nuv}$, identical to the one involved in the germi-

Table 6.1 *Effects of red and far-red light on the timing of blue light-induced cell division in the protonemata of* Adiantum capillus-veneris[a]

Light treatment	Percentage of cells in division ± standard error
B	56.0 ± 5.0
B/R	51.0 ± 1.0
B/R/FR	13.3 ± 1.7
B/R/FR/R	45.7 ± 4.2
B/FR	11.0 ± 1.5
B/FR/R	43.3 ± 1.4
B/FR/R/FR	11.0 ± 1.7
FR	0.3 ± 0.3
Dark	1.7 ± 0.7
FR/R	1.0 ± 0.6
FR/B	21.0 ± 2.5

[a] From Wada and Furuya (1972).
[b] Protonemata cultured under continuous red light for 6 days were irradiated with about 850 ergs/cm^2/s blue (B) and red light (R) and 3×10^4 ergs/cm^2/s far red light (FR) for 10 min for each treatment. Observations were made one day after the light treatment.

nation of spores has been suggested as the likely photoreceptor for the blue light effect. The exact location of this pigment in the cell has also been the subject of a certain amount of study by microbeam irradiation. As irradiation of filaments of *A. capillus-veneris* with a microbeam of blue light induces division only around the nucleus, this organelle or some other organelles closely associated with it are the prime candidates for the photoreceptive site (Wada and Furuya, 1978). This view is reinforced by another experiment in which filaments were centrifuged before microbeam irradiation. Since centrifugal stress changes the position of the nucleus basipetally, microbeam irradiation at the new spot of the nucleus, rather than at the old one, induces cell division (Kadota, Fushimi and Wada, 1986). This could not happen if the pigment is located at another site in the cell, say, for example, in the plasma membrane.

A role for blue light in enhancing mitotic activity in the germ filaments of *Pteris vittata* grown in red light has also been demonstrated (Ito, 1970, 1974). Cells of red light grown filaments are precariously balanced between the G_1 and S phases, but following a short exposure to blue light, they progress synchronously through the cell cycle, leading to an increase in the mitotic index and cell number. The effects of blue light on cell synchrony and mitotic activity are reversed by red light administered immediately after blue light. Although there are some similarities in photoresponses between germ filaments of *A. capillus-veneris* and *P. vittata*, it seems unlikely that blue light is acting in the latter through the phytochrome system. More likely, the causal relationship between blue light irradiation and mitotic activity is due to the induction of planar growth.

Interaction of light and hormones

Most experiments on hormonal effects on the growth of fern protonemata have focused on auxins. This is due to their well-known effects on cell elongation as well as due to the inability of other hormones, especially the various gibberellins to modify the pattern of filamentous growth of protonemata. Some data relevant to the possible nature of the photoreactive systems of germ filaments have come from experiments on the interaction of auxin and light quality on their growth. For example, growth of the germ filament of *Onoclea sensibilis* in red and far-red light regimens is promoted by a range of concentrations of IAA (Miller and Miller, 1965). Addition of IAA, however, does more than promote elongation, since it also produces additive effects in red and far-red light. These results may conceivably show that promotion of filament elongation in the different light regimens is not due to increased levels of endogenous auxin, as, if it were, light should inhibit filament elongation at supraoptimal levels of auxin. In contrast, auxin does not cause additive effects on filament elongation in blue light. There are also other differences between blue light-induced elongation and red and far-red light-induced elongation of the germ filament of *O. sensibilis*. One is that promotion of filament elongation in blue light almost resembles auxin-induced elongation in its requirement for an optimum concentration of sucrose, while red and far-red irradiances seem to be active regardless of the presence or absence of sucrose in the medium. Secondly, both blue light-induced and auxin-induced elongation of the germ filament in the dark are sensitive to the antiauxin, *p*-chlorophenoxyisobutyric acid (PCIB) which completely reverses the growth promotion produced. In contrast, PCIB promotes filament elongation in red and far-red light regimens. Somewhat similar results have also been reported on the effects of IAA and PCIB on the growth of the germ filament of *Athyrium filix-femina* in blue light. However, while both IAA and PCIB inhibit cell division and cell elongation of the filament in red light, the growth inhibition by the antiauxin is relieved by IAA (Bähre, 1976).

Temperature sensitivities of auxin-induced elongation and blue light-induced elongation of germ filaments of *O. sensibilis* are similar and differ from those of filaments reared in red and far-red light regimens (P. M. Miller and J. H. Miller, 1966). Thus, in both blue light and in the presence of IAA in the dark, filaments experience a steady increase in length between 24 and 29 °C whereas in red and far-red light no additional growth in length occurs at temperatures above 22 °C. From these results one is left with the impression that a type of regulatory mechanism different from that involved in red and far-red light is operating during growth of the filament in blue light. The simplest assumption seems to be the existence of a blue light absorbing pigment, although other factors unrelated to a blue light-mediated photoreaction may also account for these differences.

Another hormone which profoundly influences the growth of germ filaments of *O. sensibilis* is ethylene. Miller, Sweet and Miller (1970) found that administration of the gas promotes elongation of dark-grown filaments by a substantial amount compared to controls grown in an ethylene-free milieu. As seen in Fig. 6.5, ethylene in the range of 0.01 to 0.1 parts per million (ppm) appears to be maximally effective, although even at 1000 ppm the filaments are slightly longer than the dark controls. Since promotion of filament elongation is accompanied by inhibition of cell division, it is reasonable to conclude that the growth increment is due to the elongation of cells present before ethylene application. Ethylene also reverses the inhibition of filament elongation caused by high levels of auxin, implying that auxin-induced inhibition is not due to the production of ethylene. According to Edwards and Miller (1972*a*), cell division in light-grown filaments is also susceptible to inhibition by ethylene, although the concentration of the hormone necessary for optimum effect is nearly ten-fold higher than that required for dark-grown filaments. Based on gas-liquid chromatographic evidence for the production of ethylene by germ filaments of *O. sensibilis*, a general conclusion from these experiments is that the gas is a natural byproduct of the normal metabolism of protonemal cells. Elongation of cells of the germ filament of *Athyrium filix-femina* cultured in red light is promoted by ethylene and its effects counteracted by high concentrations of carbon dioxide or acetylcholine (Bähre, 1975*a, b*).

The key to an understanding of the interaction between light and growth hormones in the growth of the germ filament discussed here lies in the web of biophysical control mechanisms of cell elongation. From an analysis of the responses of germ filaments of *O. sensibilis* to combinations of light, hormones and cations, Cooke and Racusen (1982) have argued that cell expansion is mediated by an ethylene-Ca^{2+} system in red light and an auxin-H^{+} system in blue light. Their observations that cell expansion in red light is relatively insensitive to the inhibitory effects of Ca^{2+} and that the Ca^{2+}-chelator EGTA causes a significant decrease in cell expansion of the filament in red light are relevant to this argument. Even more significantly, cells which expand in red light do well in the presence of the ethylene absorbent potassium permagnate while the ethylene-generating agent CEPA amplifies Ca^{2+}-induced inhibition of

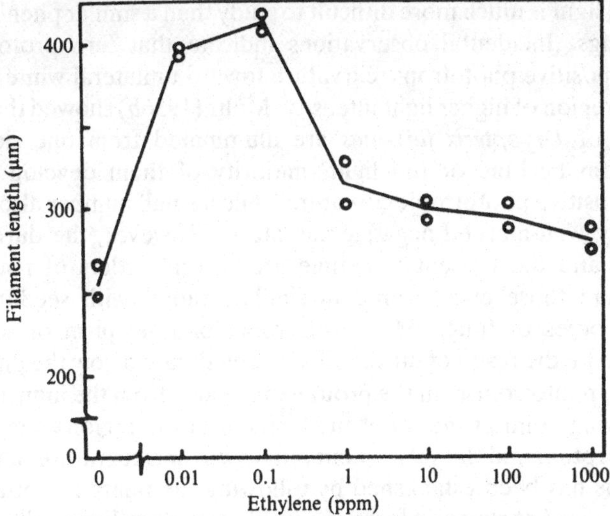

Figure 6.5. Effects of different concentrations of ethylene on the elongation of the germ filament of *Onoclea sensibilis*. (From Miller *et al.*, 1970.)

cell expansion. Somewhat different is the behavior of cells in blue light, which is sensitive to alkaline pH, PCIB, and the calcium ionophore A23187, all of which reduce cell expansion. In addition, alkaline pH also abolishes the ability of IAA to promote cell expansion. Taken together, these results have been interpreted to mean that in red light there is an enhanced, phytochrome-controled Ca^{2+} flux from the cell wall into the cytoplasm which is antagonized by ethylene. In blue light, on the other hand, there is an enhanced H^+ flux from the cytoplasm into the cell wall which is mediated by a high level of auxin activity. Electrophysiological measurements of the terminal cell of the germ filament have revealed rapid membrane depolarization in red light and hyperpolarization in blue light; alterations in membrane potential of this sort could provide a mechanism for red light-induced Ca^{2+} influx and blue light-induced H^+ efflux in the cell (Racusen and Cooke, 1982). These important findings show that physiological markers remain conspicuously stable in the germ filament following light treaments that cause long term morphological changes. The distinctiveness of these approaches is the emphasis placed on the cell wall and plasma membrane rather than on the cell wall alone.

Phototropism and polarotropism

Two additional effects of light on the growth of germ filaments that have shown interesting features are phototropism and polarotropism. As germ filaments are mostly surface-living, their curvature towards a source of unilateral light is much more difficult to study than a similar phenomenon in seedlings. Incidental observations indicate that fern protonemata exhibit a positive phototropic curvature toward unilateral white light or toward a region of higher light intensity. Mohr (1956*b*) showed that when filaments of *Dryopteris filix-mas* are illuminated from one side with monochromatic blue or red light, majority of them develop a well-marked positive phototropic curvature while a small number also exhibit an equally well-marked negative curvature. However, the duration of exposure and the presentation time are several orders of magnitude higher than those encountered in similar studies with seedlings and fruiting bodies of fungi. Moreover, while phototropism of seedlings appears to be the result of differences in growth rate below the photosensitive site, phototropism in the protonema results from the induction of a new growing point at the tip of the filament in the region of maximum light absorption and the subsequent outgrowth and curvature of the new apex. This has been established by following the migration of artificial markers placed at strategic locations on phototropically stimulated germ filaments (Etzold, 1965; Davis, 1975). Recently, Kadota *et al.* (1982) showed that localized exposure of a flank of the subapical region of the germ filament of *Adiantum capillus-veneris* with a microbeam of red light induces a curvature toward the irradiated side. It comes as no surprise that this reaction has all the hallmarks of phytochrome control such as red/far-red reversibility. Additionally, the action spectrum shows a peak at 662 nm which is close to the absorption peak of phytochrome (Kadota, Koyama, Wada and Furuya, 1984). An idea that the pigment molecules are located in the ectoplasm and/or plasma membrane of the cell has come from the observation that a phototropic curvature occurs even after microbeam irradiation of a filament centrifuged basipetally to clear much of the cytoplasm from the path of the beam (Wada, Kadota and Furuya, 1983). A further probe on this system has revealed a phototropic response to blue light, which poses a fundamental problem of accommodating phytochrome and a blue light absorbing pigment in the reaction chain (Hayami, Kadota and Wada, 1986). The fact that apart from these studies no systematic attempt has been made to identify and localize the photoreceptors in phototropism of fern protonemata indirectly indicates the inherent disadvantages of this system for phototropic investigations.

The response of germ filaments to polarized light, known as polarotropism is a variation of phototropism, as unilateral irradiation with plane polarized light causes phototropically sensitive organs either to grow in the direction of the plane of polarization or as in some cases, to grow along directions perpendicular to this plane. Bünning and Etzold (1958) made the basic observation that vertically polarized illumination applied from above orients the direction of growth of the germ filaments of *D. filix-mas* cultured on an inverted agar plate. Thus, in polarized red light the filaments grow at right angles to the plane of polarization, while a shift in the direction of polarization made with polarotropically active light causes a reorientation of growth of filaments perpendicular to the plane of vibration of the electric vector. Germ filaments of *Struthiopteris filicastrum (=Matteuccia struthiopteris)* (Hartmann, Menzel and Mohr, 1965) and *Onoclea sensibilis* (Miller and Greany, 1974) also exhibit a similar response to polarized red light. In the former, changing the plane of vibration of the light results not only in a curvature at the tip but also in the appearance of one or two additional outgrowths from the subapical region. Data on the action spectrum (Etzold, 1965; Steiner, 1967*b*, 1969*b*) and dose-response relationships at different wavelengths (Steiner, 1967*a*, 1969*a*) for polarotropism of germ filaments of *D. filix-mas* have shown that blue light is most effective, with red and UV light not far behind.

According to Etzold (1965), polarotropism of the protonema of *D. filix-mas* in the red spectral region is mediated by phytochrome which interacts in a complex way with far-red light. A major departure from classical red/far-red photoreversibility is that administration of polarized far-red light in the same plane following polarized red light does not abolish the effect of the latter, but the red light effect noticeably increases. On the other hand, if polarotropically active red light is followed by non-polarized far-red light, there is a significant reversal of the red light effect. Although these observations are difficult to reconcile, Etzold interprets them as indicating that dichroic photoreceptors, located at or very near the dome of the terminal cell of the filament, change their orientation during irradiation. This in turn leads to the displacement of the growing point to the site showing highest absorption of polarized light. There are, however, no real ultrastructural differences between normal and polarotropically activated protonemata in the disposition of organelles suggestive of recognition of a specific site in the cell for polarotropic growth. Presumably, phytochrome interferes with cell wall plasticity to induce polarotropism (Falk and Steiner, 1968).

Based on microbeam irradiation of single filaments of *A. capillus-veneris* with polarized red light, another interpretation of polarotropism is that the photoreceptor site is in the apical 5 to 15 μm region and that polarotropic growth is due to differential absorption of red light at the two flanks of the cell (Wada *et al.*, 1981). In this system, microbeam irradiation of the flank of the apical region of the protonema with polarized red light vibrating parallel to the protonemal axis is most effective in inducing polarotropic curvature which is nullified by exposure to normally vibrating far-red light (Kadota *et al.*, 1982). As there is a close relationship between phototropism and polarotropism of the filament, localization of phytochrome molecules mediating in these responses in the same region of the protonema, namely, the ectoplasm and/or plasma membrane is not at odds with it (Wada *et al.*, 1982). Physiological experiments have clearly indicated that as in the filaments of *D. filix-mas*, in *A. capillus-veneris* also, phytochrome molecules exhibit different dichroic orientations at the periphery of the cell (Wada *et al.*, 1983; Kadota, Inoue and Furuya, 1986). Other studies have led to the conclusion that differences in red light absorption and in the molecular orientation of phytochrome molecules are translated to variations in the amount of P_{fr} between the extreme tip and the subterminal region as well as between the opposite sides of the cell. Insofar as the regulation of polarotropic growth of the protonema is concerned, therefore, these differences in the amount of P_{fr} are crucial (Kadota *et al.*, 1985). It is unlikely that these will be the last words about polarotropism in fern protonema and we are certain to hear more about this in the foreseeable future.

Before leaving this admittedly engaging but complex subject, it is necessary to point out that dose-response curves at different wavelengths for polarotropism in the germ filament of *D. filix-mas* have led to the view that phytochrome action is not restricted to red and far-red irradiances, but occurs also in the blue and UV regions. One argument in favor of this view is the striking parallelism between changes in the slope with wavelength found in the dose-response curves obtained under steady state conditions for blue and UV and changes in phytochrome photoequilibrium (Steiner, 1969*b*). Dose response curves obtained by simultaneous irradiation of the germ filament with red, blue and UV regions of the spectrum, designed to alter phytochrome photoequilibrium are also consistent with the assumption of phytochrome involvement in blue and UV-mediated polarotropism (Steiner, 1970). As the polarotropic response of the germ filament in blue light is saturated at light intensities which are hardly effective in inducing planar growth, this reinforces the evidence for the existence of different pigment systems for blue light-induced polarotropism and planar growth.

Physiological gradients and polarity

Polarity is a trait that is invariably present in the germ filament from the time its initial cell is delimited in the germinating spore. The segregation of chloroplasts and other organelles to one end of the protonema initial defines the polarity of this cell. In the germ filament, polarity is implied in a linear row of cells whose most basal cell is partly enclosed in the remnants of the spore coat. An additional feature of polarity is the presence of a rhizoid on the basal cell of the filament. Although in most species, the terminal cell of the filament is free, in others it differentiates a papilla. The presumptive site of the papilla is identified by its affinity for stains such as safranin and acetocarmine and by the presence of high concentrations of RNA and proteins (Nakazawa and Ootaki, 1962). The papilla and the rhizoid share common staining reactions which are quite different from those displayed by the protonemal cells. Although the papilla appears on the fifth or sixth cell of the germ filament, it has been induced even at the three-celled stage by auxin treatment or by UV irradiation (Nakazawa, 1960*a*; Kato, 1964*a*). Nevertheless, the fate of the cytoplasmic area in the germ filament where a papilla differentiates is labile and is not rigidly fixed. For example, when filaments of *Dryopteris varia* are grown in the dark for a long time, their subsequent return to light results in the differentiation of a rhizoid in place of the papilla. A similar type of polarity reversal is observed when protonemata are cultured in media containing 10 to 100 mg/l IAA or 10^{-2}M LiCl (Nakazawa, 1960*a, c*).

As mentioned in the early part of this chapter, a conspicuous feature of polarity of the germ filament is its growth restriction to the terminal cell. However, this property is not irreversible and can be disrupted by experimental treatments such as prolonged darkness followed by exposure to light (Nakazawa and Ootaki, 1961; Ito, 1969), application of colchicine (Nakazawa, 1959), centrifugation (Ootaki, 1963; Mineyuki and Furuya, 1980), UV irradiation (Kato, 1966), and surgical operations (Ootaki, 1968; Ootaki and Furuya, 1969). These treatments generally inactivate the terminal cell and promote lateral growth and branching; the results of these experiments are described in Chapter 9.

Among other forces that may play a role in the polarity of the germ filament, some interest is attached to the pattern of regeneration of isolated cells. If individual cells of the filament of *Pteris vittata* grown in low intensity white light are isolated by amputation at one or both sides and cultured in a mineral salt medium, they display precocious functional maturity and regenerate new filaments. However, each cell exhibits an intracellular gradient in regenerative ability, the isolated terminal cell showing a greater propensity to continue apical growth than cells from the middle and basal regions of the filament. The filament itself displays an apico-basal polarity along its developmental axis by a greater tendency of cells from the apical end to grow from their anterior end and cells lower down to regenerate from the middle and posterior ends. Regenerates also appear on cells towards the apical end of the filament at progressively earlier times after culture than on cells toward the base (Ootaki, 1967).

During normal, uninhibited growth of the germ filament, an intrinsic gradient of physiological activities polarized in the apico–basal plane seems to direct the overall growth processes to the apical end. The first cell physiological gradient detected in fern protonemata is the osmotic gradient. Using the common plasmolytic method in graded solutions of sucrose, Gratzy-Wardengg (1929) showed that in the germ filaments of *Nephrodium (=Dryopteris) filix-mas* and *Struthiopteris germanica (=Matteuccia struthiopteris)*, the terminal cells have high osmotic values.

These diminish toward the basal cells. When filaments are transferred to an osmoticum, plasmolysis begins first in the basal cells which also reach full plasmolysis earlier than the terminal cell. In another work, Reuter (1953) observed marked differences between cells of the protonema of *Dryopteris (= Christella) parasitica* in the time required for deplasmolysis in urea and glycerin and concluded that the terminal cell possessed the highest permeability to these substances. In the protonemata of various ferns, the existence of apico-basal gradient has also been established for nuclear and nucleolar volumes (Ootaki, 1965, 1968; Bergfeld, 1967), chloroplast structure and morphology (Ootaki, 1968; Faivre-Baron, 1977*a*), permeability to dyes (Nakazawa and Ootaki, 1962), ability to survive after a lethal dose of radiation (Kato, 1964*a*), and polysaccharide composition of cell walls (Smith, 1972*b*). Thus each cell of the protonema appears to be unique, exhibiting cytological and physiological features slightly different from those of its immediate neighbors. In general terms we can visualize an identical stimulus programming qualitatively different changes in the terminal and basal cells of the filament because these cells occupy different positions relative to the apico-basal gradient. One is at the high point of the gradient while the other is at the low point.

Cytochemical gradients, especially those concerned with metabolic systems of cells are of importance in defining polarity of the germ filament. In the protonema of *P. vittata*, a stainable concentration of RNA is associated with the terminal cell and it progressively grades off along the filament towards its base (Nakazawa and Tanno, 1965, 1967). In a comparative study of succinic dehydrogenase activity in the germ filaments of three ferns, Nakazawa and Kimura (1964) showed that enzyme molecules are stock-piled in the subterminal region of the filament of *P. vittata* and in the basal region of the filament of *Ceratopteris thalictroides* and are uniformly scattered through the length of the filament of *D. varia*. This suggests that a gradient of enzyme activity may not be equated with polarity of the filament.

These observations constitute the basic data on the structural and metabolic aspects of polarity in the germ filament. But how they work in concert and recognize the position of each cell in a filament to define its axis is a problem for future investigations. Unfortunately, there has been a waning of interest in this area of research.

General comments

From the review of the work on the developmental physiology of fern protonemata covered in this chapter, it is clear that major advances have been possible in recent years in the analysis of the factors that control their growth. The sensitivity of protonemata to specific wavelengths of light has been plainly crucial, for it has provided a convenient handle to start or stop cell division and cell elongation at virtually any stage of growth and analyze the mechanism involved. In spite of the unquestionable differences between photosensitive organs of higher vascular plants and fern protonemata, it is likely that they share some basic mechanisms in the control of cell division, cell elongation, phototropism, and polarotropism by phytochrome.

Although seemingly simple in organization, functionally the protonema is a heterogeneous group of a limited number of cells with important structural and physiological differences. As already emphasized, the activities of the filament converge to the terminal cell. At this stage we do not know, what, if anything, this cell does to stay functionally ahead at the rest of the pack. Considering its importance in the growth in length of the germ filament and the transition of the filament to a planar gametophyte, the terminal cell should draw more attention in any future developmental studies.

7

Developmental physiology of gametophytes – induction of planar growth

In the gametophytic generation of homosporous ferns, the germ filament represents a transient phase in anticipation of a major morphogenetic event, as, sooner or later, its terminal cell characteristically divides by an oblique or longitudinal wall to initiate planar morphology. The division wall also occasionally appears in the subterminal cells; in other cases, after the terminal cell is partitioned longitudinally, most of the cells in the filament may also follow suit. For purposes of discussion in this chapter, the essential mystery of transition of the germ filament to a planar gametophyte is considered to revolve around the orientation of the mitotic spindle in the terminal cell from a position parallel to the long axis of the cell to one perpendicular to it. This and the formation of the cross-wall itself are engineered by a complex series of developmental processes of which we have only a rudimentary understanding.

Induction of planar growth in the fern protonema represents the beginning of a morphogenetic process that prepares it for the production of gametes. The major axioms governing morphogenesis in complex systems where a change in the plane of cell division initiates a change in the pattern of growth, can be analyzed with relative ease in the protonema. For this reason, as the survey in this and the next chapter shows, a sizable literature on the photomorphogenesis and physiology of planar growth has accumulated. To a large degree the questions to be addressed here are concerned with the effect of light quality on form change in the germ filament. During this discussion, we will focus our attention on the cytological changes in the terminal cell of the filament, particularly as they relate to the transduction of the light signal. In the next chapter we will consider biochemical studies to highlight the control system that relates an environmental agent with gene activation in the cell.

Morphogenesis of gametophytes

Development of the adult form is accomplished in diverse ways in the gametophytes of various homosporous ferns and involves modification of the pattern of cell division in the germ filament. Morphological investigations of the types of gametophyte development in ferns have provided valuable background material for the study of experimental control of their growth. It is therefore appropriate that we begin this section by looking at some general principles underlying the evolution of form in the gametophyte.

Origin and development of the planar gametophyte

The formation of the planar gametophyte in homosporous ferns has been attributed to the activity of a single terminal cell. As mentioned earlier, the first division of the terminal cell of the germ filament is oblique or longitudinal. This is followed by a partition at right angles to the first, generating a group of three cells. One of these, usually a wedge-shaped cell which occupies the central region at the tip of the filament, now functions as the meristematic apical cell. The formation of this cell marks the initiation of a process that produces a flat prothallus. By repeated oblique or longitudinal divisions with a left–right alternation of cell plate orientation, the apical cell gives rise to a spatulate or obovate mass of cells. At first, the apical cell appears as a small indentation at the tip of the spatulate plate, but later, as the two sides of the plate extend horizontally to assume a heart-shaped form, the meristematic cell is safely lodged in the notch between the lobes. During further expansion of the lobes, the apical cell divides transversely followed by the division of the anterior cell by two or three walls parallel to each other. Thus a row of three or four narrow cells constituting a pluricellular meristem is born. From this point onward, division of cells of the meristematic plate accounts for the growth of the gametophyte. In the leaf-like prothallus, a tissue analogous to a midrib is formed by the division of cells behind the meristem. At this stage the prothallus is considered to have attained sexual maturity. In certain ferns there is very little change in the division sequences of the germinated spore up to the stage of formation of the heart-shaped gametophyte that its changing geometry readily yields to computer simulation (Korn, 1974).

Deviations have, however, been observed infrequently, such as the selective functioning of the terminal cell as a hair and the differentiation of the apical cell from the subterminal cell as in *Coryphopteris* (Thelypteridaceae; Atkinson, 1975); differentiation of the hair from one of the cells born out of the division of the terminal cell, the sister cell functioning as the apical cell, as in *Lastrea ciliata (=Trigonospora ciliata;* Thelypteridaceae; Nayar and Chandra, 1965); and initiation of a pluricellular meristem right from the outset as in *Cyclosorus (=Christella;* Nayar and Chandra, 1963), some species of *Lastrea* (Nayar and Chandra, 1965), and other thelypteroid ferns (Loyal and Ratra, 1979). In *Dryopteris pseudomas (=D. affinis)*, the apical cell is derived by a somewhat circuitous route from the terminal cell which initially divides by a wall parallel to the long axis of the filament to produce two unequal cells. The large cell is subsequently partitioned two to four times by walls perpendicular to the first division to produce a short file of cells, the smallest of which functions as the apical cell (Dyer and King, 1979).

Other variations in the formation of the planar gametophyte have been described following the rule that (1) there are differences in the sequence of cell divisions; (2) the stage of development of the gametophyte and the spot where a meristem is established are variable and (3) adult forms of gametophytes are not alike. Among other variables involved, the contribution of the apical cell itself has come under scrutiny. The difficulties that rise in comprehending the differences described in the gametophytes of more than 200 species of ferns thus far is obvious. In order to circumvent this problem and to learn more about the evolutionary and phylogenetic implications of the pathways of gametophyte development, some investigators have attempted to classify the cell division patterns leading to the formation of the planar gametophyte in homosporous ferns. One such system of classification has recognized seven patterns of development, designated as 'Adiantum' type, 'Aspidium' type 'Ceratopteris' type, 'Drynaria' type, 'Kaulinia' type, 'Marattia' type and 'Osmunda' type (Nayar and Kaur, 1971). The Adiantum type, illustrated in Fig. 7.1, is basically similar to the general pattern described earlier in which two successive divisions of the terminal cell generate an apical cell. In Ceratopteris, Drynaria and Kaulinia types, a broad spatulate plate is formed before an apical cell is delimited; while a meristematic apical cell is formed in the Drynaria type, no organized meristematic cell or meristem is ever established in Ceratopteris and Kaulinia types. A prothallus is formed by the activity of cells of the anterior region of the plate in the Kaulinia type and by the activity of a group of cells along the margin of the plate in the Ceratopteris type. Transformation of the terminal cell of the germ filament into a hair and the formation of a planar gametophyte from the subterminal cell distinguish the Aspidium type from others. Changes in the mitotic rhythm and orientation of cell plates distort the Adiantum type of gametophyte development to produce the massive Osmunda and Marattia types, both of which are characterized by being three-dimensional in their early stages. A recent study has shown that early gametophyte development in *Gleichenia bifida,* a member of the primitive family, Gleicheniaceae, is similar to Osmunda and Marattia types in having a three-dimensional mass of cells which secondarily initiate planar growth (Haufler and Adams, 1982). As far as gametophyte morphology in the evolution of homosporous ferns is concerned, it appears that there is a distinct trend toward reduction in size of the final product.

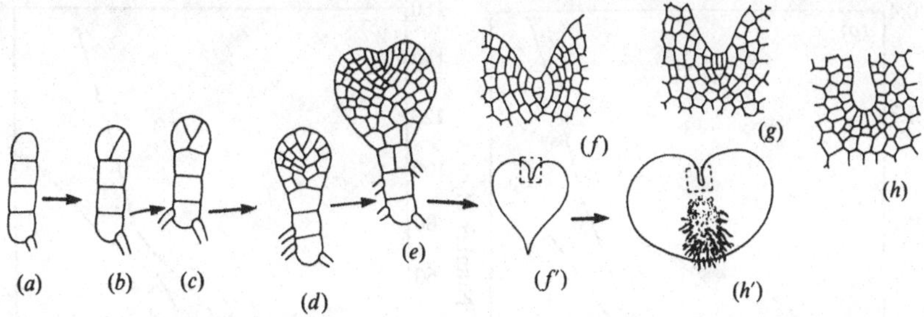

Figure 7.1. Diagrammatic representation of the Adiantum type of gametophyte development. (*a*) Germ filament. (*b*) First longitudinal division of the terminal cell of the germ filament. (*c*) Formation of a wall at right angles to the first division wall. (*d*) Spatulate plate. (*e*) Cordate prothallus. (*f*) Magnified view of the hatched square in *f′* showing the transverse division of the marginal meristem in the notch; *f′* is an outline drawing of the prothallus. (*g*) Formation of the pluricellular meristem. (*h*) Apical notch of the leaf-like prothallus, shown in outline drawing in *h′*. (From Nayar and Kaur, 1971.)

Light and planar growth of gametophytes

Since variations in light fluence produce striking form changes in fern gametophytes, regulatory effects of light on the growth of gametophytes are of pervading importance for their adjustment to the environment. In low fluence light, when the rate of cell division is low, only the filamentous form is produced. In contrast, strong light programs the filament for increased mitotic activity. The newly formed cells are integrated into a sheet-like planar form in which spatial patterning is dependent upon the successive plane of division being perpendicular to the previous one. The pronounced inhibition of elongation growth accompanied by growth in width of the planar form and the increase in length unaccompanied by increase in width of the filamentous form provide a firm basis to quantify growth data in terms of length/width ratio, designated as morphogenetic index (Mohr, 1956*b*). Miller and Miller (1961) have evaluated the effects of different fluences of white light on the growth and differentiation of germ filaments of *Onoclea sensibilis* in terms of cell number, surface area and morphogenetic index. As shown in Fig. 7.2, in weak light, the morphogenetic index of the gametophyte is high indicating the occurrence of increased cell elongation coupled with a low incidence of division. In strong light, the index is low and there is also a significant increase in the number of cells formed. The increase in mitotic activity is correlated with an increase in the surface area of the gametophyte due to planar growth.

The morphogenetic effect of light on the germ filament raises two different questions. One is concerned with the nature of light quality responsible for the transition of the germ filament to a planar

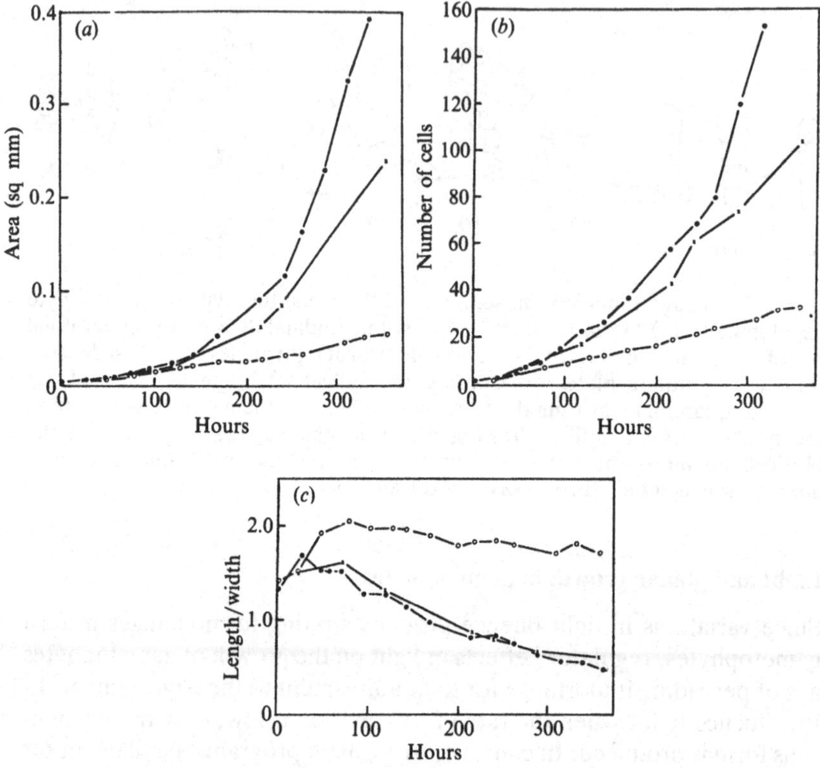

Figure 7.2. Growth of gametophytes of *Onoclea sensibilis* in different fluences of white light; ● 400 ft.-c; × 200 ft.-c; ○ 28 ft.-c. (*a*) Increase in the area of gametophytes with time. (*b*) Increase in the number of cells formed with time. (*c*) Changes in the morphogenetic index with time. (From Miller and Miller, 1961.)

gametophyte. The other question impinges on the mechanism of the light effect.

Red and blue light effects

As early as 1917, Klebs recognized that continuous exposure of protonemata of various ferns to blue light inhibited their growth in length and led to the formation of planar gametophytes. In contrast, irradiation with long wavelengths of light promoted the growth in length of the filaments with little hint of formation of new cells. Later work of other investigators like Stephan, Orth, and Teodoresco further accentuated the antagonism between short and long wavelength light in the morphogenesis of fern protonemata (see Charlton, 1938; Mohr, 1956*b*, for references).

These studies paved the way for determination of the action spectrum

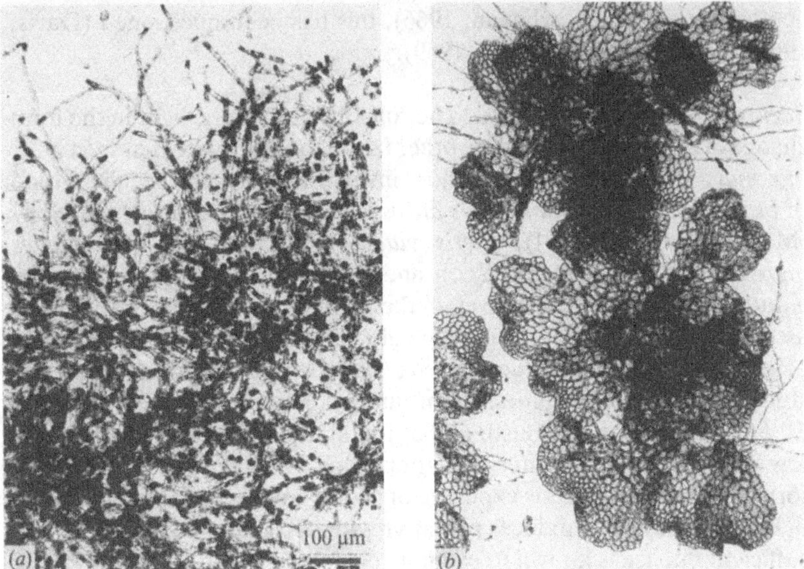

Figure 7.3. Growth patterns of gametophytes of *Asplenium nidus* in (*a*) red and (*b*) blue light. Growth periods are 50 days in both cases. Scale bar in (*a*) applies to both. (From Raghavan, 1974*c*.)

for transition of the germ filament of *Dryopteris filix-mas* to a planar gametophyte (Mohr, 1956*b*). In this work, germinated spores were exposed continuously to different wavelengths of monochromatic light. Examination of cultures at the end of six days provided the first unequivocal demonstration that planar growth occurs in the blue part (below 500 nm) of the spectrum, whereas in red light the protonema continued to grow as a filament. Fig. 7.3 shows gametophytes of *Asplenium nidus* maintained in blue and red light regimens for a period of 50 days; the distinctive effects of light quality are resoundingly clear.

From the data reviewed in the previous chapter, it is seen that elongation of the germ filament in red light is due to phytochrome action. The restriction of planar growth to the blue part of the spectrum indicates that the absorbing pigment is a yellow compound, since only a yellow substance would absorb blue light. Two flavin compounds, a flavoprotein located close to or in the cell wall with axes of maximum absorption parallel to the cell surface (Ohlenroth and Mohr, 1964; Etzold, 1965; Wada *et al.*, 1978) and a flavin-cytochrome complex with its associated electron-transport chain (Cooke, Racusen and Briggs, 1983) have been mentioned as the likely photoreceptors, but one has not been anointed yet. Although riboflavin was suggested as the pigment on the basis of its ability to reverse the inhibition of planar growth induced by nucleic acid analogs (Yeoh and Raghavan, 1966), this has been questioned (Davis, 1968*c*; Schraudolf and Legler, 1969).

Survey of red-blue light effects. The role of red and blue light in the form change of protonemata of some other ferns is analogous to that seen in *D. filix-mas* and *A. nidus*. Among those investigated, it is the peculiar virtue of protonemata of *Struthiopteris filicastrum (=Matteuccia struthiopteris)* (Mohr and Holl, 1964), *Pteris vittata* (Kato, 1965*b*), *Phymatodes nigrescens* (Dipteridaceae; Yeoh and Raghavan, 1966) and *Pteridium aquilinum* (Davis, 1968*a*) to retain the filamentous form in red light and assume planar morphology in blue light. The sensitivity of protonemata of several other ferns to blue and red light varies greatly. In a comparative study of photomorphogenesis of protonemata of *Alsophila australis* and *D. filix-mas*, Mohr and Barth (1962) found that the former require only low fluence blue light, while the latter require high fluence blue light for form change. Continuous exposure of protonemata of *A. australis* to red light for 15 days also induces planar growth. Irradiation with red light of sufficient fluence is known to promote planar growth in the protonemata of *Osmunda japonica* (Kato, 1968b) and *Pteridium aquilinum* (Sobota and Partanen, 1966). According to Kato (1967*a*, 1968*b*), protonemata of *Pteris vittata* turn planar in red light when the medium is supplemented with a carbon energy source such as sucrose, dextrose, fructose or starch, while those of *D. filix-mas* require an arbitrary combination of sucrose, yeast extract, 2,4-D, and kinetin for the acquisition of planar form in red light. Presence of PCIB in the medium along with a chlorinergic drug, pilocarpin, has been reported to inhibit cell elongation in the protonemata of *Athyrium filix-femina* grown in red light; an additional consequence of this treatment is the acquisition of planar morphology (Bähre, 1977). As pilocarpin reduces the sensitivity of gametophytes to ethylene, these results suggest that induction of planar growth is tied to the maintenance of a low level of auxin coupled with low ethylene sensitivity.

A different type of regulatory mechanism is involved in the gametophytes of *Lygodium japonicum* which assume planar morphology in both red and blue light, but remain filamentous in far-red light (Raghavan, 1973*b*; Swami and Raghavan, 1980). However, planar gametophytes generated in red light are longer than broad with narrow, rectangular cells whereas in blue light they appear broader than long with short, isodiametric cells (Fig. 7.4). Addition of 2,4-D or GA to the medium changes the morphology of the gametophytes from planar to filamentous in red light. Similarly, gametophytes growing in a medium containing 2,4-D in blue light are longer than broad, very much like those growing in the basal medium in red light. Although ABA generally retards the growth of gametophytes in both light regimens, its presence

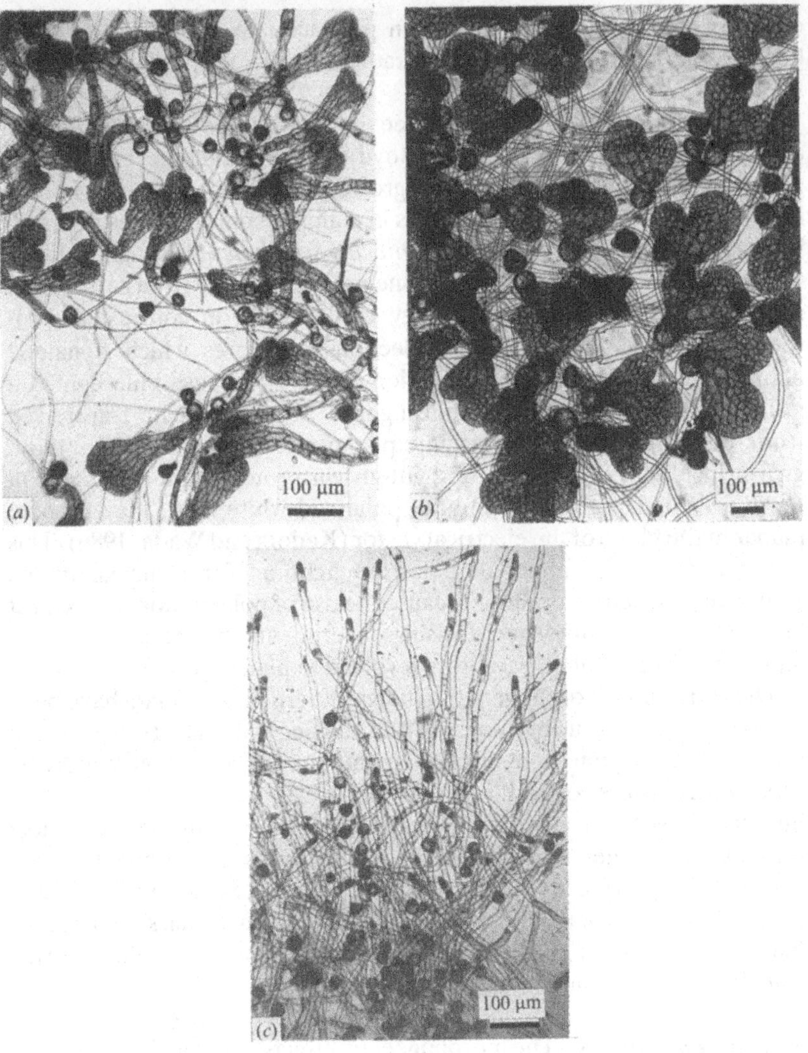

Figure 7.4. Morphological expressions of gametophytes of *Lygodium japonicum* grown in (*a*) red light, (*b*) blue light and (*c*) far-red light. Growth periods are 7 days in red and far-red and 10 days in blue light. (From Raghavan, 1973*b*.)

along with 2,4-D in red light nullifies the effect of the latter, causing gametophytes to become planar. These results suggest the possibility that light-induced form changes in the gametophytes are mediated by changes in the endogenous hormone levels. According to Grill (1987), red light-induced filamentous morphology of the protonemata of *Anemia phyllitidis* is labile. This is based on the observation that a good number of one- to three-celled red light-grown protonemata reared in the dark, become planar upon subsequent irradiation with red, yellow or green light.

By screening approximately three million spores of *P. aquilinum* exposed to γ-rays, Howland and Boyd (1974) isolated several variant gametophytes which exhibit planar growth in red light. The only known instance where planar growth occurs in complete darkness was reported in a single collection of spores of *Onoclea sensibilis* (Miller and Miller, 1970). Here, about 25% of the protonemata originating from germinated spores acquired planar morphology in the dark, in contrast to protonemata derived from other collections of spores which remained filamentous under the same condition. Finally, because blue light is a predominant component of white light, fern protonemata transferred from red light to white light acquire planar morphology similar to those grown in blue light. When red light-grown protonemata of *Adiantum capillus-veneris* are cultured under polarized white light, they become planar in the plane of the electrical vector (Kadota and Wada, 1986). This first report on the effect of polarized light on form change in the protonema is mentioned here because the use of polarization optics must be taken into account when devising experiments to determine the site and orientation of photoreceptors involved in planar growth.

Thus far, photomorphogenetic studies on fern protonemata have been largely undertaken using spores as the starting material. However, the use of tissues of gametophytic origin maintained in continuous culture by repeated transfers to fresh medium might also be rewarding for these investigations. Such a tissue, which for all intents and purposes looks like a callus, is obtained when spores of *P. aquilinum* are germinated on a solid medium enriched with yeast extract (Steeves, Sussex and Partanen, 1955). Subculture of the tissue in a liquid medium in the dark, generates a filamentous mass of cells that responds to red and blue light in a way similar to protonemata (Raghavan, 1969*a*).

Red-blue antagonism. The morphogenetic effects of red and blue light on the protonemata of ferns such as *Dryopteris filix-mas* have often been interpreted as the result of differential mitotic activity. Since protonemata develop largely by cell elongation in red light and by cell division in blue light, it was naturally thought that light regimens could act antagonistically by modulating the elongation potential or division potential of cells. Ohlenroth and Mohr (1964) found that introduction of 10% blue light into a standard red light field reduces the morphogenetic index of protonemata of *D. filix-mas* although it is significantly higher than that of protonemata grown in unadulterated blue light. Similarly, there is an increase in the morphogenetic index when 10% red light is mixed with a standard blue field. One would predict from these results that a combination of 50% blue light and 50% red light will increase the index significantly over that obtained in 50% blue alone. This is in fact the case (Fig. 7.5). These well-marked changes in morphogenetic index testify to the existence of an antagonism between red and blue light in determining the final form of gametophytes and the extent to which this is dependent upon differential cell division and cell elongation.

Transfer of gametophytes from one light regimen to another also introduces corresponding changes in their morphology in long-term cultures. Thus, a shift of gametophytes of *D. filix-mas* from red to blue light entrains them to a rapid replicative cycle resulting in planar growth, while in the reverse transfer, new growth that appears is filamentous (Mohr, 1956*b*). Transfer of gametophytes of *L. japonicum* of all ages from far-red to red or blue light induces morphological growth characteristic of the new light regimen. However, only relatively young planar

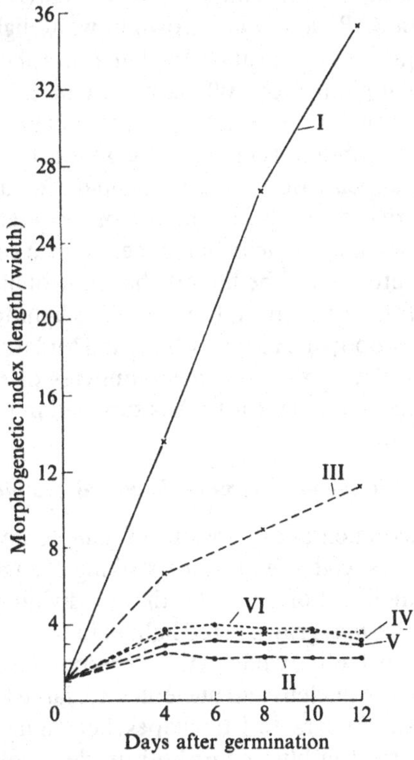

Figure 7.5. Changes in the morphogenetic index of the gametophytes of *Dryopteris filix-mas* grown in red and blue light fields and combinations of both. (I) 100% red light; (II) 100% blue light; (III) 100% red and 10% blue; (IV) 50% red and 50% blue; (V) 50% blue; (VI) 100% blue and 10% red. (From Ohlenroth and Mohr, 1964.)

forms transferred from red or blue light are able to resume filamentous growth in a subsequent period of far-red light (Raghavan, 1973*b*). This observation has an explanation based on similar experiments done with gametophytes of *Pteridium aquilinum* (Sobota, 1970). Here, the rate of reversal of planar growth to filamentous type in red light is found to decrease with increasing periods of pretreatment of protonemata in white light. For example, after 9 days in white light, less than 25% of gametophytes become filamentous during 7 days of subsequent growth in red light. After 13 days in white light only 10% revert to filamentous morphology during the next 5 days in red light. These observations underlie a progressive change in the activity of the marginal meristem which becomes more established with increasing periods of growth of gametophytes in white light. It has been postulated that filament formation is promoted through the loss of meristematic activity and accordingly, as the meristem becomes more stable, conditions which inhibit cell elongation and filament formation are favored. Davis (1971) found that when protonemata of *P. aquilinum* grown in white light are kept in red light, the time required for transition to planar morphology is correlated with the rate of their elongation. When the rate of filament elongation is low, a large percentage of protonemata become planar in a short period of time, while under conditions favoring a high rate of elongation, few, if any, planar forms appear. In summary, the main features of this system are clearly suggestive of a type of photomorphogenetic control which affects the developmental capacity of the meristem at successive stages of growth of the gametophyte. The factors that control filamentous growth in gametophytes transferred from blue or white light to dark conditions are less well-understood; one report (Miller and Miller, 1964*b*) has shown that recalcitrant white light-grown gametophytes of *Onoclea sensibilis* become filamentous in the dark if IAA is supplied in the medium.

Cytological changes in the protonemal cells in red and blue light

The primary developmental expressions of gametophytes reared in red and blue light are associated with some striking changes in the size and shape of cell organelles. For example, the nondividing basal cell of the protonema of *D. filix-mas* grown in red light or in complete darkness has a small nucleus and nucleolus, while in the corresponding cell of blue light-grown protonema, the nucleus and nucleolus are large (Bergfeld, 1963*b*). The nuclei also assume characteristic shapes, becoming spindle-shaped in red light and spherical in blue light; both nuclear volume and nuclear shape are reversible by changing the light quality. Responses of the nucleus and nucleolus of the terminal cell of the protonema to red and blue light are just the opposite of what is seen in the basal cell (Bergfeld, 1967). Although the small size of the nucleus and nucleolus of the terminal cell in blue light may reflect the high mototic activity of this cell, the burgeoning nature of these organelles in the corresponding cell of the protonema in red light is difficult to reconcile. Speculation concerning this has emphasized the possibility that red light activates the synthetic potential of the organelles leading to increase in their volume at the same time permitting only a limited transfer of macromolecules into the cytoplasm. What is appealing here is that light-induced volume changes in the nucleus and nucleolus of cells positioned in different parts of the protonema provide an ideal parameter to correlate metabolic changes with morphological expression.

No other organelle in the cell can match the chloroplast in its responses to light quality. Compared to cells of dark-grown and red light-grown protonemata, cells of protonemata nurtured in blue light have a munificient endowment of chloroplasts. Plastids in the basal cell of the protonema of *D. filix-mas* grown in red light are generally colorless and range in diameter from 0.3 to 4.8 μm while in the corresponding cell of the protonema exposed to blue light they are longer and thicker, reaching a diameter of 8.0 μm. Like nuclear volume, plastid volume is also linked to light quality and the enlargement and contraction of plastids can be controlled by changing the light field in which gametophytes are grown (Bergfeld, 1963*a*). As seen by electron microscopy, starch accumulation, along with extensive stacks of grana, accounts for the large size of plastids of the basal cell of the protonema exposed to blue light (Bergfeld, 1970). However, plastids of the terminal cell of red light-grown protonema are generally larger than those of the cell of the protonema grown in blue light. The former are also characterized by an array of vesicle-like protrusions extending from their surface. It has been suggested that the most swollen plastids carry the major burden for macromolecule synthesis, the vesicles functioning to relieve topological restraints in the transfer of biosynthetic products into the cytoplasm.

Rapid effects related to planar growth

Although the first longitudinal division of the terminal cell of the germ filament occurs in about 12 to 24 hours in blue light, a tangle of events all competing in the same region is detected even at earlier times following exposure to actinic radiation. Recent work by Cooke *et al.* (1983) has called attention to modifications in membrane properties as a critical initial step in the growth response of *Onoclea sensibilis* filaments to blue light. The import of this work is that a massive hyperpolarization of the membrane occurs in the terminal cell of the filament within seconds after blue light exposure and that inhibitors like SHAM and sodium vanadate block this. Based on the activity of SHAM as an inhibitor of flavin-cytochrome complex and of vanadate as an inhibitor of proton-pumping

ATPase in the plasma membrane, it seems probable that the photoreaction is followed by plasmalemma hyperpolarization as a result of enhanced activity of ATP-dependent proton pumps. Regardless of the possible mechanism, the essential point is that some really impressive electrical changes occur very quickly in the membrane at the business end of the protonema in blue light.

But how could these changes promote morphogenesis? One possibility is that a high osmotic pressure or turgor built up in the terminal cell pushes the cell wall outward resulting in a swelling at the apex and inhibition of elongation of the cell, thus setting the stage for the first longitudinal division. Several investigators have noted that a broadening of the terminal cell (apical swelling) is the earliest visible change that occurs after exposure of red light-grown filaments to blue or white light (Mohr, 1965; Mohr and Holl, 1964; Drumm and Mohr, 1967*a*; Davis, 1969; Davis *et al.*, 1974; Howland, 1972); Fig. 7.6 displays this in a representative case. For protonema of *D. filix-mas* shifted from red light to blue light, the response time for apical swelling is of the order of three hours (Mohr, 1965; Drumm and Mohr, 1967*a*). Time-lapse photography has shown that the swelling is due to the incorporation of newly synthesized wall material on the flanks as well as on the dome of the terminal cell and not by the softening of cell wall originally synthesized in red light (Howland, 1972). A response time of one to two hours is typical of the protonema of *P. aquilinum* transferred from red light to white light (Davis, 1969; Davis *et al.*, 1974). According to Davis *et al.* (1974) who mapped the relative positions of ion exchange beads adhering to the terminal cell of the protonema during the early hours of its growth in white light, the swelling which occurs in the dome and in the proximal region of this cell, may well involve recycling of the wall material produced in red light. This conclusion is at variance with that of Howland (1972) referred to earlier, and although the contribution of the wall material produced in red light may be quantitatively trivial, it renders likely a role for white light in triggering the synthesis of a wall macerating enzyme. The most vivid demonstration of expansion of the subterminal wall is given by Stockwell and Miller (1974) who observed a pronounced hypertrophy of cells when dark-grown protonemata of *O. sensibilis* were transferred to 0.15% colchicine in white light. Comparison of the diameter of the terminal cell at different points from its dome before and after colchicine treatment has shown that increase in cell diameter occurs behind the dome. It will be interesting to follow this up by electron microscopy to examine the role of microtubules in the swelling phenomenon.

While prepattions for apical swelling are in progress, a different type of change is also initiated in the terminal cell and that is the inhibition of its axial elongation. Because of the variability in length of germ filaments, it

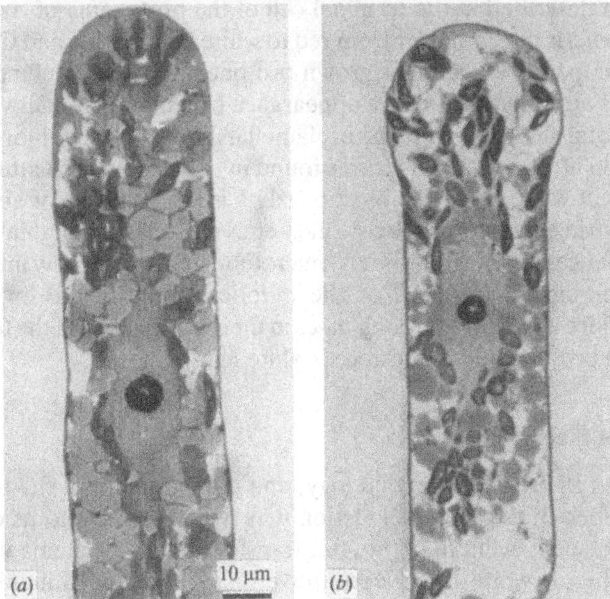

Figure 7.6. Apical swelling of the protonema of *Adiantum capillus-veneris* (*a*) before and (*b*) 3 hours after transfer from red light to blue light. Scale bar in (*a*) applies to both. (From Wada *et al.*, 1978; photographs supplied by Dr M. Wada.)

is difficult to determine the response time for this effect with any degree of accuracy. Data based on time-lapse photography (Howland, 1972; Davis *et al.*, 1974) indicate that filament elongation does not continue in blue light or in white light beyond the time when the first signs of swelling appear. Thus, it is possible that both swelling and inhibition of elongation involve the concerted action of the same sets of microfibrils in the terminal and subterminal zones of the cell.

Not all investigations of the terminal cell have focused on its growth parameters. Significant ultrastructural changes occur in this cell in response to actinic radiation, although their role in signal transduction is not understood. Cran and Dyer (1975) have described some modifications of chloroplast structure as early as one hour after transfer of the protonema of *D. borreri* (=*D. affinis*) to blue light. The modified chloroplast has numerous small structures resembling prolamellar bodies associated with certain undefined circular bodies in the region of the thylakoid. We have been alerted to the possibility that the prolamellar bodies may function as sites of accumulation of chlorophyll precursors, but this is purely speculative. Certainly, these bodies are something to be watched for in the protonemal cells of other ferns exposed to short periods of blue light. The most clear-cut changes in organelle disposition have been described in the terminal cell of the protonema of *Adiantum capillus-veneris* after transfer from red to white light (Wada and O'Brien, 1975). Compared to red light-grown protonema, apical swelling of the filament is accompanied by the appearance of numerous small vacuoles and the total disappearance of prolamellar bodies in the chloroplasts. Among the ultrastructural features found in the terminal cell after about ten hours in white light, the most puzzling is the presence of abnormal mitochondria characterized by swollen ends connected by a thin tube or by the absence of cristae. Although microtubules are obviously important in determining the position of the mitotic spindle in the impending mitosis, there are no dramatic changes in their distribution in the terminal cell of the protonema transferred to white light.

Division of the terminal cell

Changes in the size, shape, cytology, and ultrastructure of the terminal cell described in the previous section may well be critical to its division which is the next event to occur after transfer of the protonema from red light to blue or white light. Descriptively, increase in cell number and DNA content are the most characteristic events registered when red light-grown protonemata are transferred to blue or white light (Drumm and Mohr, 1967*b*). As seen in *Pteris vittata,* the first significant wave of mitotic activity occurs with incredible synchrony as early as 13 hours after transfer of protonemata from red to blue light, then declines, to be followed by a smaller wave about 10 hours later (Fig. 7.7). These results lend strong support to the conclusion that nuclei of red light-grown filaments are retained essentially in the G_1 phase of the cell cycle. Even if the filaments are exposed to blue light for only a short period during this time, the cells progress synchronously through the S and G_2 phases causing increased cell production (Ito, 1970, 1974).

After exposure of protonemata to blue or white light, there are two planes in which the first division wall is engineered in the terminal cell. In one case, the initial one or two walls are formed in the oblique or longitudinal plane. This seems to be a logical enough means to set off the protonema in the planar way almost immediately. However, the most general form of cytokinesis is one in which the first one or two walls are laid down perpendicular to the long axis of the terminal cell followed by a second or third in the oblique or longitudinal plane initiating planar growth. We might expect that in both cases, the plane of wall formation will be absolutely predictable if it were controlled by red-blue (or white) light interactions, and in fact it is. As shown elegantly in the protonema of *A. capillus-veneris,* cell plate formation for the first transverse division occurs between 14 to 18 hours after transfer of red light-grown protonema to white light. Under the same condition, the second transverse division

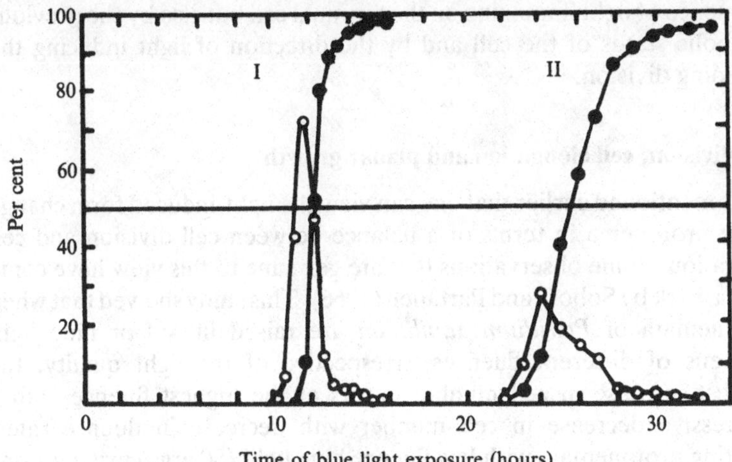

Figure 7.7. Changes in the percentage of cells in mitosis (mitotic index, ○) and the percentage of divided cells (●) for the first and second divisions of the protonemata of *Pteris vittata* grown initially in red light and then transferred to blue light. (From Ito, 1970.)

occurs within 24 hours, followed by the first longitudinal division in the terminal cell in 72 hours (Wada and Furuya, 1970). There are a few means of achieving a controllable delay in the orientation of the cell plate for the longitudinal division, the most precise one being the administration of red light between the first and second or between the second and third divisions. To complicate matters further, exposure of the protonema to red light for more than 100 hours after the first transverse division or for more than 42 hours after the second transverse division even changes the orientation of the cell plate for the third division during a subsequent period in white light, from the longitudinal to the transverse plane. Similar temporal and spatial changes in the alignment of the cell plate for the longitudinal division are seen when white light is administered to the tip of the protonema from different directions. Thus, when white light for the induction of the second and third divisions is administered from the same direction, the cell plate for the third division is laid down in most filaments in the longitudinal plane. On the other hand, if the direction of light for the third division is at right angles to that for the second, the percentage of protonemata with a longitudinal cell plate is diminished with a concomitant increase in the number of filaments with a third transverse cell plate in the terminal cell (Wada and Furuya, 1971). These experiments show the complexity of the interactions between contrasting environmental signals that occur in the terminal cell of the protonema preparatory to planar growth. All told, it would appear that the orientation of the cell plate initiating planar growth is controlled not only by the fluence of light impinging on the terminal cell but also by the previous metabolic status of the cell and by the direction of light inducing the preceding division.

Cell division, cell elongation and planar growth

It was mentioned earlier that one can view the light-induced form change in the protonema in terms of a balance between cell division and cell elongation. Some observations that are germane to this view have come from a work by Sobota and Partanen (1966). This study showed that when protonemata of *Pteridium aquilinum* are raised in red or blue light regimens of different fluences, irrespective of the light quality, the greatest increase in cell number occurs at the highest fluence with a progressive decrease in cell number with decrease in fluence rates. Treating protonemata with low fluence blue light (50 ergs/cm^2/s) for one hour every 12 hours results in their continued filamentous growth, while those exposed to high fluence red light (7000 ergs/cm^2/s) for 12 hours every day become planar like those given high fluence blue light. To explain these results, an interpretation of the development of form in fern gametophytes as a function of the rate of cell division which in turn influences the plane of division has been found to be useful. The idea of a causal relationship between mitotic activity and the orientation of the mitotic spindle has been attributed to an influential hypothesis by Stebbins (1967), who suggested that when the frequency of cell division is low, cells normally tend to grow between divisions, and that in such elongating cells wall formation will occur in a transverse plane. However, as mitotic activity increases, the interval between divisions does not make up for the halving of cell size, thus effectively limiting cell length and increasing the probability that the mitotic spindle will be oriented in a different plane at each division. In other words, according to this hypothesis, in organisms which exhibit a regularly shifting synchrony between cell division and cell elongation as a way of life, the plane of division of the cell will depend upon its length.

This is logically a very attractive hypothesis. If the form of the gametophyte is controlled by the relative frequency of cell division and cell elongation, on any grounds we would anticipate that substances promoting cell elongation or inhibiting cell division would prevent planar growth and that substances that promote cell division relative to cell elongation would speed up this transition. Some experiments have provided a body of evidence to confirm this. For example, Miller (1968*b*) found that addition of 10^{-7}M IAA to the medium inhibits planar growth in the protonemata of *Onoclea sensibilis* and causes them to grow as filaments in white light beyond the stage at which protonemata grown in the basal medium become planar. An impressive promotion of cell division relative to cell elongation is obtained by the application of 2-chloroethylmethylammonium chloride (CCC). Because of its inhibitory action on cell elongation in various angiosperms, this compound is considered as a growth retardant, and in this sense, its effects are opposite to those induced by GA. When protonemata of *Pteridium aquilinum* are grown in white light in a medium supplemented with 10^{-3}M CCC, there is a four- to five-fold increase in cell number. The production of new cells is of such magnitude as to commit the filament to an enhanced transition to planar morphology (Kelley and Postlethwait, 1962). At a first glance, therefore, it appears that the relationship between the rate of cell division, plane of cell division, and planar growth in the protonema is justified. However, other properties of the protonema, such as the geometry of the terminal cell and its metabolic status may also influence or even determine the induction of planar morphology.

In searching for a model to interpret the division of the terminal cell of the protonema, the biophysical properties of the wall of a dividing cell have come under close scrutiny. The general situation is summarized by the statement that the ability of the cell plate in the dividing terminal cell to respond to thermodynamic constraints of minimal surface area will determine the pattern of its growth (Cooke and Paolillo, 1980*b*). In an operational sense, one can envisage that a terminal cell geometry that makes the transverse division plane the position of minimal surface area will perpetuate filamentous growth, whereas planar growth occurs if the longitudinal division plane becomes the position of minimal surface area. Time-lapse photographic observations of growth of the protonema of *O. sensibilis* in various light regimens as well as in chemical milieu and actual measurements of wall areas of the protonema have indeed shown that in each case the new division wall assumes the position dictated by the minimal surface area. While these observations are compatible with the hypothesis which is thus an accurate predictor of the plane of division in the terminal cell, it does not provide a mechanism to account for the phenomenon. Based on a concept developed by Lintilhac (1974) to explain the plane of wall orientation in the cells of a tissue, Miller (1980*b*) has proposed that division of the terminal cell of the protonema is due to the local pattern of stress imposed at the site of the nucleus during mitosis and cytokinesis. This has been found to be the case in cells in which unequal stress was generated in the longitudinal or transverse planes by deplasmolysis. In each case, division of the cell occurs in the stress-free plane.

Evidence of another kind seems to favor the view that acquisition of morphogenetic potency by the protonemal cells is an important control of planar growth; for purposes of this discussion, morphogenetic potency is viewed as a series of biochemical events concerned with the synthesis of specific macromolecules. Fig. 7.8 shows an experiment in which the time

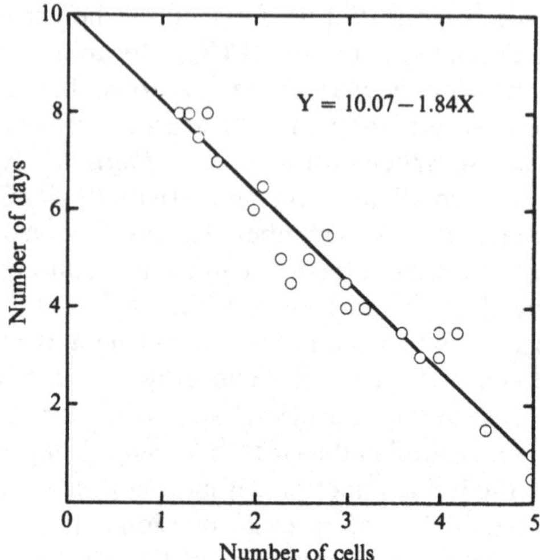

Figure 7.8. Graph showing the number of days required for planar growth in 50% of germ filaments of *Asplenium nidus* with different initial cell numbers. The regression line drawn is calculated according to the equation given. (From Raghavan, 1974*a*).

for 50% transition to planar morphology of protonemata of *Asplenium nidus* with different initial cell numbers was determined (Raghavan, 1974*a*). It is seen that protonemata with a fewer number of cells to begin with, require a longer time in white light to become planar than those with more cells; implication of this is that the division of the protonema initial to form more cells is part of a process which endows morphogenetic potency under the light conditions employed. As supplementation of the medium with sucrose does not result in an appreciable increase in the percentage of planar forms in the protonemata with fewer cell numbers, the contribution of additional cells of the filament does not appear to be photosynthetic. More direct evidence to the point is derived from biochemical analyses of culture filtrates of protonemata grown under light conditions promoting planar morphology, showing the presence of promotory and inhibitory compounds (Smith *et al.*, 1973). There is no reason to suppose that either group of compounds confers morphogenetic potency on the protonema, although nothing excludes that possibility.

At one time it was assumed that induction of planar growth is correlated with the formation of a fixed number of cells in the protonema. This assumption is no longer true, as protonemata attain planar morphology after generating fewer and fewer numbers of cells when germinated spores are exposed to increasing photoperiods or increasing fluences of light. As shown in Fig. 7.9, under a 24-hour photoperiod provided by

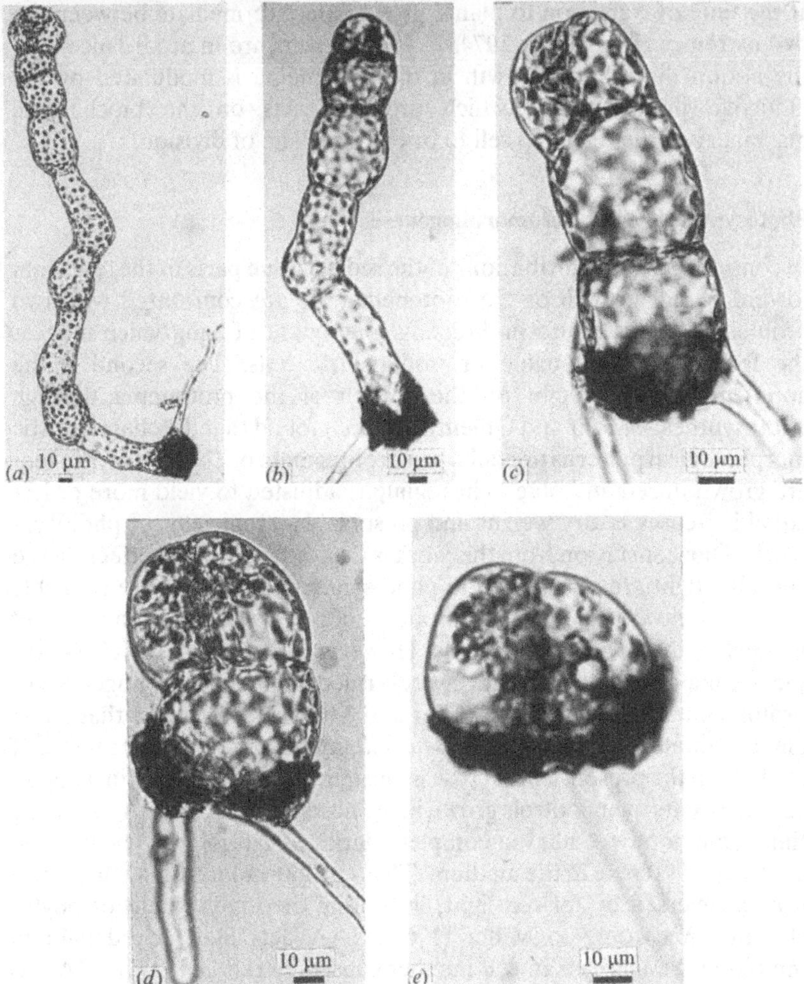

Figure 7.9. Representative gametophytes of *Asplenium nidus* grown in different photoperiods, showing the longitudinal division of the terminal cell. (*a*) Photoperiod of 3 hours for 26 days. (*b*) Photoperiod of 6 hours for 16 days. (*c*) Photoperiod of 12 hours for 10 days. (*d*) Photoperiod of 24 hours for 6 days; all of the above at 1000 µW/cm²/s. (*e*) Photoperiod of 24 hours (1400 µW²/s) for 5 days. (From Raghavan, 1974*a*.)

1400 µW/cm²/s white light, even the protonema initial of *A. nidus* is partitioned longitudinally to initiate planar morphology. In this light regimen, 50% of protonemata become planar when they have an average of 1.8 cells. In contrast, 50% of protonemata grown in a 3-hour photoperiod do not become planar until the average cell number is 9. In intermediate photoperiods, the critical cell numbers in the protonemata at the time of transition to planar growth are intermediate between the two extremes (Raghavan, 1974*a*). These results are in accordance with the notion that planar growth in the protonema is modulated by the inductive light regimen which probably acts on the biochemical machinery of the terminal cell to orient the plane of division.

Photosynthesis and photomorphogenesis

In considering the contributions of the red and blue parts of the spectrum toward planar growth of the protonema, we are confronted with two problems. One is the form-directing component of light which induces the formation of elongate or isodiametric cells. The second is the apparent effect of light on the growth of the protonema through photosynthesis. Mohr and Ohlenroth (1962) found that the characteristic morphological patterns prevail when protonemata of *Dryopteris filix-mas* are grown in red and blue light regimens adjusted to yield more or less equal increases in dry weight and presumably equal rates of photosynthesis. One conclusion from this work is that the differential effects of red and blue light are not related to photosynthetic rates, but are probably due to the flow of the primary products of photosynthesis in a different manner in the two light regimens. However, an effect of the photosynthetic component of light is not entirely ruled out in the morphogenesis of protonemata of other ferns. Miller and Miller (1961) found that when planar gametophytes of *Onoclea sensibilis* are exposed to a low fluence of red light in the presence of sucrose in the medium, a three-fold increase in growth occurs over controls grown in the absence of sucrose. At the same time, gametophytes held in complete darkness fail to grow even in the presence of sucrose in the medium. The red light-induced growth promotion is reversed by far-red light, indicating the involvement of phytochrome. According to Miller (1961), IAA acts like far-red light in inhibiting the increase in cell number caused by red light. Since IAA is practically inactive in the absence of sucrose in the medium, the photosynthetic component of light which provides the carbon energy source seems to be somehow involved in phytochrome action and in the photomimicking action of auxin.

The role of photosynthesis in photomorphogenesis has been demonstrated more directly in experiments in which the photosynthetic rates of gametophytes of *Pteridium aquilinum* were determined under different light conditions (Donaher and Partanen, 1971). Although photosynthesis continues uncurbed during the initial stages of growth of the protonemata in red and blue light, there is a later decline in activity in protonemata grown in red light. Since the form of gametophytes as filaments in red light or as planar structures in blue light is established during the period of their high photosynthetic activity, it has been concluded that specificities of form in the two light regimens are due to differences in the photoreactions. It was also shown that although blue light provides the stimulus for form change, it nonetheless modifies both growth and form. The form-directing function of blue light is satisfied at low fluence levels when the typical isodiametric cell shape is attained; however, this level of blue light even in conjunction with sucrose is unable to support continued growth, indicating a requirement for high energy blue light to sustain vigorous photosynthesis. The final outcome of morphogenesis is therefore dependent upon the quality of light which determines the form as well as its quantity which synthesizes substrates for growth. The effects of the latter are so little comprehended and variable in gametophytes of different ferns that it is difficult to speak of them in general terms. Nonetheless, they exist and cannot be ignored.

General comments

Light signals modulate the growth and development of fern gametophytes leading to the evolution of two morphological patterns – a filament in red light and a planar form in blue light. The potential to undergo rapid mitoses distinguishes the planar form from the filament in which cells display a tendency for elongation. How the protonema initial of the germinated spore responds to different light signals depends upon the informational state of the cell as determined by nucleocytoplasmic traffic following a photoreaction.

Photomorphogenesis of the fern gametophyte is a multifaceted process; several steps of control exist to sustain filamentous and planar types of growth, but the real question is how light signals cause the mitotic spindle to align in one plane rather than in the other in the target cell. A key discovery, such as differences in the number of microtubules fabricated or in the directionality of movement of tubulin or its precursors in the terminal cell of the protonema grown in red and blue light could make all the difference in the world in our understanding of form change in this system. Thanks to technical advances such as high resolution electron microscopy, immunofluorescence techniques and *in situ* localization methods, we are now in a position to examine the cells of the fern protonema more closely than before.

8

Role of protein and nucleic acid synthesis in planar growth

We have seen in the previous chapter that planar growth of the protonema in blue light is due to a change in the orientation of the plane of division of the terminal cell from transverse to oblique or longitudinal. This response to blue light may be the microscopically visible manifestation of a sequence of complex biochemical processes in the cell subsequent to the photochemical reaction. These processes are initiated by the transcription of a particular set of polysomal mRNA which moves into the cytoplasm to direct the synthesis of new proteins. The general view is that the pattern of nucleic acid and protein synthesis in the protonema might provide the operational criterion for gene transcription during planar growth. We will use this as a frame of reference for discussion in this chapter on the reprogramming of the cellular machinery of the gametophyte during photomorphogenesis. The approach chosen here is to introduce the various facets of earlier investigations and to focus on those concepts which have received renewed experimental attention in recent years. This subject is also considered in the reviews of Miller (1968*a*), DeMaggio and Raghavan (1973), and Raghavan (1974*c*).

Protein synthesis

A change in the rate synthesis of enzymatic, regulatory and structural proteins is the basic attribute of an organism embarking on a specific mode of growth. Generally speaking, the higher the rate of cell elongation or cell division, the more active is protein synthesis. Thus, the rate of protein synthesis and less frequently, the rate of protein accumulation are often used as an index of the change in the pattern of growth.

Attempts to relate the changes in protein content of the germ filament to the pattern of its growth have led to contradictory interpretations. In the first investigations on this problem, it was found that during growth of the gametophyte of *Dryopteris erythrosora*, there is a rise in the absolute amount of protein nitrogen, which is particularly marked at the time of transition from filamentous to planar morphology (Hotta and Osawa, 1958). If the protein nitrogen data are expressed in terms of mg dry weight of gametophytic tissue, the curve follows the same trend as for the absolute amount. A study of the amino acid composition of the proteinaceous fraction of gametophytes of different ages suggested that proteins synthesized during planar growth are qualitatively different from those made during the filamentous phase (Hotta, 1960*a*). It seems likely, therefore, that the change in form of the gametophyte is associated with and is possibly caused by a rapid synthesis of proteins. As increased cell division is the principal activity of the gametophyte during planar growth, many of the proteins made are probably concerned with membrane biogenesis, chromosome replication and mitosis.

The case for a causal role for protein synthesis in planar growth was however considerably weakened by the linear relationship obtained when the logarithms of values for absolute nitrogen content are plotted against time (Bell and Zafar, 1961). This work also showed that if the mean organic nitrogen contents of individual gametophytes of the related *D. borreri* (= *D. affinis*) are plotted logarithmically, the rate of increase in nitrogen content is constant as one would expect in an organism growing exponentially in a medium providing unlimited nutrition. It has been reasoned from these results that during transition of the filament to a planar form, there is no fluctuation in the rate of protein synthesis. An interesting post-script to this controversy is the observation that when the protein content of the gametophyte of *Pteridium aquilinum* is expressed per unit dry weight, there is an increase associated with planar growth; however, on a per cell basis, cells of the filamentous gametophyte have a higher protein content than those of the planar form (Sobota and Partanen, 1967).

Later studies by Mohr and associates (Ohlenroth and Mohr, 1963, 1964; Kasemir and Mohr, 1965; Deimling and Mohr, 1967; Payer, 1969; Payer and Mohr, 1969; Payer, Sotriffer and Mohr, 1969) on the gametophytes of *D. filix-mas* growing at nearly the same rates of photosynthetic dry matter accumulation in red and blue light regimens have unequivocally shown that increased protein synthesis occurs during planar growth. As shown in Fig. 8.1, there is little difference in the absolute protein content of gametophytes growing in red and blue light until at least four days after irradiation and often until even later. Thereafter the protein content of gametophytes growing in blue light clearly outstrips that of gametophytes growing in red light. The differences in the relative protein content of gametophytes in red and blue light are maintained when the dry matter accumulation is modified by increasing or decreasing CO_2 tension in the medium. Moreover, when gametophytes are exposed to increasing fluences of red or blue light, the expected surge in protein nitrogen is greater in blue than in red light (Ohlenroth and Mohr, 1963, 1964). These results are clearly consistent with the general view that blue light is an enriched environment and red light is an impoverished environment for the growth of gametophytes. Transfer of gametophytes from the enriched to the impoverished environment not only causes filamentous growth but also leads to a precipitous decrease in protein content, while the reverse transfer initiates planar growth with a concomitant increase in protein content. There is no evidence to suggest that blue light induces increased protein accumulation through increased amino acid synthesis, for if it does, there should be an increase in the pool size of amino acids of gametophytes grown in blue light. This is not the case and hence, the suggestion that the control of protein synthesis in blue light is at the level of polypeptide assembly is probably tenable (Deimling and Mohr, 1967).

Further studies have characterized the pattern of protein synthesis during photomorphogenesis in gametophytes of *D. filix-mas* by incorporation of radioactive label. In one experiment it was found that gametophytes allowed to photosynthesize in blue light in an atmosphere of $^{14}CO_2$ incorporate the label into proteins at a faster rate than those exposed to red light (Payer *et al.*, 1969). Another work showed that during a 40 min administration of $^{14}CO_2$, the rate of photosynthesis and the rate of label incorporation into the amino acid pool of gametophytes are similar in the two light regimens (Payer and Mohr, 1969). However, protein synthetic activity increases about one hour after transfer of

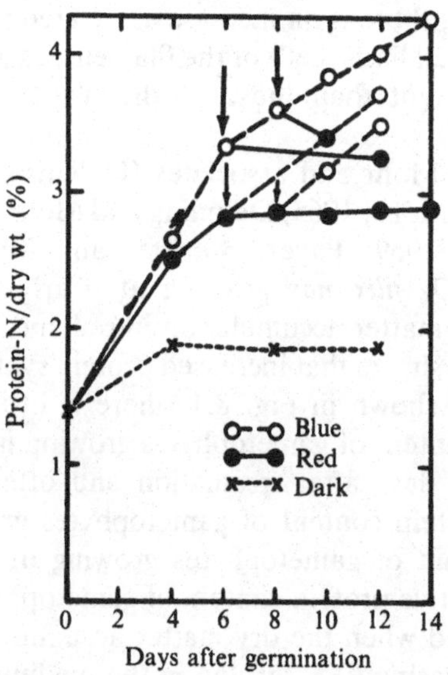

Figure 8.1. Changes in protein nitrogen content per dry weight of gametophytes of *Dryopteris filix-mas* during growth in complete darkness, in red light and in blue light, and during transfer from one light regimen to another. Arrows indicate the times during the growth period when transfers were made. (From Ohlenroth and Mohr, 1963.)

gametophytes from red light to blue light, the increase continuing up to 16 hours at which time the incorporation of the label catches up with that of gametophytes grown in continuous blue light. Data on the specific activity of free alanine in gametophytes exposed to $^{14}CO_2$ in red and blue light regimens also agree well with the pattern of incorporation of radiocarbon into proteins (Payer, 1969). Generally, an increase in the specific activity of alanine occurs after exposure of gametophytes to blue light for one hour. When the incorporation reaches a steady rate after about 4 hours, the specific activity of ^{14}C in alanine is about 1.8 times higher in blue light than in red light. On the face of it, one might conclude from these results that protein synthesis in blue light is correlated with the swelling of the apical cell of the filament which begins at about 3 hours. Conceivably, these early formed proteins have a high turnover rate and constitute the enzymes which oversee the reactions initiating morphogenetic change. A puzzling feature of the enzyme factor in gametophyte morphogenesis is that it has been intractable to analysis so far and we know so little about the relevant enzymatic activities.

Chloroplast protein synthesis

We will now discuss the nature of the large fraction of proteins synthesized by the gametophyte in blue light. From an analysis of the relationship between chlorophyll content, total protein content and dry weight of gametophytes of *D. filix-mas* grown in red and blue light, Kasemir and Mohr (1965) concluded that the increase in protein content in blue light is due to the synthesis of structural proteins which are later stockpiled in the chloroplasts. Partly as a result of this analysis, chloroplasts of fern gametophytes grown in the two light regimens have been the focus of considerable cytological and biochemical research. The comparatively large size of chloroplasts of gametophytes grown in blue light attests to their increased protein content. This seems to be consistent with the observation that 5-methyltryptophan, a general inhibitor of protein synthesis, not only blocks planar growth, but also prevents the enlargement of chloroplasts in blue light (Bergfeld, 1964*a*). In contrast, chloramphenicol, an inhibitor of mitochondrial and chloroplast protein synthesis, does not curb planar growth of the protonema, but blocks the swelling of the plastids (Bergfeld, 1968). Thus, significant as is the accumulation of proteins in the plastids, its connection to form change in the protonema appears tenuous.

Other approaches have verified that the plastids function as sites for the synthesis and accumulation of proteins during the growth of gametophytes. Determinations of protein contents of particulate fractions of gametophytes of *Pteridium aquilinum* obtained by differential centrifugation have shown that a considerable proportion of the total proteins is found in the chloroplasts (Raghavan, 1968*b*). Using ^{3}H-leucine as a precursor, a relatively high level of protein synthesis is found in the chloroplast-rich fraction. Generally, chloroplasts isolated from blue light-grown gametophytes exhibit a higher protein content and label incorporation than a corresponding fraction isolated from red light-grown gametophytes, suggesting that plastids of the same dry weight are more efficient in blue light than in red light (Raghavan, 1968*a*). Additional evidence for this view was obtained by the demonstration that if gametophytes raised in red light are exposed to blue light, there is nearly a five-fold increase in the amino acid incorporation activity of isolated chloroplasts. The enhanced activity is maintained only during the first 48 to 72 hours after transfer of gametophytes to blue light, after which it attains the level of activity of chloroplasts isolated from gametophytes exposed to continuous red light (Raghavan and DeMaggio, 1971*a*). As gametophytes are maintained under conditions of equal dry weight increase in red and blue light, a strong case can be made here for the existence of a photosynthesis-independent effect of blue light on protein synthesis in the plastids. Beyond this general statement, we do not know much about the role of plastid proteins, which remains quite mysterious.

Further investigations are necessary before we can write off chloroplast protein synthesis as being inconsequential in the form change of the protonema in blue light. On the basis of some limited studies, plastids of fern gametophytes appear to be somewhat unusual in the pattern of their division (Gantt and Arnott, 1963) and in their spartan requirements for amino acid incorporation into proteins (Raghavan and DeMaggio, 1971*a, b*). It would indeed be surprising if we do not find a coordination between membrane properties of chloroplasts, protein synthesis and planar growth of the protonema.

RNA synthesis

Studies on RNA synthesis during fern gametophyte development have provoked two questions worthy of discussion. The first one, based on various experimental approaches, relates to the nature of specific, newly synthesized RNA which is involved in the form change of the protonema. The second question is speculative and concerns how the synthesis of a specific RNA is coupled to the photochemical reaction and the altered orientation of the mitotic spindle initiating planar growth.

An early investigation by Hotta, Osawa and Sakaki (1959) provided much of the inferential evidence for a role for RNA synthesis in the induction of planar growth in fern gametophytes. It was found that RNA content of the protonema of *Dryopteris erythrosora* increases slowly during filamentous growth and somewhat steeply at the time of and following planar growth. In gametophytes of the same age, RNA content is found to increase earlier than proteins and later runs downhill, while the protein content continues to increase. A conspicuous part of this work was also addressed to the differences in base composition of RNA between filamentous and planar forms of the gametophyte. RNA of the former is found to be high in guanine and uracil and low in adenine and cytosine while that of the planar form has high cytosine, low uracil, and intermediate amounts of adenine and guanine. These experiments make a strong case for the conclusion that planar growth of the protonema requires the synthesis of RNA qualitatively different from that which is necessary for filamentous growth. The two morphological forms of the gametophyte apparently draw on separate nucleotide pools, and judging from the magnitude of changes in the base composition of RNA, induction of planar growth appears to be signaled by a stable RNA rather than by an unstable mRNA.

Following this work, other reports have shown that gametophytes growing as planar forms in blue light synthesize RNA more actively than those growing as filaments in red light (Drumm and Mohr, 1967*a*; Raghavan, 1968*a*). Using ^{14}C-uridine as a precursor of RNA synthesis, Drumm and Mohr (1967*a*) found that the specific activity of the label in RNA which increases after the onset of blue light is detected as rapidly as the swelling of the apex of the germ filament of *D. filix-mas*. Studies of ^{3}H-uridine incorporation in the subcellular fractions of gametophytes of *P. aquilinum* cultured in red and blue light have indicated that the chloroplast-rich fraction is labeled much sooner and more extensively than other fractions (Raghavan, 1968*a*). However, radioactivity in all fractions isolated from blue light-grown gametophytes is more pronounced than in corresponding fractions of red light-grown gametophytes to rule out a possible blue light effect exclusively on the chloroplast ribosomes.

Although the role of chloroplast RNA in the growth of the gametophyte in blue light is unresolved, it is clear that chloroplast RNA is not causally connected with form change. Relevant to this statement are two investigations which merit a brief discussion here. Burns and Ingle (1970) examined by one-dimensional gel electrophoresis the kinetics of formation of stable RNA species in the gametophyte of *D. borreri (= D. affinis)* during growth in blue light and found that there is no apparent change in the quality or quantity of chloroplast and cytoplasmic ribosomal RNA (rRNA) components accompanying planar growth. This is probably due to the fact that while the method can detect quantitative changes in major RNA components, it is not sensitive enough to identify small changes in major or especially minor RNA species. This work also revealed that the proportion of chloroplast rRNA to the total rRNA remains surprisingly constant during a growth period of more than seven days in blue light, thus ruling out the possibility of a preferential synthesis of chloroplast rRNA during planar growth. Somewhat different results have come from the work of Howland (1972) who found that there is an increase in the incorporation of ^{32}P into chloroplast rRNA in the gametophytes of *D. filix-mas* after about 15 to 20 hours in blue light. However, this does not appear to be a biochemical turning point connected with form change, since the accelerated incorporation of the label lags several hours behind the first sign of morphological change in the apex of the germ filament.

An interesting experiment on the possible role of specific RNA species in the form change of gametophytes was done by Sobota (1972), who studied RNA metabolism in the gametophytes of *P. aquilinum* transferred from white light to red light. In many ways, the transformation of rapidly dividing cells of the gametophyte grown in white light to a cell population with a low propensity for division in red light is analogous to the step-down growth syndrome described in bacterial cultures. The initial response of the gametophyte to a step-down condition is a rapid decrease in RNA synthesis. This is followed by a slightly increased synthesis of RNA, which stabilizes to a new rate lower than that occurring in white light. As regards the base composition, RNA extracted from gametophytes 12 hours after shift to red light has a low guanine-cytosine ratio and displays a particularly high adenosine monophosphate (AMP) content compared with gametophytes grown in white light. If the RNA is run on an acrylamide gel, a broad peak following rRNA and trailing into the rRNA region of the gel dominates the profile (Fig. 8.2). In all essential respects, this RNA is clearly of the heterogenous nuclear (hnRNA) class, and its appearance on the gel is related to the new mode of growth of the gametophyte. There is no indication as to what fraction of this RNA represents mRNA.

In summary, this brings us to the heart of the matter of synthesis of specific RNAs connected with planar growth of the gametophyte in blue light. The main question is: to what extent do newly synthesized RNAs direct the morphological pattern of growth of the protonema in blue light? It seems that an mRNA is synthesized during planar growth of the protonema, but as will be seen below, evidence for this is fragmentary and is based on the effects of actinomycin D.

Effects of inhibitors of RNA and protein synthesis

Some persuasive, if not conclusive evidence for a role for RNA and protein synthesis in the planar growth of the protonema has come from its response to inhibitors of RNA and protein synthesis. It is now known that apart from blocking a specific step in a biochemical pathway, some inhibitors interfere with several metabolic pathways and thus produce undesirable side effects. Naturally, this complicates the interpretation of

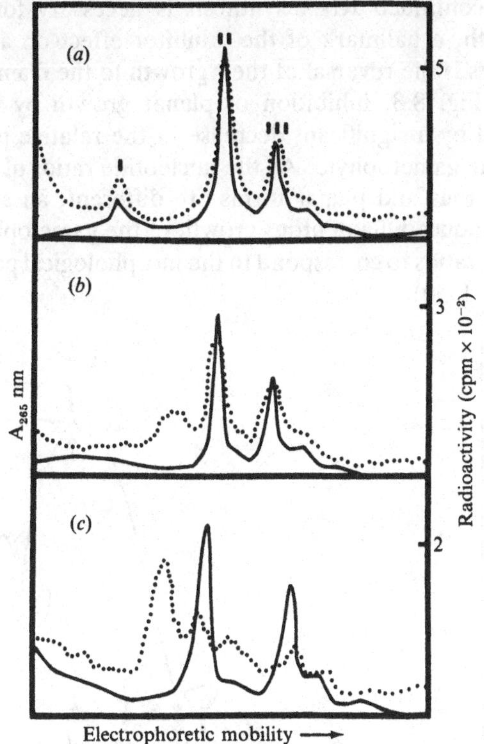

Figure 8.2. Acrylamide gel electrophoretic separation of nucleic acids of the gametophytes of *Pteridium aquilinum* during step-down growth. (*a*) Gametophytes grown in white light for 4 days and labeled with ^{32}P for 7 hours. (*b*) Gametophytes grown in white light for 4 days and labeled for 2 hours. (*c*) Gametophytes grown in white light for 4 days and transferred to red light for 12 hours and labeled for 2 hours. In (*b*) and (*c*) samples were digested with deoxyribonuclease prior to fractionation. Solid line, absorption at 265 nm; dotted line, radioactivity. (From Sobota, 1972.)

inhibitor experiments to the point that they may not offer anything but circumstantial evidence. Additional problems often ignored in the use of inhibitors are those relating to their permeability and the high concentrations required.

With this precautionary note on the real and potential hazards of the use of metabolic inhibitors, it is informative to examine their effects on form change in the fern protonema. In the first use of inhibitors in this system, Hotta and Osawa (1958) found that in the presence of a low concentration of 8-azaguanine, an inhibitor of RNA synthesis, germinating spores and early stage protonemata of *Dryopteris erythrosora* continue to grow as filaments for as long as 50 days, although spores germinated in the basal medium had long since become planar. To the extent that continued RNA synthesis is necessary for uninterrupted planar growth, a hallmark of the inhibitor effect on advanced planar gametophytes is the reversal of their growth to the filamentous pattern. As seen in Fig. 8.3, inhibition of planar growth by 8-azaguanine is accompanied by a significant decrease in the relative protein nitrogen content of the gametophyte. As the nucleotide ratios of RNA extracted from filamentous and planar forms are different, an antecedent of 8-azaguanine-induced filamentous growth of the gametophyte is a change in RNA base ratios to correspond to the morphological pattern of growth (Hotta *et al.*, 1959).

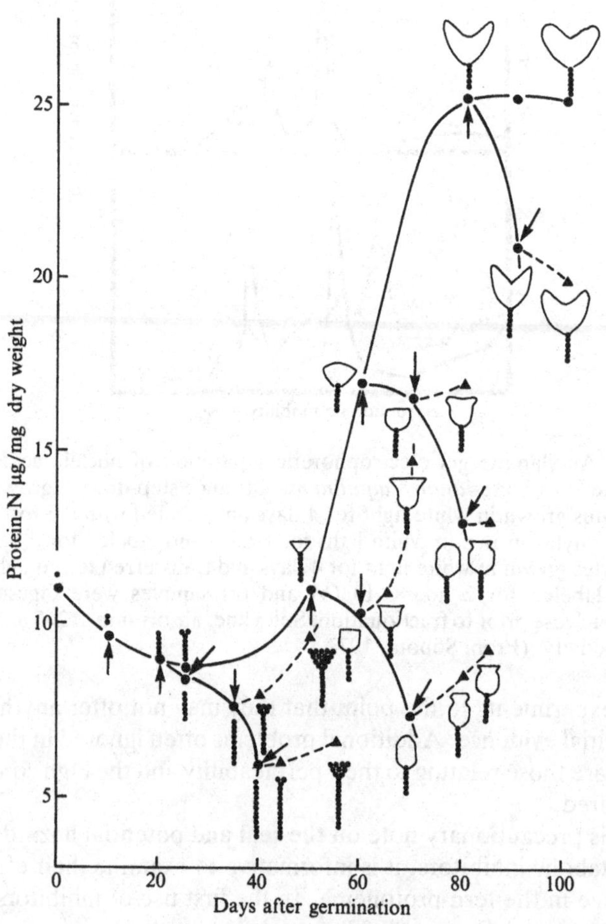

Figure 8.3. The effects of 8-azaguanine on the morphology (schematically represented) and protein nitrogen content (black curve) of the gametophytes of *Dryopteris erythrosora.* Dashed curves show the protein nitrogen content of gametophytes treated first with 8-azaguanine at points indicated by arrows pointing upward and later with guanine at points indicated by arrows pointing downward. (From Hotta and Osawa, 1958.)

Essentially similar to the inhibition by 8-azaguanine are the effects of amino acid analogs with known routes of action such as ethionine and 5-methyltryptophan, which compete with methionine and tryptophan, respectively, in the synthesis of proteins (Hotta and Osawa, 1958). These results raise the possibility that formation of the planar gametophyte is under transcriptional control. Apparently, interference with RNA synthesis affects the translation of specific proteins and impairs the ability of cells to divide longitudinally and perpetuate planar growth.

Following this work, there appeared a spate of articles (for example, Raghavan, 1964, 1965*b*, 1968*d*; Raghavan and Tung, 1967; Bergfeld, 1964*b*; 1965; Davis, 1968*b*; Faivre-Baron, 1980) on the qualitative and quantitative effects of inhibitors of nucleic acid and protein synthesis on the planar growth of gametophytes of various homosporous ferns. In one investigation it was found that if spores of *Asplenium nidus* are allowed to germinate and grow in nontoxic concentrations of a range of analogs of all four RNA bases, instead of a planar gametophyte, a single file of slowly dividing cells, akin to a filament appears (Raghavan, 1964, 1965*b*). Yet, if the filament is transferred to the mineral salt medium after several weeks in the inhibitor, it resumes planar growth. In contrast to RNA base analogs, DNA base analogs induce nonspecific growth inhibition without tampering with the basic planar morphology. Although ribonuclease (RNase) is not an inhibitor of RNA synthesis, its target is cellular RNA. It is therefore not surprising that addition of RNase to the medium promotes filamentous growth of gametophytes of *A. nidus* (Raghavan, 1969*b*).

Results on the reversal of inhibition of planar growth of the gametophytes of *A. nidus* by purine and pyrimidine analogs have shown that the antagonism of each analog is reversed by the corresponding base or its ribotide or riboside (Raghavan, 1968*d*). Thus, adenine and guanine and their ribosides and ribotides are more effective than cytosine, uracil and thymine and their derivatives in canceling the inhibition caused by 8-azaadenine and 8-azaguanine. Likewise, the inhibitory effects of 2-thiocytosine, 5-fluorouracil (FLU), and 2-thiouracil are overcome by the pyrimidines and their derivatives, but not usually by the purines. As shown in Fig. 8.4, when a purine analog and a pyrimidine analog are used in combination, the inhibition of planar growth is completely reversed in the presence of both natural bases. These results help to assign to both purines and pyrimidines of RNA a central role in the induction of planar growth in *A. nidus* gametophytes.

Actinomycin D effects

Planar growth of the fern protonema is sensitive to inhibitors of RNA synthesis because form change possibly relies on the continued synthesis of RNA. This raises the question as to which molecular species of RNA is

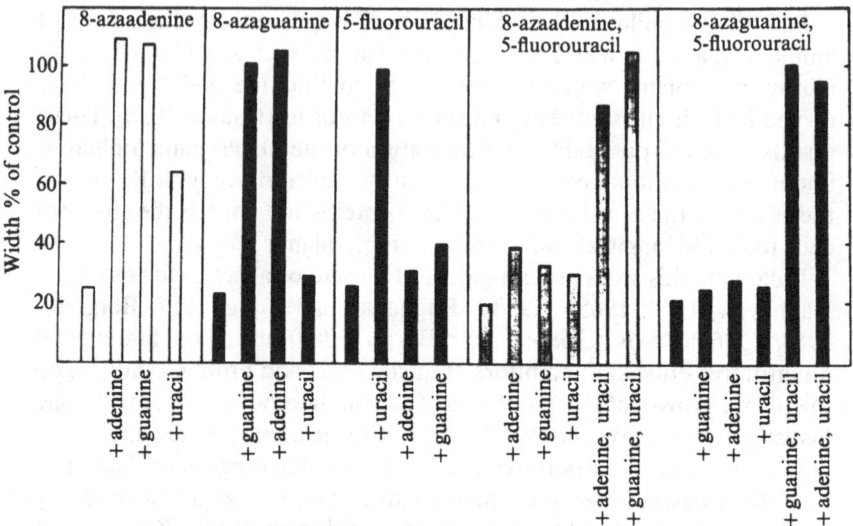

Figure 8.4. The inhibition of planar growth of gametophytes of *Asplenium nidus* by 8-azaadenine, 8-azaguanine and 5-fluorouracil, separately and in combination and its reversal by the respective bases. Growth is expressed in terms of the width of the gametophytes as a percentage of control (basal medium). (From Raghavan, 1968*d*.)

involved in inducing planar growth. It is unlikely that rRNA and transfer RNA (tRNA) possess the specificity for promoting morphogenetic changes, which leaves mRNA solely in this role. One way to determine, albeit indirectly, the role of mRNA in planar growth is to block transcription in the protonema using actinomycin D. Such experiments (Raghavan, 1965*a*; Nakazawa and Tanno, 1967) have shown that spores of certain ferns allowed to germinate and grow in the presence of the drug under light conditions favoring planar growth remain essentially as filaments. In the gametophytes of *Dryopteris filix-mas*, actinomycin D inhibits planar growth in blue light, but allows continued filamentous growth in red light (Mohr, 1965). This is to be expected if the latter is independent of continued transcription. Fig. 8.5 illustrates a different manifestation of growth inhibition observed when gametophytes of *Gymnogramme (= Pityrogramma) calomelanos* are exposed to the drug; here growth of the original apex is greatly disturbed and new spurts of meristematic activity are initiated at other parts of the gametophyte (Faivre-Baron, 1980). Both autoradiographic (Raghavan and Tung, 1967) and biochemical analyses (Raghavan, 1968*c*) have shown that inhibition of planar growth is paralleled by decreased synthesis of RNA in the gametophytes, but the question whether this includes inhibition of mRNA synthesis has not been elucidated. If actinomycin D acts as an inihibitor of mRNA synthesis in the fern gametophyte, these results are

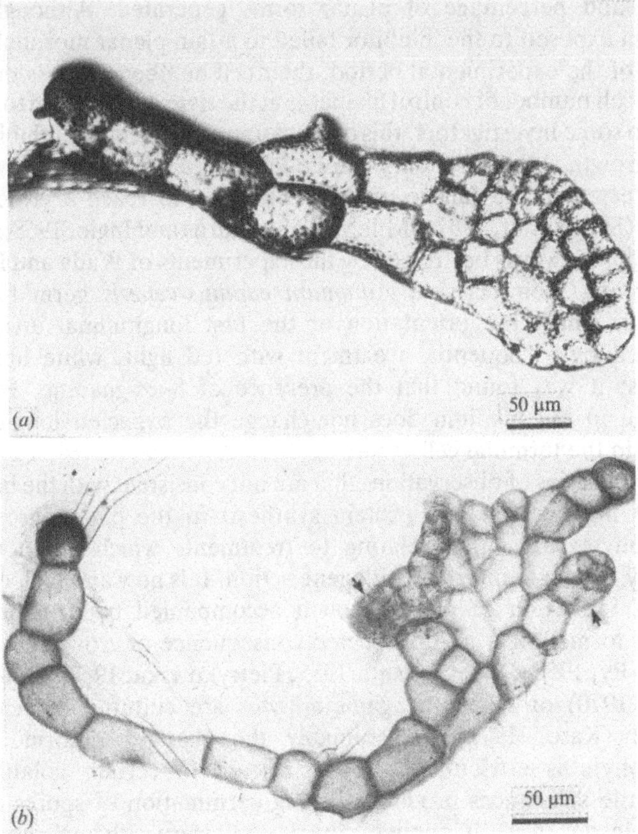

Figure 8.5. Effets of actinomycin D on the gametophytes of *Gymnogramme* (= *Pityrogramma*) *calomelanos*. (*a*) Gametophyte after growth for 10 days in the basal medium. (*b*) Gametophyte after growth for 10 days in a medium containing actinomycin D (1.2×10^{-6}M). Arrows point to the new pockets of meristematic activity. (From Faivre-Baron, 1980; photographs supplied by Dr M. Faivre-Baron.)

in harmony with the view that induction of planar growth in this system requires a newly synthesized mRNA. Since light of a specific wavelength is an essential prerequisite for this morphogenetic change, the primary role of the photoreaction is apparently to trigger the synthesis of mRNA.

Critique

In much of the published work on the effects of inhibitors on planar growth of fern gametophytes, germinated spores were allowed to grow in a medium containing a fixed concentration of the inhibitor in blue or white light and determinations were made of growth in length, cell number and percentage of planar forms generated. Although protonemata exposed to the inhibitor failed to attain planar morphology at the end of the experimental period, their cell number was less than the average cell number of control filaments at the time of transition to planar form. To some investigators, this observation suggested that inhibition of planar growth is the secondary consequence of a general growth inhibition reflected in the failure of the protonema to reach a critical cell number (Schraudolf, 1967*a*; Miller, 1968*b*; Burns and Ingle, 1968; Davis, 1970). This view was buttressed by the experiments of Wada and Furuya (1973) using their classical *Adiantum capillus-veneris* germ filament system in which the orientation of the first longitudinal division is controlled by a sequential treatment with red light, white light and darkness; it was found that the presence of 8-azaguanine, FLU or ethionine in the medium does not change the expected longitudinal division in the terminal cell.

Another series of observations that are not consistent with the involvement of nucleic acid and protein synthesis in the planar growth of gametophytes are those relating to treatments which are not even remotely known to interfere with gene action. It is now amply documented that inhibition of planar growth accompanied by promotion of filament formation is an unexpected consequence of crowded cultures (Hurely-Py, 1950*a*; Steeves *et al.*, 1955; Pietrykowska, 1962*b*; Smith and Rogan, 1970) or when the gametophytes are cultured on activated charcoal (Kato, 1973*b*). Presumably the change in form of the gametophyte is attributable to the release of certain volatile and nonvolatile substances produced during germination of spores (Smith and Robinson, 1971). It is unlikely tha they interfere with nucleic acid and protein metabolism of the gametophyte, any more than low temperature (Kawasaki, 1954*a*), mineral ions (Parès, 1958), NaSCN (Nakazawa, 1960*c*), mannitol (Kato, 1964*b*, 1965*b*), IAA (Miller, 1968*b*), reduced oxygen tension (Kato, 1969*a*), ethanol (Smith and Robinson, 1969), phenylboric acid (Caruso, 1973), unidentified phenols and alkaloids of plant extracts (Wadhwani, Bhardwaja and Mahna, 1981), colchicine (Singh and Roy, 1984*a*), chloral hydrate (Singh and Roy, 1984*b*), acidic pH (Otto *et al.*, 1984), and salinity stress (Lloyd and Buckley, 1986) which have also been reported to block planar growth in gametophytes of diverse ferns grown under otherwise favorable conditions. Despite the lack of information on the biochemical pathways blocked by these compounds, one could go so far as to say that planar growth opens up a broad spectrum of metabolic pathways in the cells of the germ filament rather than solely activating genes.

Before a final verdict is reached on the interpretation of inhibitor effects on the planar growth of gametophytes, there are two questions worthy of further discussion. For example, are there age-dependent changes in the sensitivity of the protonema to the inhibitor? Do changes in the metabolic status of the protonema alter its responses to a given concentration of the inhibitor? Answers to these questions have been sought in a study of the effects of FLU on the transition of the protonema of *Asplenium nidus* to a planar gametophyte (Raghavan, 1974*b*). Under a constant fluence of light and a constant photoperiod, progressively older protonemata require increasingly higher concentrations of the inhibitor for blocking planar growth. The optimum concentration of FLU required to inhibit planar growth in five- to six-celled, 15-day old filaments poised on the threshold of planar growth is 25 to 30 times higher than the optimum which inhibits planar growth in the protonemata of just germinated spores. Similarly, when protonemata of a certain age are exposed to increasing photoperiods or light fluences, the concentration of FLU required to inhibit planar growth increases, suggesting that the effects of the inhibitor are counteracted by light-induced acquisition of morphogenetic potency in the cells. This also agrees with the observation that by transferring germinated spores to increasingly higher concentrations of FLU, it is possible to maintain the gametophytes as filaments beyond the stage when planar growth is normally initiated in them. Summarily, these results indicate that when the responses of a developing multicellular system to an exogenous chemical are evaluated, possible age-dependent changes in the sensitivity of the system to the added chemical should be taken into account. Failure to recognize this principle may explain the apparent conflicts in the literature on the specificity of the effects of inhibitors of nucleic acid and protein synthesis on the form change in the protonema.

A model and some perspectives

The generally accepted dogma that DNA of an individual cell harbors the genetic blueprint characteristic of the whole organism of which it is a part has led to the premise that differential gene activation accounts for functional differences between cells of the same organism. The transition of the fern gametophyte growing as a filament in red light to a planar form in blue light obviously involves a change in the direction of differentiation and hence in the functioning of particular sets of genes. According to this view, first proposed by Ohlenroth and Mohr (1964), in the protonema growing in red light most genes are silenced except the active 'housekeeping' genes which are transcribed to provide proteins for sustaining filamentous growth. Since protonemata of many ferns grow as filaments at a reduced rate in complete darkness, red light probably stimulates low levels of gene expression already under way. On the other hand, blue light derepresses certain 'potentially active' genes to transcribe proteins necessary for planar growth. The model implies that differences in the nature of signals produced by the two photoreactions serve as cues to select the set of genes that are programmed in a particular light regimen. On the whole, biochemical evidence reviewed earlier concerning enhanced RNA and protein synthesis in the gametophyte in blue light is consistent with the essential features of this model. However, we do not have any notion about the nature of the intermediate steps between light absorption and gene activation, and the chain of enzymatic events set in motion in the protonemal cell subsequent to gene activation escapes us completely.

The basic question is how blue light orients the mitotic spindle in the terminal cell of the filament from a position parallel to the longitudinal axis of the cell to one perpendicular to it. The work of Pickett-Heaps and Northcote (1966) on the division of epidermal cells of graminean plants suggested that the orientation of the mitotic spindle is anticipated by certain changes occurring in the spatial location of microtubules in the cell before the nucleus enters prophase. It was found that although the microtubules are segregated in the outer cytoplasm of a resting cell, entry of the nucleus into prophase is preceded by the appearance of a new band of microtubules at right angles to the direction of the mitotic spindle and coincident with the region of the parent cell wall where the future cell plate dividing the daughter nuclei would join the wall. Based on this, one might envisage that microtubules are the target of light action in the fern protonema, that is, light quality might control their assignment to ensure the orientation of the division wall at right angles to or parallel to the long axis of the terminal cell of the protonema on the threshold of division. How this might come about is not clear, but it is tempting to speculate that gene activation in the protonemal cell in the different light regimens is linked to polymerization of microtubule precursor proteins. However, until we learn how light regulates microtubule precursor protein traffic in any particular direction in the cell, we are left with the feeling that light is somehow involved in fern gametophyte morphogenesis, without knowing how.

General comments

At a certain stage in its ontogeny, under the influence of blue light, the filamentous fern protonema is triggered to replicate DNA, transcribe RNA and synthesize proteins for a rapid burst of mitotic activity resulting in a planar gametophyte. Although some details about the timing of macromolecule synthesis during form change in the protonema are currently known, they are really a prolegomenon to a study of the molecular basis of gene expression during photomorphogenesis. The development of the concept that a reorientation of the mitotic spindle in the terminal cell of the germ filament is the fundamental basis for morphogenesis has brought about a change in the perspective from which induction of planar growth is analyzed.

It is clear that the great advancements in molecular biology are yet to illuminate the events of planar growth in the fern protonema in a satisfactory way. Further work to establish the validity of the transcription–translation theory of form change requires new approaches like characterization of mRNAs involved, preparation of hybridization probes specific for individual mRNAs and their analysis by kinetics of hybridization, translational efficiencies and rates of synthesis. However, an even more important question concerns the identification of the proteins that are synthesized in blue light. Among the possible candidates, attention should be paid to ribulose-1,5-biphosphate carboxylase, ribosomal proteins and structural proteins like extensin and tubulin. Analysis of the proteins differentially expressed in red and blue light regimens will become a cornerstone in the molecular biology of fern gametophyte morphogenesis.

9

Vegetative and reconstitutive growth of gametophytes

With the description of the landmark stages in the early development of fern gametophytes behind us, we are now in a position to consider the factors that regulate their growth and maturation. Studies in this area have been largely aided by the ease with which gametophytes of homosporous ferns can be raised from spores in large numbers under aseptic conditions. For the most part of its growth, the gametophyte is only one cell layer thick, a property which makes determination of cell numbers a less formidable task; the structural simplicity of the gametophyte also makes it possible to study a range of morphological expression of cells which seem to possess unlimited potentialities. It is, therefore, no wonder that the effects of a staggering variety of chemicals on the growth of gametophytes of a number of ferns have been tested, with a view to determine the optimum nutrient or chemical environment for their growth. Surprisingly, in these studies there were relatively few examinations for the changes of biochemical constituents or hormone levels during growth. This was because it is so much easier to follow growth of gametophytes in a given nutrient milieu in terms of a few growth parameters like increase in surface area, increase in cell number or change in the dimensionality of growth (such as from filamentous to planar) than process the material for arduous biochemical determinations. Nonetheless, a knowledge of the chemical and environmental conditions for growth of gametophytes is important from both physiological and ecological perspectives.

Regeneration of gametophytes is tied to their vegetative growth as the capacity of well-developed fern prothalli to break off and form new entities is widespread in nature. Gametophytes of some ferns also survive almost indefinitely by the production of vegetative propagules. The developmental biology of reconstitutive growth of gametophytes poses important questions inasmuch as small groups of cells retain their growth potential and proliferate into adult forms in full sexuality and structure. The term reconstitutive growth is used in this book to denote both regenerative and reorganizational types of growth resulting in the formation of new phenotypes, an ability bestowed by the totipotency of cells.

As Sinnott (1960) has stated, reconstitution is virtually synonymous with regeneration in the sense used by animal embryologists and so admittedly, there is a little ambiguity in the use of these terms in the book.

Vegetative growth of gametophytes

We will begin this section with some comments about the physiological ecology of fern gametophytes. Although ferns produce an enormous quantity of spores, only a few reach the stage of mature gametophytes. This is primarily due to the fact that spores are disseminated by wind to places unfavorable for their germination and for the subsequent growth of gametophytes. There are also a great variety of stressful conditions such as variations in temperature, humidity, and light fluences, highly acidic and alkaline soils, and nutritional deficiency to which gametophytes are exposed daily and which limit their growth. Because gametophytes are devoid of elaborate protective devices, such as a thick cuticle on their epidermal cells, they are sensitive to these changes and succumb easily. Based on peroxisomal enzyme activity, it appears that gametophytes are capable of photorespiration; this is bound to limit their photosynthetic production (DeMaggio and Krasnoff, 1980). However, gametophytes of some ferns appear to be much hardier than their sporophytes. They remain viable and regenerate new growths even after desiccation to a low water content or after freezing to liquid N_2 temperature; some even develop and survive at acid pH as low as 2.2 (Evans and Bozzone, 1977; Sato and Sakai, 1980, 1981; Quirk and Chambers, 1981). Evidently, the lower limits of other factors for active growth of gametophytes have not been determined and there is much scope for research. The ability of gametophytes to withstand extremes of environments is obviously important in the establishment and survival of species in ecological niches not otherwise colonized by ferns.

We shall discuss here the morphology of mature fern gametophytes, followed by an analysis of the genetics of their growth. Since almost all of the work in this area has been done with gametophytes of homosporous ferns, they form the focus of this account.

Growth and maturation of gametophytes

Following the initiation of planar morphology, the gametophyte increases in length and width by the activity of the apical cell. A widely cordate structure (cordate-type) is the most ubiquitous form of the mature gametophyte found among homosporous ferns. Descriptively, it is a flat, leaf-like photosynthetic unit with a midrib several layers of cells thick and normally-proportioned one-cell layer thick symmetrical laminar extensions on either side. Development of rhizoids from the ventral side of the gametophyte is heavy and continuous. There is a strong convergence of cells toward the anterior notched end of the prothallus, but at this stage, instead of an apical cell, a meristematic cell plate, stabilized in a state which permits continued divisions, is in place in the notch. As shown in Fig. 9.1, the final shape of the gametophyte is clearly the result of a precise control of the rhythm and location of cell division and cell expansion activities.

Attempts to interpret the mature form of gametophytes growing without an apical cell of sufficient stature have brought to light some variations of the cordate form. This is no doubt due to the location and timing of activity of a marginal meristem which takes over the function of the apical cell. If the meristem is lateral, the result is a prothallus exhibiting varying degrees of asymmetry, as seen in several species of *Anemia* (Atkinson, 1962; Pray, 1971; Nester and Schedlbauer, 1981) and *Pellaea* (Pray, 1968). In *Elaphoglossum villosum* (Aspleniaceae), the marginal meristem is almost apical from the beginning and the final product is a long cordate structure which even becomes lobed. In some members of Hymenophyllaceae and Adiantaceae including species of *Vittaria*, pockets of meristematic cells dot the margins of early stage gametophytes and the result is the formation of cordate, branching prothalli (Stokey, 1951*a*; Stokey and Atkinson, 1957; Farrar, 1974, 1978).

In *Rhipidopteris (=Peltapteris) peltata* (Aspleniaceae) and several species of *Elaphoglossum*, Stokey and Atkinson (1957) have described longer than broad unbranched prothalli (strap-like). Although an apical cell is differentiated late in the ontogeny of the gametophyte, it gives way to a meristematic plate of cells lodged in the notched apex of the gametophyte. Extension growth of the strap-like prothallus is programmed by this meristem. In some ferns, in which an apical cell or a marginal meristem is invariably absent in the gametophytes (Kaulinia type), the latter continue to grow indefinitely as narrow elongate structures (ribbon-like), bearing profuse lateral branches. Here, cells of the entire anterior region are considered the chief motivative force for the lengthening of the gametophyte. Ribbon-like prothalli are seen in members of Grammitidaceae, Adiantaceae and some Hymenophyllaceae and Polypodiaceae (Nayar and Kaur, 1971). These observations make a compelling case to conclude that prothallial growth in ferns in relation to an apical cell makes only part of the story.

Patterns of cell division and cell expansion are developmental properties that not only contribute to the form of the prothallus, but also have some evolutionary and phylogenetic significance. In the gametophytes of primitive species such as those included in Ophioglossaceae (Bierhorst, 1958; Whittier, 1972, 1983) as well as in *Actinostachys* (Schizaeaceae) (Bierhorst, 1966, 1968*a*) and *Stromatopteris* (Stromatopteridaceae) (Bierhorst, 1968*b*), cell division is often random and disorganized while

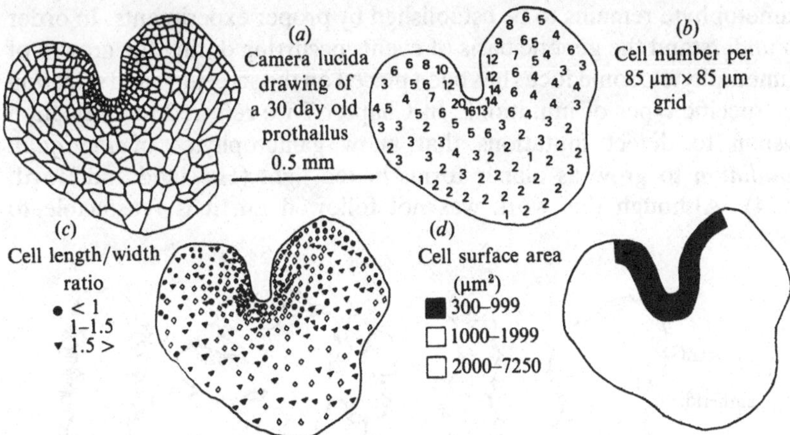

Figure 9.1. Distribution of cell member, and of cells of different sizes and shapes in a 30-day old cordate prothallus of *Todea barbara*. The greatest concentration of cells is at the apical notch of the prothallus where cells are isodiametric and have much less surface area than those in the basal region. (From von Aderkas and Cutter, 1983*a*.)

cell expansion continues unabated. Associated with this, a remarkable transformation that occurs in the gametophytes is the appearance of subterranean, often mycorrhizal, cylindrical dorsiventral structures composed of parenchymatous cells (tuberous type). Another mature form of the prothallus is the branched filamentous type found in some species of *Schizaea* and *Trichomanes* (Bierhorst, 1968*a*; Nayar and Kaur, 1971). Representative examples of these variations in prothallial morphology are illustrated in Fig. 9.2.

Bierhorst (1967) has described some very remarkable features in the gametophyte of *Schizaea dichotoma*. Although the mature gametophyte is subterranean, fleshy, mycorrhizal, and nongreen like a tuberous form, it is basically a bundle of filaments with progressively thinner branches that terminate in single filaments. From the ontogenetic point of view, the entire gametophyte is constructed on a body plan in which packets of cells generated by the activity of apical cells with a single cutting face are stacked in linear rows. These findings have in some ways modified the oversimplified interpretation of the phylogenetic relationships of Schizaeaceae.

Genetics of gametophyte growth

It goes without saying that virtually every step in the growth of the gametophyte from its single-celled beginning is under genetic control. But the reality of gene expression and gene function during growth of the gametophyte remains to be established by proper experiments. In order to understand the genetic basis of events occurring during the growth of gametophytes, some focus has been placed on the screening and selection for specific types of mutations. In Chapter 7 we referred to a selection system to detect mutations that allow gametophytes of *Pteridium aquilinum* to grow as planar forms in red light (Howland and Boyd, 1974). Although this work was not followed up, it is reasonable to

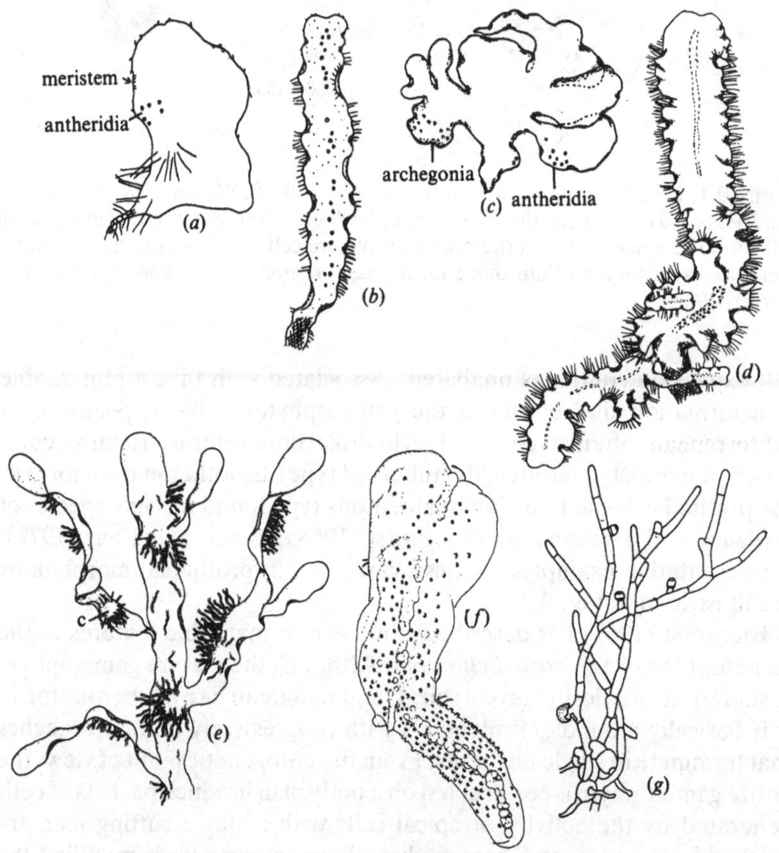

Figure 9.2. Variations in prothallial morphology in homosporous ferns. (*a*) An asymmetrical prothallus of *Anemia rotundifolia*. (From Atkinson, 1962). (*b*) Long, cordate prothallus of *Elaphoglossum villosum*. (From Stokey and Atkinson, 1957). (*c*) Cordate, branching prothallus of *Vittaria elongata*. (From Stokey, 1951*a*). (*d*) Strap-shaped prothallus of *Rhipidopteris* (=*Peltapteris*) *peltata*. (From Stokey and Atkinson, 1957). (*e*) Ribbon-like prothallus of *Paraleptochilus decurrens*; c, superficial multicellular cushion. (From Nayar, 1963*b*). (*f*) Tuberous prothallus of *Botrychium virginianum*. (From Bierhorst, 1958). (*g*) Branched, filamentous prothallus of *Schizaea pusilla*. (From Atkinson and Stokey, 1964.)

conclude that the mutations presumably prepared the gametophyte for an altered photomorphogenetic reaction. Germinating spores are particularly sensitive to various toxic substances found in polluted environments whose effects show up as mutational damage. A high incidence of such damage has been documented in the gametophytes of *Osmunda regalis* growing in a river heavily polluted with industrial waste. The mutant phenotypes attain only a fraction of the size of the wild type or succumb to unorganized callus growth (Klekowski, 1978; Klekowski and Davis, 1977; Klekowski and Klekowski, 1982). The susceptibility of gametophytes appears to be due to post-zygotic mutations in the sporophyte which are transmitted to the spore. Variant gametophytes with reduced number of cells have also been obtained from spores treated with chemical mutagens (Carlson, 1969). Another example of mutation which affects the growth of gametophytes is tumor formation induced by exposure of spores to ionizing and ultraviolet radiations; these will be discussed in more detail in a later section of this chapter (p. 192).

Some mutations from radiobiological experiments on gametophytes are seen as aberrations in chloroplast morphology (Knudson, 1940; Maly, 1951; Breslavets, 1951, 1952; Howard and Haigh, 1968; Mehra and Palta, 1969; Haigh and Howard, 1970). Because the abnormal chloroplasts revert to normal morphology, different types of plastids in the same cell of the gametophyte as well as chimeric gametophytes have been frequently observed; this has cast considerable doubt on the genetic nature of the aberrations. Only in *O. regalis* has it been established by breeding experiments that a nuclear gene mutation is responsible for the appearance of variant prothalli in which chloroplasts aggregate on one side of the cell ('bar' mutation; Haigh and Howard, 1970).

From the perspective of genetic control of growth, recent studies on the isolation of ABA, herbicide (paraquat), and NaCl resistant mutants of *Ceratopteris richardii* gametophytes are of interest (see Hickok, Warne and Slocum, 1987 for review). Protocols used in these investigations include sowing X-irradiated spores and isolating variants that grow better than the wild type in high concentrations of the screening agent. Determination of the segregation patterns in gametophytes produced by hybrid sporophytes has established that mutants selected for enhanced tolerance to the chemical stress have single nuclear gene inheritance patterns (Hickok, 1985; Hickok and Schwarz, 1986*a, b*; Warne and Hickok, 1987*a*). Availability of mutants as well as information on the mode of action of the selective agents used should help to elucidate the physiological mechanisms and the biochemical lesions of the stress.

Gene dosage effects on the growth of gametophytes. Increase in chromosome number of cells induced by colchicine and other chemicals generally inhibits the growth of fern prothalli. Other side effects of polyploidy are swelling of the surviving cells and irregular branching of gametophytes (Rosendahl, 1940; Mehra, 1952; Mehra and Loyal, 1956; Yamasaki, 1954; Döpp, 1955*b*). In the gametophytes of *Lygodium japonicum* treated with colchicine, the increase in cell size is associated with an increase in the organelle population and change in their ultrastructural morphology (Trivedi and Bajpai, 1977).

Changes in the ploidy level of cells of the gametophyte also occur as a natural consequence of apospory (see Chapter 14). This developmental anomaly allows the regeneration from sporophyte of gametophytes of higher ploidy levels, without intervening stages of meiosis or spore formation. In general, although diploid and tetraploid gametophytes are morphologically similar in appearance to the haploid, they exhibit considerable variations in size and in growth rate (Partanen, 1961, 1965; Kott and Peterson, 1974). In a polyploid series of gametophytes of *Todea barbara*, increasing complement of nuclear genes causes increase in cell size, dry weight, chloroplast size, nuclear volume, and peroxidase activity (DeMaggio and Stetler, 1971; DeMaggio and Lambrukos, 1974; Hannaford and DeMaggio, 1975). These results indicate that gene dosage of the cells probably determines their morphological and metabolic activities.

Growth requirements of gametophytes in culture

While the composition of the nutrient medium is of paramount importance for the growth of gametophytes, conditions of culture such as light or its absence thereof, temperature, liquid *versus* solid medium and density of the population, among others, also assume importance as major factors to be considered. In some of the early studies, attention was paid to the establishment of optimum temperature conditions, direction of illumination of cultures, protocol for sterilization, and culture of spores and the development of complicated glass support for effective nutrition and growth of fern prothalli in culture. The considerable volume of literature on these and other aspects of gametophyte culture has been integrated into a useful review by Dyer (1979).

Among the various components of the medium necessary for prothallial growth, far and away the best analyzed are mineral salts, carbohydrates and hormones.

Mineral salts

Although spores of several ferns germinate freely in distilled water, the gametophyte requires a variety of mineral elements of the type found in a balanced nutrient solution for continued growth, in the absence of which characteristic deficiency symptoms surface. According to Schwabe (1951), for the growth of gametophytes of *Pteridium aquilinum*, a supply of phosphorus in the medium is most essential. Lack of this element leads to the appearance of pale green and transparent gametophytes which cease to grow further. Feeble growth is also characteristic of gametophytes of *Gymnogramme (=Pityrogramma) calomelanos* nurtured in a medium deficient in phosphorus and potassium (Sossountzov, 1956, 1957*b*) and of *Blechnum brasiliense* deprived of potassium and sulfur (Laurent and Lefebvre, 1980). In N-deficient media, gametophytes of diverse ferns exhibit an array of deficiency symptoms, chiefly, reduced growth, bleaching, lack of organized meristem, increased starch content of cells, cessation of archegonial growth, and appearance of antheridia (Czaja, 1921; Schwabe, 1951; Laurent and Lefebvre, 1980). Indications are that the N-requirement for a luxurious growth of gametophytes can be met by NO_3^- or NH_4^+ salts or by a mixture of both (Parès, 1958; Hotta, 1960*a*).

Having established that N is essential for the growth of gametophytes, it is natural to determine whether organic nitrogen compounds can duplicate the effects of NO_3^- or NH_4^+ salts. It is somewhat surprising to note that to date comprehensive studies on the effects of organic nitrogen compounds have been undertaken with gametophytes of only *G. (=P.) calomelanos*. From these investigations, which involved a range of compounds tested at different concentrations during a growth period extending over several months, and scores of publications, it is difficult to identify a single organic nitrogen compound as the key metabolite essential to sustain growth of gametophytes, as well as NO_3^- or NH_4^+ salts. Moreover, some of the work was done under insufficiently critical conditions to warrant detailed treatment here. Nonetheless, the following general conclusions appear to be noteworthy. The substances tested can be divided into two classes, based on their ability to support proliferative growth of gametophytes through varying numbers of transfers each of 45 days duration. Glycine, leucine, and serine and aspartic acid, glutamic acid, and their amides (Sossountzov, 1954) and methylamine (Sossountzov, 1957*a*) belong to a class of compounds which are nearly as effective as nitrate in maintaining continued growth and increase in fresh weight of gametophytes through six transfers. In the second group of compounds which includes alanine, phenylalanine, and valine (Sossountzov, 1955), the colonies appear to suffer some sort of irreversible damage that effectively limits their growth to three transfers before necrosis sets in. Naturally, any compound that drastically affects the growth of gametophytes may be expected to cause repercussions on their nitrogen metabolism but the specific way by which alanine, phenylalanine, and valine inhibit growth is not apparent.

An extension of this work came from experiments on the effects of mixtures of two amino acids on the growth of gametophytes. It needs no emphasis that in experiments of this kind some combinations of amino acids might produce synergistic effects while others act antagonistically. The synergistic interaction between toxic and nontoxic amino acids such as between valine and glycine, alanine and leucine, alanine and serine, and alanine and glycine (Sossountzov, 1955, 1959) or between two toxic amino acids such as alanine and phenylalanine are mainly evident in the ability of gametophytes to survive a large number of transfers, in a decreased level of necrosis in them, in increased growth and in generally increased level of tolerance to the toxic component of the mixture. A combination of low concentrations of glycine, alanine and leucine is even effective in restoring the growth of gametophytes to the level obtained on a medium containing NO_3^- (Sossountzov, 1958). These clearly defined effects of amino acid mixtures pose a dilemma in interpretation as it is not known if growth modifications are associated with a repatterning of enzyme activities in the gametophytic cells.

Carbohydrates

Some investigations also bear on the question of whether addition of sugars to the medium stimulates the growth of gametophytes. The apparent ability of fern gametophytes to grow in the absence of light presents a fundamental problem in establishing whether exogenous sugars substitute for the primary products of photosynthesis or merely serve as a combustible source of energy. Generally, when the intent is to obtain massive growth of the prothallus, addition of a sugar such as glucose or sucrose at a concentration of 1 to 2% is beneficial. Fragments of the colony so formed can be used as inocula to initiate new cultures so that in a few weeks' time, enormous quantities of prothalli can be obtained. Other sugars which have supported the growth of gametophytes are fructose, maltose, ribose, and xylose (Hurel-Py, 1955*a*).

In the above studies, prothalli were grown on the supplemented nutrient media and subjected to different light-dark regimes. In physiological terms, we can ask whether this variable affects the utilization of exogenous sugars by prothalli. An investigation on the effects of photoperiodic conditions on the utilization of sugars by prothallial cultures of *Lygodium japonicum* seems to offer a partial answer to this question (Hurel-Py, 1969). Basically, this work showed that poor growth of the gametophytes in short days in a medium containing fructose is correlated with a high soluble sugar content in their cells, while a high growth rate in long days is associated with a low sugar content. These results are consistent with the view that a long photoperiod provides a more compatible environment for assimilation of fructose by prothalli than short days. Assimilation of glucose by *L. japonicum* gametophytes is not, however, dependent upon the duration of the light period.

Characterization of the morphogenetic effects of sugars on fern gametophytes made very little headway by these purely physiological studies. Wetmore, DeMaggio and Morel (1963) nurtured single prothalli of *Onoclea sensibilis, Osmunda cinnamomea, O. regalis* and *Todea barbara* on media containing various levels of sucrose and found that if sucrose concentration of the medium is high, prothalli grow as tall, upright, sturdy structures. A concentration of sucrose at 1% is optimal although levels of sucrose as high as 5% do not do much damage. Lowering the sucrose level in the medium leads to the formation of membranous prothalli. An intriguing question here is whether sucrose functions as a morphogenetic determinant or as a source of energy for the increased mitotic activity necessary for determining the shape of the gametophyte. In this connection, a finding of great interest is that in cultures of *T. barbara* gametophytes, addition of sucrose induces the formation of more or less complete or incomplete strands of xylem (DeMaggio, Wetmore and Morel, 1963). Since the cells of the gametophyte are clearly of the uncommitted parenchymatous type, the trigger provided by a simple carbohydrate in the medium for xylogenesis *in vitro* has implications with respect to lack of vascular tissue differentiation in fern gametophytes *in vivo*. As shown in a later study (DeMaggio, 1972), growth hormones such as IAA, NAA, GA, kinetin, and benzyladenine supplied individually to the medium do not surpass the effect of sucrose in inducing xylogenesis.

Of the many studies pertaining to the effect of complex carbohydrates on the growth of fern gametophytes, only that of Kato (1964*b*, 1965*b*) acknowledges a role for these compounds in promoting planar growth. When spores of *Pteris vittata* are germinated in white light, induction of planar growth is preceded by the formation of a two- to five-celled protonema. However, upon addition of 0.5 to 1.0% soluble starch to the medium, the germinated spores virtually bypass the protonemal phase and form massive planar gametophytes. Evidence also indicates that gametophytes of *Polypodium vulgare* grow more rapidly in a medium containing starch than on the basal medium; maltose generated from extracellular amylase activity has been postulated to account for this growth promotion (Smith, 1973). Growth of gametophytes in the starch-containing medium lags 24 to 48 hours behind growth in the maltose-containing medium; this fits in with the observation that the enzyme becomes functional in about 36 hours after gametophytes establish contact with the substrate.

In reflecting upon the effects of carbohydrates, we can say that except in isolated instances, the final form of the fern gametophyte owes much more to processes mediated by endogenously synthesized energy sources than to exogenous supplies.

Growth hormones

Representatives of all five major groups of plant hormones have been identified in fern spores or gametophytes; included in the list are IAA from *Pteris longifolia* (Albaum, 1938*b*) and *Dryopteris erythrosora* (Hotta, 1960*b*), ethylene from *Pteridium aquilinum* (Elmore and Whittier, 1973; Tittle, 1987), and GA (Schraudolf, 1966*d*; Weinberg and Voeller, 1969*b*), cytokinin (Schraudolf and Fischer, 1979), and ABA (Cheng and Schraudolf, 1974; Bürcky, 1977*c*) from *Anemia phyllitidis*. It was noted in an earlier chapter that under certain light conditions, growth of fern protonemata is promoted by auxins. However, auxins are generally inhibitory or only marginally effective in promoting growth of older gametophytes (Hurel-Py, 1943; Hickok and Kiriluk, 1984). In an interesting series of experiments, Reinert (1952) showed that a high concentration of 3,4-benzpyrene leads to a decrease in the extractable auxin and in the inhibition of growth of the gametophytes of *Stenochlaena palustris* (Blechnaceae), while lower concentrations of the chemical which increase the auxin content of the gametophytes promote their growth. GA is generally inhibitory or only slightly promotory for growth of gametophytes (Kato, 1955; von Witsch and Rintelen, 1962; Schraudolf, 1962). With regard to the effects of cytokinins, it has been noted that although kinetin does not have any immediate effect on the form change of gametophytes of *Gymnogramme (=Pityrogramma) calomelanos*, low concentrations of the hormone, nonetheless, stimulate the rate of cell division during their subsequent growth (Faivre, 1969*a, b*). Somewhat similar results have been reported when gametophytes of *Pteris longifolia* are exposed to kinetin action (Durand-Rivierès and Fillon, 1968). Benzimidazole, a naturally occurring azole which functions as a growth regulator in certain plants, has been shown to promote the growth of gametophytes of *Thelypteris felix-mas (= Dryopteris filix-mas)* by increased mitotic activity (Dyar and Shade, 1974).

Changes observed in gametophytes grown in the presence of maleic hydrazide and triiodobenzoic acid in the medium follow a pattern that is consistent with their well-known property as anti-auxins. In the prothalli of *G. (=P.) calomelanos* (Sossountzov, 1953) and *P. longifolia* (Durand-Rivierès and Fillon, 1968), for example, the syndrome of inhibitory effects of these compounds includes cessation of growth, formation of necrotic spots, and regeneration of filamentous or plate-like adventitious outgrowths. The slow tempo of growth of prothalli of *Cheilanthes farinosa* and *Dryopteris cochleata* exposed to varying concentrations of maleic hydrazide results in the regeneration of filamentous outgrowths at low concentrations and in the transformation of the prothallus into a cushion-like tissue at high concentrations (Khare and Roy, 1977; Khare, 1981). Whether these manifestations of anti-auxin effect are based on the enzymatic removal of auxin required for prothallial growth has not been determined. On the whole, the scanty information available indicates that the promise of a regulatory role for hormones in the growth of fern prothalli predicted by their occurrence in the system has not been realized.

Allelopathy

Before leaving the subject, we should note that many natural plant products leached into the soil may strongly influence the growth of fern gametophytes in nature. The compounds produced by other plants are almost always detrimental to the growth of gametophytes; this chemical inhibition of growth and development of one organism by another is known as allelopathy. Although it has been reported that extracts of angiosperms growing in the immediate vicinity of the fern inhibit the growth of gametophytes of the latter (Wadhwani *et al.*, 1981), the few instances of allelopathy in ferns on record involve interaction between the sporophyte of one species and the gametophyte of the same or different species, or between gametophytes of different species. One well-studied case is that of *Thelypteris normalis* whose roots and fronds produce two allelopathic chemicals, identified as indole derivatives and christened as thelypterin A and B. These compounds retard not only the growth of gametophytes of *T. normalis*, but also those of *Pteris longifolia* and *Phlebodium aureum* (Davidonis and Ruddat, 1973, 1974). The use of culture filtrates of gametophytes of *Dryopteris intermedia* and *Osmunda cinnamomea* has demonstrated reciprocal allelopathy, in which gametophytes of two antagonistic species inhibit each other's growth (Petersen and Fairbrothers, 1980). By stifling competition, allelopathic mechanisms probably help in controlling population size of gametophytes in nature.

Growth of gametophytes of Ophioglossaceae

The problems associated with identifying the growth requirements of gametophytes of Ophioglossaceae are not so much with regard to their carbon and nitrogen nutrition or growth hormone needs, as they are in establishing gametophytes in axenic culture in the absence of the endophytic fungus. The main stumbling blocks in this work are the natural dormancy of spores that is not easily overcome and, related to this, an inordinately long time, extending in some cases to several years, for spore germination, and the slow growth of gametophytes in culture. However, in recent years, prothalli of several species of *Botrychium* (Whittier, 1972, 1981; Gifford and Brandon, 1978) and *Ophioglossum* (Whittier, 1981) have been successfully cultured to maturity from spores in a nutrient medium containing sucrose but without the endophyte; of these the morphology of the prothalli of *B. dissectum* (Whittier, 1972) and *O. engelmannii* (Whittier, 1983) is essentially the same as that seen in nature.

The only study on the organic nutrition of gametophytes of Ophioglossaceae is that by Whittier (1984) who has shown that increasing the sucrose concentration accelerates the maturity of gametophytes of *B. dissectum*. Glucose, fructose, trehalose, mannose, and glycerol have also proved to be effective carbon sources for supporting the growth of mature prothalli. At the present time we have to assume, based primarily on these observations, that the carbon energy needs of other members of Ophioglossaceae are basically similar in nature to those of *B. dissectum*.

Regenerative growth of gametophytes

The term regeneration as commonly applied to plants includes processes that are initiated after the loss of a tissue or an organ, or elicited by releasing nongrowing parts from correlative inhibition of the more dominant growing regions. Often, regeneration in plants is not limited to the formation of a replica of the lost tissue or organ, but what is reformed is a whole new array of tissues, organs or even a whole new organism. The ability of ferns to reconstitute a normal functional whole from a part of the original is nowhere better illustrated than in their gametophytes. In nature, by compensating for nonfunctional parts through renewed growth, regeneration serves to restore the functional competence of the gametophyte.

Before we consider experimentally induced regeneration, it should be pointed out that spontaneous regeneration is well-known in natural populations of fern gametophytes. This faculty depends upon growth conditions that lead to nutritional deficiency, starvation or derangement of the normal physiological balance between the meristem and the older cells (Mottier, 1927; Stokey and Atkinson, 1956*a, b*; Parès, 1958; Faivre-Baron, 1978*b*). The new proliferations commonly arise from the margins of the prothallus although it is not uncommon to find regenerates on the midrib or the region of the apical meristem. Thus, as long as its cells are capable of mitotic activity, the gametophyte can attain immortality. Irrespective of their origin, there are no differences in the pattern of growth of the new gametophytic structures which go through a filamentous phase before they become planar.

Regeneration of cut prothalli

Regeneration episodes from pieces cut from the prothallus of *Pteris aquilina* (=*Pteridium aquilinum*) and *Pteris longifolia* described by

Albaum (1938*a*) are typical of most cases studied. Despite the simplicity of body plan of the prothallus, a clear difference is seen between the regenerative behavior of its apical and basal parts. When the prothallus is cut transversely into three pieces, the apical segment containing the meristem grows maximally by the addition of new cells. The middle piece grows only slightly while hardly any growth is noticed in the basal region during a 14-day experimental period. Most interesting is the observation that an isolated nongrowing basal part of the prothallus produces numerous adventitious gametophytes from the cut margin. Rarely, as seen in *Platyzoma microphylla*, both apical and basal halves of the cut gametophyte produce regenerates freely, indicating a lack of morphogenetic control by the meristem (Duckett and Pang, 1984). Experiments with tissue punches extracted from gametophytes of *Gymnogramme (= Pityrogramma) calomelanos* (Soyerman, 1963, 1964) have shown that fragments from all parts of the prothallus frequently exhibit renewed growth and regeneration. Apparently, a release from the physiological dominance of neighboring cells is the signal that sets off the dormant and unexpressed cells of the gametophyte in the morphogenetic pathway.

It is clear that certain structural and physiological changes occur in the cells of the prothallus destined for regeneration. Light microscopic studies of regeneration from excised fragments of prothalli of *G.(=P.) calomelanos* (Vaudois, 1963), *Lygodium scandens* (Vaudois, 1964), and *L. japonicum* (Vaudois, 1969) appear to indicate that starch accumulation in the chloroplasts and their migration to the vicinity of the nucleus are important changes that prepare the cells to divide. At the physiological level, a demonstration that the meristematic region of the gametophyte synthesizes auxin has led to the conclusion that the inhibition of regenerative overtures of cells in the basal region during normal growth is caused by the basipetal migration of auxin. In support of this thesis it has been shown that when IAA mixed in lanolin is applied to the cut basal half of the gametophyte of *P. longifolia*, formation of adventitious outgrowths is completely inhibited (Albaum, 1938*b*).

Another study has indicated that a substance diffusing from the meristematic region of the prothallus of *Anemia phyllitidis* actively inhibits adventitious growth from its meristemless half when both are cultured in the same medium (Reynolds, 1979). In this species, moderate levels of IAA promote rather than inhibit regeneration of adventitious gametophytes from the meristemless half (Reynolds and Corson, 1979). An additional interesting feature is that ABA inhibits regeneration from the meristemless half and that IAA greatly increases the inhibitory effect of ABA (Reynolds, 1981). There are some vague indications in these results of an interaction between IAA and ABA in the maintenance of form in the intact gametophyte, but additional data are required to confirm this.

Maintenance of normal growth correlations in the prothallus depends upon communication between its different parts. The channels of communication operate in both directions, from apex to base and from base to apex of the prothallus. The phenomenon of intracellular communication involving exchange of hormones and metabolites in the prothallus has been attributed to plasmodesmata which maintain cytoplasmic continuity between cells through their walls. As plasmolytic conditions lead to the breakage of plasmodesmata, it is not surprising that a plasmolyzed prothallus exposed to an isotonic milieu experiences a renewed cycle of divisions in scattered cells which thus behave as if they are not in organic connection with the rest of the prothallus (Nakazawa, 1963; Schraudolf, 1966*c*).

Regeneration of single cells and isolated protoplasts

In the experiments on regeneration from cut halves and fragments of gametophytes described above, a general influence of neighboring cells has to be considered, as there is no guarantee that a given regenerate originates in fact from a single cell. The basic experiment to demonstrate complete totipotency of a cell is to isolate it and nurture it in culture into a regenerate. Ito (1960, 1962) isolated single cells from the prothallus of *Pteris vittata* by pricking the neighboring cells with fine glass needles and cultured them in a mineral salt medium. Physiologically released, the cells acquire independent morphogenetic potencies and develop into individuals like the parent. When cells isolated from the margin of the prothallus are cultured, they exhibit the developmental repertoire of a germinated spore by dividing initially to form the rhizoid and protonema, the latter giving rise to a planar structure. Generally speaking, cells isolated from the basal region of the prothallus which are older, regenerate more rapidly than newly formed cells in the vicinity of the meristematic notch; cells from the margin also fare better than those from the meristematic region. Certain fortuitous occurrences such as accidental death of most of the cells of the prothallus (Meyer, 1953), lethality by high doses of ionizing radiations (Kato, 1964*a*, 1967*b*), and freezing injury by ultra-low temperatures (Sato and Sakai, 1980, 1981) can lead to the same end result that one obtains with glass needles, namely, isolated single cells; when the cells that survive death and destruction are isolated and cultured, they regenerate normal gametophytes.

From isolated single cells it is one further step to enzymatically divest cells of their walls and follow the isolated protoplasts through steps involving reformation of the cell wall, growth and division of the reconstituted cell and regeneration of a new gametophyte. To date, successful reports of isolation and culture of protoplasts from cells of the gametophytes of a mutant phenotype of *Pteridium aquilinum* (Partanen,

Power and Cocking, 1980) and *Lygodium japonicum* (Maeda and Ito, 1981) in media of high osmolarity and formation of organized structures from isolated protoplasts have appeared. A critical step in the differentiation of the reconstituted cells or cell aggregates into gametophytes is a reduction in the osmotic value of the medium. Depite these successful reports, the fact remains that difficulties inherent in ensuring survival of a high percentage of protoplasts have not been overcome.

In summary, it seems clear that the developmental potency of cells to replay the program of a haploid spore persists during continued growth of the fern gametophyte. The regenerative behavior of prothallial fragments and of cells isolated from different parts of the prothallus suggests that the expression of totipotency is dependent upon the age of the cells as well as on their internal chemical milieu. It is perhaps more than a coincidence that the older cells which regenerate readily are also placed further from the source of hormones and other metabolites.

Regeneration in filamentous protonemata

Cells of the filamentous fern protonema seldom regenerate spontaneously. A long-standing explanation is that the terminal cell holds a sway over the other cells of the filament. The ability of cells of the protonema whose terminal cell is inactivated by chemical stimuli, by surgical isolation or by environmental or other stresses, to regenerate new growth has supported this interpretation. In experiments with *Dryopteris varia*, most of the test substances are found to modify the polarity of the filament, while colchicine treatment in addition unmasks the regenerative potential of individual cells. In the protonema transferred to 0.05% colchicine, failure of cell division is associated with abnormalities such as lateral expansion of cells, cellular hypertrophy and formation of giant, globular terminal cells. In a high proportion of the hypertrophied protonemata transferred to a medium devoid of the drug, one of the cells enters a dedifferentiative phase and forms a new gametophyte (Nakazawa, 1959). In another experiment, two- or three-celled protonemata of *Pteris vittata* are plasmolyzed in calcium chloride and transferred to isotonic conditions. In about a week's time, regenerative growth is initiated in all but the terminal cell of the filaments (Nakazawa and Kimura, 1964). In both treatments, regeneration is caused by interruption of correlations between the terminal and basal cells of the filament. This is also dramatically seen in experiments involving surgical removal of the terminal cell of a five-celled protonema of *P. vittata.* This operation prompts one or more of the basal cells, thus freed from apical dominance, to divide and produce new branches (Ootaki and Furuya, 1969). There is no sign of regeneration of side branches from the terminal cell of the filament when the basal cells are amputated; this is also the case when the terminal cell alone is isolated and grown (Ootaki, 1967). Cells progressively down the axis of a filamentous protonema show an increasing tendency for regeneration of side branches. This observation reinforces the notion that the regenerative potential of the cell is dictated by its age.

Experiments on decapitated protonemata have provided a window through which one can discern a subtle nuclear control over the regeneration process. For example, as early as 24 hours after amputation of the terminal cell of the protonema of *P. vittata,* an increase in nuclear and nucleolar volumes of the basal cell is seen (Ootaki, 1968; Ootaki and Furuya, 1969). It continues steadily as the nucleus enters the mitotic cycle and is largely complete in about 40 hours. Other impressive changes noted in the regenerating cell are the uniform dispersal of chloroplasts as opposed to their crowding in the anterior end, concentration of chloroplasts around the nucleus, and a change in the shape of plastids and their division. The ultrastructural and metabolic changes observed during regeneration of the protonema of *Gymnogramme (=Pityrogramma) calomelanos* are also quite similar to those seen in *P. vittata* (Faivre-Baron, 1977*b*). The period leading to the division of the basal cell of the protonema is characterized not only by nuclear activity such as DNA synthesis, but also by an increase in the number of chloroplasts and by starch accumulation, DNA synthesis, and protein synthesis in the plastids. Cells of an intact protonema maintain a low structural and synthetic profile; consequently, these cells have a greater need for organelles and for metabolic activity before they can regenerate. Compared to cells of the decapitated protonema, those of the newly regenerated filament have a greater affinity for precursors of DNA and protein synthesis (Faivre-Baron, 1977*a, c*).

The complex nature of regeneration of cells of the fern protonema would lead us to suppose that many different metabolic reactions are involved in the process, but it is not easy to decide which reactions are critical. Because some important enzymes of cell division and carbohydrate metabolism require the integrity of SH-groups, their role in the regenerative growth of protonemata of *G.(=P.) calomelanos* has been explored (Faivre-Baron, 1979*b*). As far as can be seen from histochemical and ultrastructural observations, dictyosome-derived vesicles rich in SH-groups abound in the vicinity of the plasma membrane of the cell induced to regenerate. If protein-bound SH-groups increase during the preliminary stage of regeneration of the protonemal cell, it is reasonable to assume that sulfur-containing amino acids such as cysteine should promote regeneration. This has been found to be the case. Changes in the concentration of SH-groups in the cells poised for regeneration may have a broader meaning in the assembly of the mitotic apparatus and in the activity of SH-enzymes in mitosis.

Effects of light and other factors. It has been known for some time that the branching pattern in the protonemata of certain ferns is modified by the quality and intensity of light. As shown by Ootaki (1965), protonemata of *P. vittata* grown in moderate fluence white light (750 lux) attain planar morphology without regenerating adventitious branches. When such gametophytes are transferred to high fluence white light (2500 lux), continuation of planar growth is associated with the regeneration of a branch from the distal cell of the filamentous part. Transfer of planar gametophytes growing in moderate fluence white light to red light or to darkness for a few weeks followed by reactivation in white light of the same fluence also induces regeneration of branches (Fig. 9.3). Regenerative growth from the protonemal cells of *D. pseudo-mas (=D. affinis)* is induced by culturing them at a temperature range of 27.5 to 30 °C in white or red light (Dyer and King, 1979). It is significant that the various treatments apparently destroy the apico–basal gradient of physiological activities of the protonema and lead to the establishment of new metabolic and growth centers in certain cells which subsequently regenerate.

Questions concerning the metabolic activity of the terminal cell are of

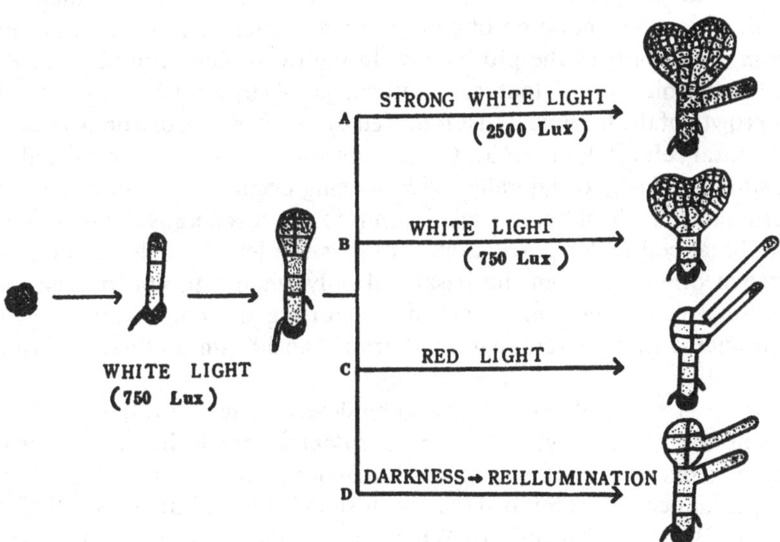

Figure 9.3. Patterns of regeneration of protonemata of *Pteris vittata* under different light–dark conditions. Gametophytes were transferred to each condition on day 10 after sowing. Dots show localization of chloroplasts. (*a*) Regeneration from the most apical cell 2 weeks after transfer to high fluence white light. (*b*) Absence of regeneration 2 weeks after transfer to moderate fluence white light. (*c*) Regeneration one week after transfer to red light. (*d*) Regeneration after 3 weeks in darkness and one week in moderate fluence white light. (From Ootaki, 1965.)

particular interest in the analysis of regeneration, since this cell is at the crossroads of differentiative pathways of a protonema. When the protonema of *G. (=P.) calomelanos* is transferred from a light regimen of moderate fluence to complete darkness, the activity of the terminal cell comes to a standstill. However, regeneration of a side branch from the cell adjacent to the terminal cell of the filament occurs when it is brought back to light. Transfer of a filament to complete darkness after amputation of its terminal cell also prompts regeneration of a side branch from the next cell. From experiments in other systems described previously, it would appear that an intact protonemal filament never regenerates a branch when it is nurtured in a light regimen allowing moderate metabolic activity, but in *G. (=P.) calomelanos*, an intact protonema regenerates a branch in light when the apical zone is shielded by aluminium foil (Faivre-Baron, 1978*b*). Thus, the problem of regeneration of cells of the protonema appears to center around the metabolic activity of the terminal cell and the ability of the regenerating cell to deflect metabolites from this cell. The effects of light indicate that the reactions may have a photochemical basis.

Centrifugation of the protonema causes a pattern of regeneration similar to that induced by light-dark treatments. Here also there is evidence for regeneration of cells into which metabolites are deflected from other parts of the protonema. In the two-celled protonema of *P. vittata* in which cell contents are stratified basally by centrifugation, arrest of growth of the terminal cell is followed by regeneration of a branch from the basal cell (Ootaki, 1963). On the contrary, when the terminal cell is stratified apically or laterally, no branching occurs. One feature of the terminal cell which bears a relationship to the regenerative potential of the basal cell under gravitational stress is the length of the former. A branch originates from the basal cell only when a protonema with a terminal cell longer than 300 μm is centrifuged, while a centrifuged protonema with the terminal cell shorter than 300 μm continues to grow normally.

The last series of experiments to be described here are those undertaken with gametophytes of *G. (=P.) calomelanos* which are in an early stage of planar growth and in which regeneration from the filamentous part is induced by excision of the spoon-shaped plate (Faivre, 1970, 1975; Faivre-Baron, 1978*a*, 1979*a*). When parts of the planar plate excluding the incipient apical meristem are excised, no regeneration occurs from the cells of the filament; nonetheless, removal of the meristem alone provokes regeneration (Fig. 9.4). At the metabolic level, as the cells in the filamentous part begin to regenerate, substantial increases in their microbody population and catalase activity are evident. Although the young gametophyte exhibits the hallmarks of apical dominance, it has

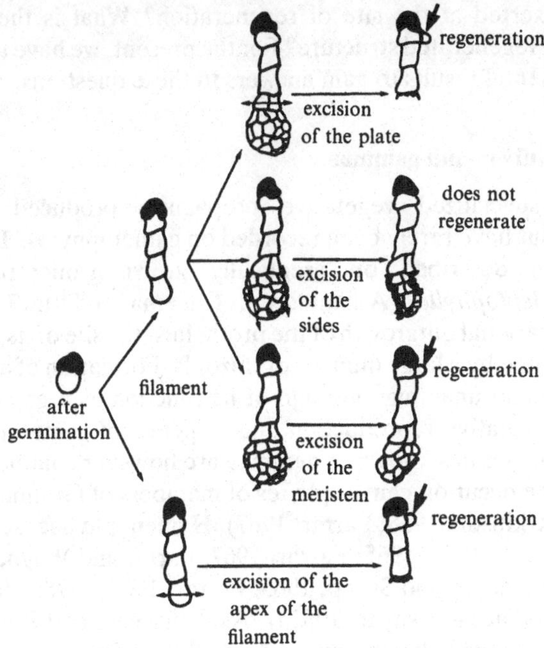

Figure 9.4. Regeneration from cells of the filamentous part of planar gametophytes of *Gymnogramme* (=*Pityrogramma*) *calomelanos* after surgical treatments. Regeneration of the most basal cell of the filament is indicated by an arrow. (*a*) Excision of the entire planar plate. (*b*) Excision of the sides of the planar plate. (*c*) Excision of the meristem. (*d*) Excision of the terminal cell of a filamentous protonema. (From Faivre-Baron, 1978*a*.)

been suggested that cells of the filamentous part of the gametophyte are prevented from expressing their innate morphogenetic potential by diversion of metabolites to the growing part, rather than by a direct intervention of auxin as in apical dominance. If this were true, then removal of the planar plate and substitution of auxin on the decapitated stump should not affect the pattern of branching in the cells of the filament. This has been found to be the case (Faivre-Baron, 1981). Moreover, as shown by autoradiography of ^{14}C-IAA incorporation, acropetal movement of exogenous IAA seems to be an intrinsic property of the gametophyte and the label taken up by its basal cells invariably accumulates in the regenerating cells during a chase period. This could not have happened if IAA is acting in an inhibitory role in apical dominance of the early planar gametophyte.

The examples of regeneration discussed here raise several questions. What initiates regeneration in the cells of an amputated protonema? Is it wound stimulus, physiological isolation or a combination of both? How is the control exerted at the site of regeneration? What is the basis of polarity of the regenerated structure? For the present, we have to wait for more experimental results to gain answers to these questions.

Formation of tubers and gemmae

Tubers are specialized vegetative propagules produced by fern sporophytes but have rarely been recorded on gametophytes. They have recently been described on aseptically grown gametophytes of *Anogramma leptophylla* (Adiantaceae; Cheema, 1980). The tuber appears as a marginal outgrowth of the prothallus, the site of its inception being demarcated by a large number of rhizoids. Formation of antheridia on the tuber is an unambiguous sign of its function as a gametophytic propagule. Vegetative reproduction and dispersal of gametophytes by specialized regenerates known as gemmae are however common among ferns. Gemmae occur on gametophytes of members of Grammitidaceae (Stokey and Atkinson, 1958; Farrar, 1967), Hymenophyllaceae (Stokey, 1940, 1948; Stone, 1958, 1965; Farrar, 1967; Farrar and Wagner, 1968), Adiantaceae (Wagner and Sharp, 1963; Farrar, 1967, 1978; Masuyama, 1975*c*), Polypodiaceae (Nayar, 1963*a*), Aspleniaceae, and Blechnaceae (Page, 1979*a*). In considering gemma formation in ferns, several points merit emphasis. A gemma does not arise by the direct transformation of the prothallial cell but appears on a short stalk cell or gemmifer[1] which is an outgrowth of a prothallial cell. In fact, a single prothallial cell may give rise to many gemmifers each of which generates one or more gemmae. As the gemmifer grows, a transverse constriction appears at its base cutting off a globular terminal cell which functions as the mother cell of the gemma. This cell divides by transverse walls to form a row of three to five cells or two to four rows of 20 to 40 cells (Fig. 9.5). Occasionally, as seen in *Mecodium flabellatum* and *M. rarum* (Hymenophyllaceae; Stone, 1965), a wedge-shaped apical cell which remains active for several generations is formed by the division of the terminal cell of the group.

Farrar (1974, 1978) has made some interesting observations on gemma formation on gametophytes of several species of *Vittaria* maintained on their natural substrate in the laboratory. In *V. lineata*, the adult gemma plan is blocked out by initial divisions of the gemmifer followed by a division of the daughter cell which results in the formation of two gemma primordia. Then, by a series of transverse divisions these cells form a pair of gemmae of 5 to 12 cells. Presumptive rhizoid primordia appear shortly thereafter, one on each terminal cell or occasionally one on a central cell. In other *Vittaria* gametophytes, the gemmifer may give rise to second and

[1] Although this cell is known as the sterigmata in the early literature, the word shares usage with a cell on which a basidium is born. To avoid confusion, Farrar (1974) has advocated the use of the term 'gemmifer', which the present author has accepted.

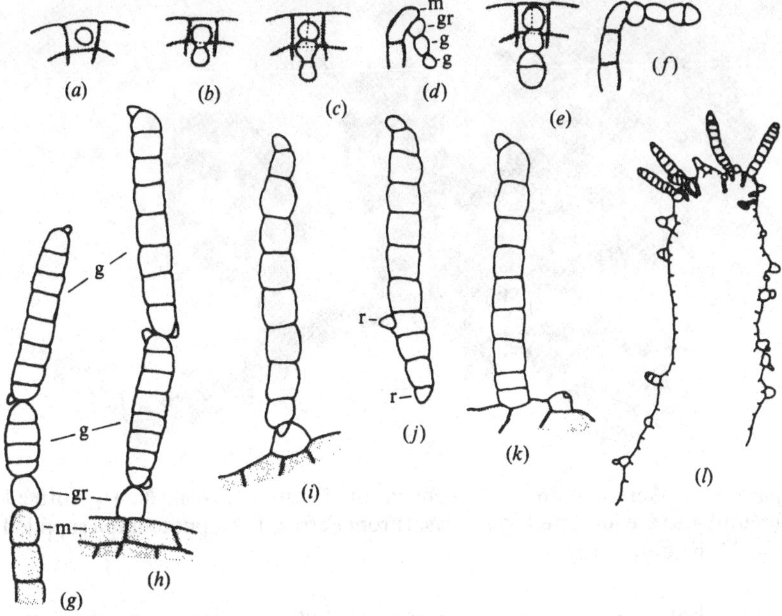

Figure 9.5. Gemma formation on the gametophyte of *Vittaria lineata*. (*a*)–(*j*) Diagrammatic representation of the development of gemmae. (*k*) A mature gemma. (*l*) Part of a prothallus with gemmae in various stages of development. g, gemma; gr, gemmifer; m, margin of the gametophyte; r, rhizoid. (From Farrar, 1974; print supplied by Dr D. R. Farrar.)

third generation gemmifers to the extent that a gemmifer might support 15 or more gemmae. Gametophytes of certain species of *Vittaria* which produce an overabundance of gemmae have been found to occur in North Carolina and in the Appalachian mountains of the southeastern United States. Although sex organs are present on the gametophytes, viable sporophytes have never been found in the region. This has led to the view that the dominant sporophytic phase has essentially been erased from the life cycle of these ferns which thus exist only as gametophytes (Wagner and Sharp, 1963; Farrar, 1967, 1978).

Gemmae detach from the gametophyte and germinate as a prelude to an independent existence. Germination is by an outgrowth of cells at either end or occasionally from the central cells and results in the formation of a new gametophyte (Fig. 9.6). The occurrence of antheridia (Farrar, 1974) on germinating gemmae or on young gametophytes produced from gemmae, and the demonstration of antheridia formation on gemmae by the antheridiogen system from mature gametophytes or by GA (Emigh and Farrar, 1977) confirm the sexual competence of gemmae and their derivatives.

In the scenario involved in the formation and functioning of gemmae

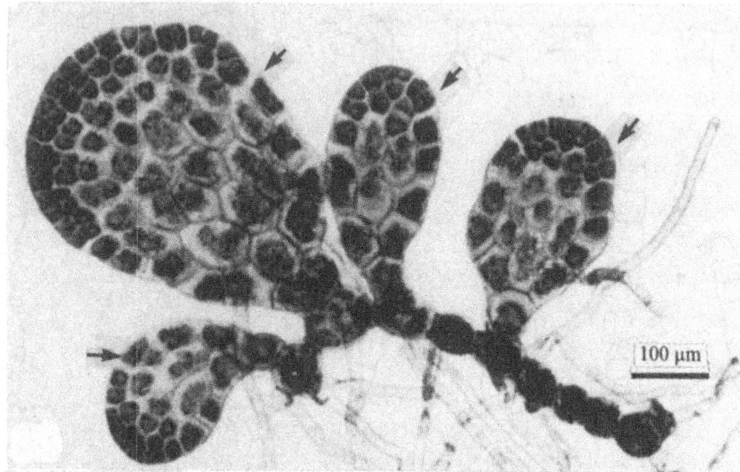

Figure 9.6. Germination of a gemma of *Vittaria lineata*. Newly formed gametophytes are indicated by arrows. (From Farrar, 1974; photograph supplied by Dr D. R. Farrar.)

we see that certain cells of the gametophyte spontaneously display a specific pattern of morphogenesis. These cells probably acquire their developmental potential during the growth of the gametophyte. As a consequence, cells become highly specialized to channel their resources efficiently and indirectly regenerate gametophytes similar to the ones of which they were a part. Cell specialization relies on communication between adjacent cells and with the outside world for signal transduction that deflects the developmental program of the gametophytic cells in the gemma-forming pathway. However, there is need for more information before we can understand the mechanics of gemma formation from the moment a progenitor cell is carved out, to the germination of the propagule.

Neoplastic growth and regeneration of fern prothalli

In the studies described above, some types of abnormal development were frequently noted, but in the investigators' fascination with the extraordinary regenerative powers of fern prothalli, these were often ignored. Later work has shown that certain cultural conditions and ionizing radiations induce the formation of abnormal and tumorous types of growth on the prothallus. These tissues are isolated and cultured as more or less permanent cell lines capable of regenerating normal gametophytes. In some cases, like explants of angiosperm tissues or organs, the prothallus forms a callus which has the potential to regenerate both gametophytes and sporophytes.

Growth of abnormal tissues

The most comprehensive observations have been made on the morphology and powers of regeneration of abnormal proliferations on prothalli of *Pteridium aquilinum* cultured on a medium containing 2% glucose and 0.05% yeast extract (Steeves *et al.*, 1955). The total range of abnormalities observed during a three year culture period has been conveniently placed in two groups. One group consisting of filamentous pseudocallus and parenchymatous callus types is characterized by the absence of any prothallial features and the inability to form sex organs; the filamentous pseudocallus also loses its capacity to revert to normal prothallial growth with time (Partanen, 1972). Under certain conditions, the abnormal proliferations in this group are interconvertible, giving the impression that they are two faces of the same coin. The other group includes morphological types ranging from dense masses of filaments and more compact hemispherical filamentous aggregates to branched club-like processes composed of true tissues which possess seemingly functional sex organs and undiminished potentiality for normal prothallial expression for a long period of time. There can be little doubt that these growth expressions are a function of the particular conditions of culture, although it is unlikely that culture conditions alone can account for their survival in this form through repeated subcultures.

Among the questions that relate to neoplastic growth of the prothalli, one that has received some attention is the degree of polyploidy associated with the modified growth patterns. Using the chromosome configuration and DNA content of normal prothallial cells as a standard of reference, it has been established that in the normal prothallial cells the mode is precisely determined at the haploid level with 1C DNA amount, as one would expect, with an occasional diploid number in older cells (Partanen, Sussex and Steeves, 1955; Partanen, 1956, 1959). No deviations from the normal haploid number are also found in the cells of essentially prothallial configuration represented by the types included in the second group. In contrast, a wide disparity in chromosome numbers from an apparently normal haploid condition to a state of variable aneuploidy in the 3N or 4N range with a simultaneous increase in DNA content by endoreduplication is found in the filamentous pseudocallus and the parenchymatous callus types. The extent to which a recently derived pseudocallus retains its capacity for regeneration of a normal prothallus is also related to its cytological condition, as the progressive loss of capacity for regeneration is correlated with an increase in chromosome number. These observations are obviously in line with the view that in cultured fern prothalli the balance of competing genotypes shifts the morphogenetic potential in favor of deviations from the normal growth pattern. Yet, these data do not indicate that quantitative nuclear changes usher in the neoplastic transformation of prothalli of *P. aquilinum*. Perhaps one could envision the existence of a plurality of mechanisms at the cellular level to explain the occurrence of abnormal types of growth in cultured prothalli.

Tumor induction by ionizing radiations

As a sequel to the work described above, other investigations have shown that abnormal proliferations comparable to spontaneously occurring parenchymatous pseudocallus type are produced in high frequency when spores or young gametophytes of *P. aquilinum* are exposed to X-rays, γ-rays (Partanen and Steeves, 1956) or UV-irradiation (Partanen and Nelson, 1961). Dose-response relationships of spores to treatment with X-rays and UV-rays are linear with increasing radiation up to a point when spore viability becomes an issue; in terms of actual numbers of tumors, at X-ray dose of 16 000 r, there are 26.5 tumors/10 spores at six weeks as opposed to 0.2 tumors for the same number of unirradiated spores – in other words – a nearly 150 fold increase; the figures are even more striking with UV-rays (Partanen, 1958; Partanen and Nelson, 1961).

What is the mechanism that regulates the growth of tumors on prothalli generated from irradiated spores? An irreversible lesion in the division mechanism of the spore may commit it to an autonomous proliferation. Biochemical or nutritional deficiency or variants arising as mutations might also be expected to show such abnormalities. If we are dealing with nutritionally impaired variants it should be possible to overcome the defect by supplying the missing component in the medium and restore normal phenotypic expression in the prothallus. When irradiated spores of *P. aquilinum* are grown in a medium supplemented with casein hydrolyzate, the frequency of tumor induction on the gametophyte is reduced to 34% of the control growing in the unsupplemented medium (Partanen, 1960*a*). The effect of casein hydrolyzate in the medium is obviously linked to the deficiency of some specific amino acids; in line with this reasoning, it was found that a mixture of certain amino acids corresponding to a synthetic approximation of casein hydrolyzate is as effective as the commercial preparation in depressing the incidence of tumor growth. Single amino acids also reduce the frequency of tumors induced by both X-rays and UV-rays, the most effective ones in this respect being *l*-methionine and *l*-lysine (Partanen, 1960*b*; Partanen and Nelson, 1961).

These studies open up some important questions. Are we dealing with mutants in the genetically accepted sense of the word, or are we dealing with some genetic variants exhibiting defined nutritional deficiency? Supporting the case for the mutant nature of the tumor is the observation that cells dissociated from the tumor maintain the altered morphology for prolonged periods of culture under conditions which preserve normal gametophytic morphology in the wild type (Partanen, 1972). Arguing against the true mutant nature of these outgrowths is the fact that in no case has the trait been shown to be transmitted through sexual crosses in a typical Mendelian fashion. Moreover, it has not been established whether the tumorous growth is an epigenetic mechanism due to stabilization of genes that program a tumorous state.

Tumorous outgrowths have also been induced by ionizing radiations in the gametophytes of *Osmunda cinnamomea* (Partanen and Nelson, 1961), *O. regalis* (Haigh and Howard, 1973*b*), *Onoclea sensibilis* (Estes, 1963), and *Pteris vittata* (Palta and Mehra, 1973). A remarkable feature of old prothalli of *O. sensibilis* exposed to lethal doses of UV-rays is the regeneration of cells of the midrib region into normal gametophytes (Estes, 1963). In *O. regalis* tumors appear as spherical or ovoid masses of cells some of which later differentiate into normal planar gametophytes with functional sex organs. They are generally associated with dead cells on the meristematic notch of the prothallus, and are rarely observed on irradiated prothallus lacking an apical meristem. From these observations it has been suggested that cell death and mitotic arrest rather than gene mutation might account for the ability of the irradiated prothallus to respond to systemic factors regulating their growth (Haigh and Howard, 1973*b*). A postulated inhibition of auxin biosynthesis in the apical meristem of the gametophyte by irradiation being the cause of tumorization, could not be confirmed (Howard and Haigh, 1974).

Growth of the callus

The first demonstration that the fern gametophyte can be transformed into an undifferentiated callus stems from the observations of Morel (Morel, 1950; Morel and Wetmore, 1951) who found that a small number of gametophytes in a population of spores of *O. cinnamomea* germinated in a mineral salt medium spontaneously regenerated a callus. Upon its transfer to a medium enriched with certain B vitamins, the callus was found to grow into a moderate-sized granular and friable tissue composed of parenchymatous cells. Later work with other fern species has shown that besides the usual mineral salts and a carbon energy source such as glucose or sucrose, the essential recipe for callus induction is continued maintenance of germinated spores in a medium containing varying concentrations of 2,4-D, yeast extract, or coconut milk (Mehra and Sulklyan, 1969; Kato, 1963, 1969*a*; Padhya and Mehta, 1981).

In the transformation of undifferentiated gametophytic callus into normal planar gametophytes, light or its lack thereof and a reduced concentration of sucrose or auxin or their total absence in the medium

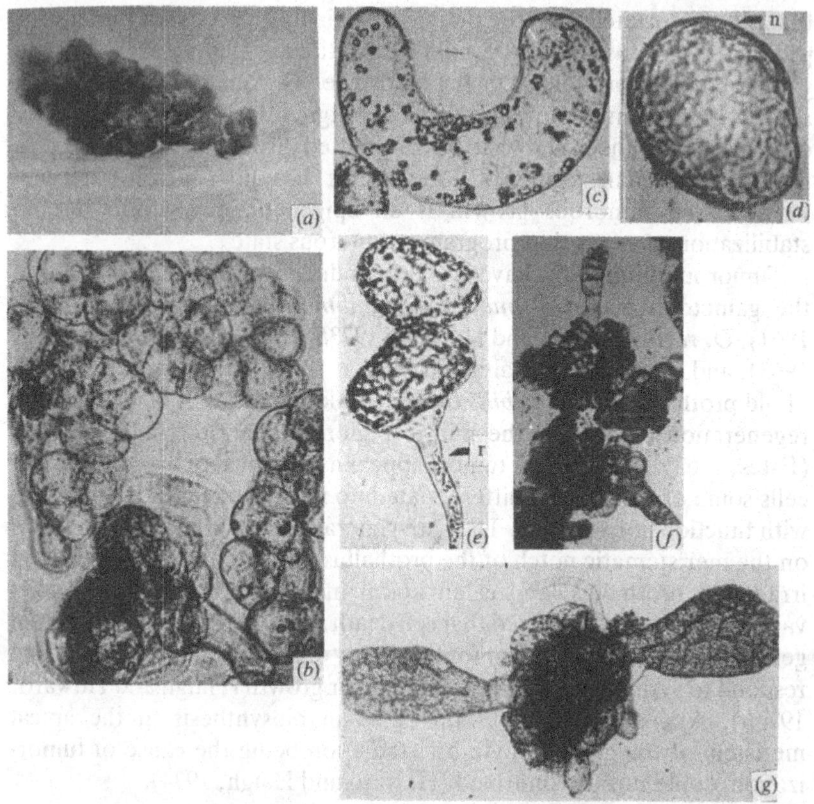

Figure 9.7. Callus formation on the gametophytes of *Pteris vittata* and regeneration of new gametophytes from single cells of the callus. (*a*) A compact callus. (*b*) Typical groups of cells in suspension culture. (*c*), (*d*) Examples of isolated single cells. (*e*) Rhizoid formation from a single cell. (*f*) Formation of filamentous protonemata from a cell cluster. (*g*) Formation of planar gametophytes from a cell cluster. n, nucleus; r, rhizoid. (From Kato, 1964*c*.)

play a decisive role (Mehra and Sulklyan, 1969; Partanen, 1972; Padhya and Mehta, 1981). Perhaps the best example illustrating a high degree of plasticity in the regeneration of gametophytes from a callus is afforded by *P. vittata*. In this system the callus becomes friable after a short period of culture in a medium containing sucrose and yeast extract. Abundant reformation of gametophytes is obtained when the friable tissue consisting of small groups of cells and single cells is transferred to a medium containing only mineral salts (Kato, 1963, 1964*c*). In the conversion of single cells which merely grow and divide in culture, into cells with the ability to reconstitute gametophytes, the first step appears to be surprisingly identical to the germination of the spore, namely, formation of a rhizoid (Fig. 9.7). The free cells obtained in shake culture exhibit a variety of growth patterns to suggest that the gametophytic callus is a complex of different cell types more than one of which may be capable of regenerating normal gametophytes (Kato, 1968*a*, 1969*b*).

In summary, these studies show that manipulation of the nutritional and hormonal substances in the medium is a powerful means of controlling the dedifferentiative growth of the fern gametophyte cells and to achieve its clonal multiplication. The regeneration of sporophytes from the gametophytic callus is considered in Chapter 13.

General comments

The plasticity of fern gametophytes has made them useful objects to study their normal and regenerative growth. In the true sense of its autotrophic nature, the gametophyte hardly responds to exogenous carbohydrate, nitrogen or hormonal supplements in the medium. The regenerative ability of the gametophyte has provided important information for our understanding of the internal relations of its cells and the reciprocal interaction of its parts. The experiments reviewed in this chapter indicate that both intracellular and extracellular factors regulate regenerative and reconstitutive growth of fern gametophytes. Ultimately, regeneration of a new gametophyte coud involve activation of specific genes – a switching-on of the morphogenetic program in response to chemicals or to surgical operation. Questions connected with this could guide future research on the regeneration of fern gametophytes.

Part III

Reproductive strategies

10

Control of differentiation of sex organs on gametophytes

Gametophytes of homosporous ferns acquire the potential to form antheridia and archegonia during a period of growth and maturation. A striking aspect of sexuality in fern gametophytes is the complexity of the division sequences giving rise to sexual cells and the simplicity of the final products. Initiation of sex organs (gametangia) on the gametophyte thus poses important developmental questions inasmuch as certain cells in a homogeneous population respond to reprogramming cues and differentiate into gametes adapted for sexual recombination. What is the trigger that starts off cells of the gametophyte on a particular course of metabolism and behavior which will turn them into antheridia and archegonia? Analysis of this question has been an important thrust in the developmental biology of ferns and as a result there is strong evidence to show that antheridium formation on the gametophyte occurs in response to hormonal signals. The controlling factors in the initiation of the archegonium have as yet been hardly considered, but the challenge is great.

In this chapter we emphasize the ontogeny of the antheridium and archegonium and follow it up with a discussion about the physiological control of their differentiation. Beginning with a survey by Näf (1962*b*) which was partially devoted to the physiology of antheridium formation in fern gametophytes, new knowledge gained in the field has been incorporated into periodic reviews (Näf, 1963, 1969, 1979; Näf, Nakanishi and Endo, 1975; Voeller, 1964*a*; Voeller and Weinberg, 1969). To the interested reader, these articles should be of value in synthesizing the wealth of information available on the control mechanisms involved in antheridium initiation on fern gametophytes.

Antheridium development

Mature fern prothalli behave as centers of growth and morphogenesis from which antheridia and archegonia differentiate. One feature common to homosporous ferns is that their prothalli differentiate antheridia earlier than archegonia. However, there is some variability in the lag period between the appearance of antheridia and archegonia and it is not uncommon to find prothalli in which antheridial function is partially or completely terminated before archegonia are formed as well as those in which both sex organs occur side by side. When antheridia precede archegonia, they are confined to the ventral surface behind the apical notch of the prothallus. Antheridia may also be scattered over the entire prothallial surface, confined to the margins of the prothallus or found laterally on filaments. In a prothallus harboring both sex organs at the same time, competition for space and resources restricts antheridia to the midrib region in the posterior half and archegonia to the anterior region of the midrib. As we will see in the next chapter, the distribution of sex organs on the gametophyte has an adaptive value in favoring inter-gametophytic fertilization and heterozygosity.

Ontogeny of the antheridium

Two models of antheridium development that we will describe here have been arbitrarily assigned to the less specialized families of Ophioglossaceae and Marattiaceae and the advanced filicalean ferns. In the primitive families the antheridium is sunken in the parenchymatous tissue of the gametophyte. It is formed from a superficial cell of the prothallus (antheridium initial) which is partitioned transversely into a lower spermatogenous cell and an upper cover initial. In the mature antheridium, a mass of sperm produced by repeated divisions of the spermatogenous cell is ensheathed in a jacket layer of cells cut off from the surrounding prothallial cells. In the final act of antheridium formation, the cover initial divides to form an incipient cushion of cells of which a centrally placed cell functions as an opercular cell. Discharge of sperm is facilitated by this cell which flips open like a cap (Nayar and Kaur, 1971; Whittier, 1983).

In the advanced filicalean ferns, cellular makeup of the antheridial wall is numerically different and accordingly, antheridia with many cells in the wall and those with a wall composed of only three or four cells have been recognized. The antheridium with a many-celled wall develops from a wedge-shaped initial cell. Although subsequent divisions of this cell are chaotic in the different species investigated, the final product in all cases is a large spermatogenous cell enveloped by a multicellular wall. By the division of a wall cell towards the anterior end of the antheridium, one or two cap cells constituting an operculum are also differentiated. Antheridia with a multicellular wall attain massive size and have been described in *Osmunda, Cyathea* and *Sticherus palmatus* (Gleicheniaceae), among others (Atkinson and Stokey, 1964).

Antheridia with a limited number of cells in the wall are common in polypodiaceous ferns. The typical antheridium of this type has a central spermatogenous cell enveloped in a jacket of three cells. The formation of the antheridium centers around the establishment of the jacket cells which are unique in their size and shape; starting from the base there is a cup-shaped basal cell, a doughnut-shaped ring cell and a disc-shaped cap cell. Although this basic configuration of cells in the mature antheridium has been accepted in all quarters for well over a century, there has been continuing debate about the genesis of these cells. According to the classical concept, the first division of the antheridium initial cell in the prothallus is either transverse, concave or funnel-shaped. The second division is hemispherical or periclinal while the third division is also funnel-shaped. Variations of this pattern have been proposed by Davie (1951), Verma and Khullar (1966), and Leung and Näf (1979). The points of contention among the investigators have been with regard to the orientation of the phragmoplast, the deposition of the cell wall and the plane of the new cell wall during each division giving rise to the jacket cells (see Kotenko, 1985 for review). A recent study on the cytology of antheridium formation in *Onoclea sensibilis* has cleared much of the confusion surrounding the earlier interpretations and has convincingly shown that the origin and disposition of each of the three cells of the jacket follow the classical concept (Kotenko, 1985, 1986). According to this work, an asymmetric division of a prothallial cell giving rise to a small antheridium initial marks the beginning of antheridium formation. During the first division of the antheridium initial, the phragmoplast undergoes a change in shape that causes it to sink and develop into a funnel-shaped wall delimiting a basal cell and an upper cell. The second division occurs in the upper cell by a wall oriented parallel to its outer curved wall. At this stage the antheridium is composed of a basal cell, an upper dome-shaped cell and a central spermatogenous cell. The next division occurs in the dome-shaped cell by a funnel-shaped wall resulting in a ring cell and a cap cell (Fig. 10.1). The morphogenetic importance of the divisions that yield the jacket cells is foreshadowed by localized cytoplasmic accumulations which ensure that one daughter cell is vacuolate and the other enriched with cytoplasm. That some type of supercellular agency is at work in coordinating these divisions is suggested by the observation that the cytoplasmic accumulation is rich in RNA (Leung and Näf, 1979). Although light and electron microscopic investigations of antheridium formation in other ferns have confirmed the occurrence of primary funnel shaped walls during the first division of the antheridium initial (Schraudolf, 1968; Schraudolf and Richter, 1978; Stone, 1962) and during the formation of the ring and cap cells (Bierhorst, 1975*a*; Schraudolf, 1963, 1968; Stone, 1961, 1962), Kotenko's (1985) work ranks as the first one in which a complete picture of these complex events has been fully traced. The prolonged controversy surrounding the genesis of the wall cells of the antheridium of polypodiaceous ferns serves to underscore the difficulties inherent in portraying the events in a cell that

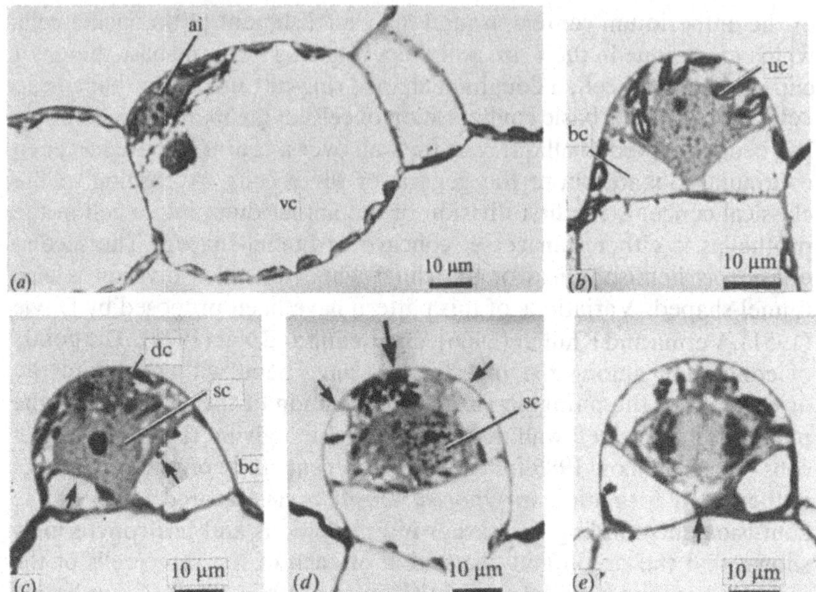

Figure 10.1. Stages in antheridium development on the gametophyte of *Onoclea sensibilis*. (*a*) Division of a prothallial cell to form the antheridium initial (ai) within a vacuolate vegetative cell (vc). (*b*) Two-celled stage showing the vacuolate basal cell (bc) and a densely cytoplasmic upper cell (uc) separated by a funnel-shaped wall. (*c*) Three-celled stage showing the curved periclinal wall (arrows) separating an upper dome-shaped cell (dc) from a central spermatogenous cell (sc); bc, basal cell. (*d*) Section through the dome-shaped cell showing the formation of the cell wall (small arrows) around the cap cell nucleus (large arrow); sc, spermatogenous cell. (*e*) Three-celled stage showing the nucleus of the spermatogenous cell in division. Arrow points to the phragmoplast. (From Kotenko, 1985; photographs supplied by Dr J. L. Kotenko.)

lie at some depth beneath other cells unless they are followed in thin serial sections.

With this brief background about the ontogeny of the antheridia, we will examine the factors that control their differentiation.

Discovery of antheridiogens

The history of the discovery of factors controlling antheridium differentiation on fern gametophytes divides rather strikingly into two periods – the first up to 1950 and the second dating from 1950 to the present. In the pre-1950 epoch, antheridium intiation was observed on gametophytes of diverse ferns growing at reduced light intensities, in starvation conditions, and in crowded cultures. The implication of these observations was that genes for antheridium formation become functional when the normal vegetative growth of the gametophyte slows down. In other observations, the occurrence of unisexual male gametophytes in certain ferns was found to be enhanced by deficiency of nutrient substances in the culture medium or by changing the environmental conditions of growth of the gametophytes. This generated the idea that an array of mineral nutrients or light conditions may be an efficient means of perpetuating male sexuality in fern gametophytes and led to occasional bursts of interest in the experimental control of sexuality. Although these observations resulted in the publication of nearly two dozen articles, they did not implicate a triggering mechanism in the differentiation of antheridia on gametophytes growing under normal conditions in the laboratory or in the field.

The first breakthrough in the post-1950 period occurred as a result of an accidental observation by Döpp (1950) that in mixed cultures of adult and juvenile gametophytes of *Pteridium aquilinum*, only the latter bear antheridia. This led him to suspect that induction of male sexuality on the young gametophytes was regulated by a substance made by the old gametophytes. This was indeed found to be the case. By careful experiments it was shown that addition of an agar medium on which gametophytes of *P. aquilinum* were grown to maturity or an aqueous extract of mature gametophytes to the culture of young gametophytes hastens antheridium formation on the latter by a few days. Aware of the claims of previous investigators that antheridium initiation occurred as a result of nonspecific inhibition of growth of the gametophyte, Döpp (1950, 1959, 1962) ruled out such a possibility and convincingly showed that even at extremely low dilutions of the extract which produced no growth inhibition whatsoever of the gametophyte, a few antheridia still appeared. Another noteworthy observation from this work was that the antheridium-inducing extract lacks species-specificity. Thus, an extract of *P. aquilinum* gametophytes could induce antheridium formation on the gametophytes of *Dryopteris filix-mas*; reciprocally, an infusion of *D. filix-mas* gametophytes was active on gametophytes of *P. aquilinum*. The sum of these studies, which constitute one of the most significant advances in the developmental biology of fern gametophytes, immediately suggested that, although antheridium induction is a seemingly simple response of certain cells of the gametophyte to internal biochemical turmoil, it is actually triggered by a specific organ-inducing chemical substance. This work indeed set the stage for much of the contemporary analysis on the control of sexuality on fern gametophytes.

Subsequent work by Näf (1956) not only led to a broad confirmation of these results, but also helped to reinforce the view that a natural hormone causing antheridium formation is present in the gametophytes of *P. aquilinum*. This biologically active factor was assayed against gametophytes of *Onoclea sensibilis* which do not form antheridia spontaneously under certain conditions of culture. In this system, a liquid medium that supports the growth of *P. aquilinum* gametophytes for seven weeks or an extract of gametophytes of the same age induces the formation of antheridia at a dilution of 1:6250 to 1:31 250. Still later investigations (Pringle, Näf and Braun, 1960) showed that a purified substance isolated from the culture filtrate of *P. aquilinum* is effective against *O. sensibilis* gametophytes at a dilution as low as one part in ten billion. The prevailing light regimen which alters the morphology of gametophytes has little effect on their response to the antheridium-inducing substance, since planar gametophytes formed in white light and filamentous forms produced in red light generate abundant antheridia in the presence of the extract. Pringle (1961) proposed the name antheridogen for the substance isolated from gametophytes of *P. aquilinum*, although for reasons advanced by Näf, Sullivan and Cummins (1969), the preferred term used in this book for the chemical inducer of male sex organs on fern gametophytes is antheridiogen.

Thus far, besides *P. aquilinum* and *D. filix-mas*, native antheridiogen activity has been established only in a limited number of ferns. Among those examined, it is the peculiar virtue of gametophytes of *Blechnum gibbum, Thelypteris* sp., *T. hexagonoptera* (Näf, 1956), *Anemia phyllitidis* (Näf, 1959), *A. hirsuta* (Zanno *et al.*, 1972), *A. mexicana* (Nester and Schedlbauer, 1982), *Lygodium japonicum* (Näf, 1960), *O. sensibilis* (Näf, 1961*b*; Näf *et al.*, 1969), *Microlepia speluncae* (Dennstaedtiaceae; Fellenberg-Kressel, 1969), *Ceratopteris thalictroides* (Schedlbauer and Klekowski, 1972), *C. richardii* (Hickok, 1983), *Vittaria lineata* (Emigh and Farrar, 1977), *Bommeria ehrenbergiana, B. hispida* (Adiantaceae; Haufler and Gastony, 1978), *Athyrium filix-femina* (Schneller, 1979), *Pityrogramma calomelanos* (Dubey and Roy, 1985), *Cystopteris protrusa* (Aspleniaceae; Haufler and Ranker, 1985), and *Pteris vittata* (Gemmrich, 1986*a*) to elaborate this substance in their cells and release it into the medium to induce antheridia formation in the neighboring gametophytes.

A recent investigation (Rubin and Paolillo, 1983) has questioned earlier claims (Näf, 1956; Näf *et al.*, 1969) that an antheridiogen is required for the development of male sexuality on gametophytes of *O. sensibilis*. It has been shown that whereas agar used as a substratum in the earlier studies suppresses antheridia formation on the gametophytes, those nurtured on unsterilized soil or ashed soil develop antheridia spontaneously. This result would seem to weigh against antheridiogen involvement in the fundamental control of male sexuality on the gametophytes of this fern. Yet, there are some convincing physiological studies, to be reviewed later, which indicate that under certain defined conditions of culture there is no substitute for antheridiogen for antheridium induction on the gametophytes of *O. sensibilis*. Apparently there is something in the agar medium that makes gametophytes sensitive to the hormonal signal as there is something in the soil that makes them insensitive to the same signal. As an experimental system, *O. sensibilis* seems favorable enough for further investigations to reconcile these conflicting results.

Activity spectrum of antheridiogens

Some antheridiogens show a great deal of interspecific activity, while others are relatively species-specific. For convenience, Näf (1969) has proposed abbreviations for the different antheridiogens, which reflect the genus from which they are derived.

***Antheridiogens from* Pteridium aquilinum *and* Pteris vittata.** To date, the best studied activity spectrum is of the antheridiogen isolated from gametophytes of *P. aquilinum* (A_{Pt}). As seen in Table 10.1, positive responses to A_{Pt} have been recorded in gametophytes of 38 species belonging to 25 genera. While the Table does not list the genera and species which fail to respond to A_{Pt}, some examples of this type are worthy of comment. The least sensitive species are probably found in Polypodiaceae. The classic example is *Aglaomorpha meyeniana* (Polypodiaceae) whose gametophytes require a titer of A_{Pt} nearly 10 000 times more than that which is effective on *O. sensibilis* gametophytes (Näf, 1966), while other members of Polypodiaceae tested such as *Drynaria quercifolia, D. rigidula, Polypodium polycarpon (= Microsorium punctatum), P. subauriculatum, P. (=Niphidium) crassifolium* (Voeller, 1964*a*), and *P. (=Phlebodium) aureum* (Näf, 1956) completely fail to respond to the stimulus. Moreover, within the same genera such as *Dryopteris, Tectaria* (Aspleniaceae), *Blechnum, Dennstaedtia* (Dennstaedtiaceae) and *Pteris* both sensitive and insensitive species are encountered (Voeller, 1964*a*). These observations cast some doubt on the usefulness of the activity spectrum of A_{Pt} to monitor the taxonomic relationships and phylogenetic affinities of homosporous ferns.

In many ways, the recently investigated activity spectrum of antheridiogen of *Pteris vittata* (A_{Ps}) appears to be similar to that of *Pteridium aquilinum* with the added attraction that several species of *Pteris* are sensitive to the hormone (Gemmrich, 1986*b*).

Activity of antheridiogen from other ferns. An unexpected outcome of studies on the specificity of A_{Pt} was the lead that resulted in the discovery of additional antheridiogens. Näf (1959) found that although gametophytes of *Anemia phyllitidis* are particularly recalcitrant and fail to respond even to high concentrations of A_{Pt}, they nonetheless elaborate

Table 10.1 *Activity spectrum of antheridiogen from* Pteridium aquilinum[a]

Family	Subfamily	Genera and species	Reference
Adiantaceae	Adiantoideae	*Adiantum pedatum*[b]	Voeller, 1964*a*
		Bommeria ehrenbergiana[e]	
		B. hispida	Haufler and Gastony, 1978
		B. subpaleacea	
		B. pedata	
	Cryptogramma crispa	Döpp, 1959	
		Hemionitis arifolia[b]	Voeller, 1964*a*
		Notholaena sinuata	
		N. distans	Döpp, 1959
		N. vellea (= *Cheilanthes catanensis*)	
		Pellaea viridis (= *Cheilanthes viridis*)	Döpp, 1959
	Pteridoideae	*Pteris longifolia*[b]	Voeller, 1964*a*
		P. vittata	Gemmrich, 1986*b*
Aspleniaceae	Athyrioideae	*Athyrium thelypterioides*[c]	Näf, 1956
		Cystopteris protrusa	Haufler and Ranker, 1965
		Gymnocarpium robertianum	Döpp, 1959
		Matteuccia struthiopteris[c]	Näf, 1956; Döpp, 1959, 1962
		Onoclea sensibilis	Näf, 1956
		Woodsia obtusa[d]	Näf, 1959
	Dryopteridoideae	*Dryopteris filix-mas*	Döpp, 1950
		Polystichum acrostichoides[c]	Näf, 1956
		P. tsus-simense[d]	Näf, 1966
	Tectarioideae	*Tectaria incisa*[c]	Näf *et al.*, 1975
Blechnaceae		*Blechnum brasiliense*[b]	Voeller, 1964*a*
		Doodia media[b]	Voeller, 1964*a*
		Woodwardia areolata[c]	Näf, 1956
		W. virginica[b]	Voeller, 1964*a*
Davalliaceae	Oleandroideae	*Nephrolepis hirsutula*[c]	Näf, 1960
		N. cordifolia[b]	Voeller, 1964*a*
Dennstaedtiaceae	Dennstaedtioideae	*Dennstaedtia punctilobula*[d]	Näf, 1959
		Microlepia speluncae	Fellenberg-Kressel, 1969
		Pteridium aquilinum	Döpp, 1950
Polypodiaceae	Drynarioideae	*Aglaomorpha meyeniana*[d]	Näf, 1966

Table 10.1 (*cont.*)

Family	Subfamily	Genera and species	Reference
Thelypteridaceae		*Thelypteris hexagonoptera*[b]	Näf, 1956
		Thelypteris sp.	
Thyrsopteridaceae		*Cibotium barometz*[b]	Voeller, 1964*a*

[a] Data from Voeller (1964*a*) and Näf *et al.* (1975) with additions.
[b] The purified extract was applied at a concentration more than 100 times greater than that required to induce antheridia formation in *Onoclea sensibilis*.
[c] Antheridiogen containing preparation was applied at a concentration 200 times greater than that required to induce antheridia formation in O. *sensibilis*.
[d] Antheridiogen containing preparation used at concentrations expressed as mutiples of the concentration effective in O. *sensibilis* are: *P. tsus-simense*, 3000; *T. incisa*, 1000-3000; *W. obtusa*, 25; *D. punctilobula*, 125; *A. meyeniana*, 10000.
[e] *B. ehrenbergiana* and *B. subpaleacea* required greater concentrations of antheridiogen than did *B. hispida* and *B. pedata*.

and secrete into the medium their own antheridiogen (A_{An}). Based on its failure in the *Onoclea* bioassay system, and certain physiological properties, A_{An} appears to be different from A_{Pt}. As revealed by chromatography, A_{An} apparently consists of two separate components, A_{An1} and A_{An2}, which reach their peak activity about 40 to 50 days after spore germination (Schraudolf, 1972; Bürcky, 1977*a*).

Investigations on the control of antheridium differentiation in *Lygodium japonicum* led to the discovery of yet another antheridiogen (A_{Ly}) (Näf, 1960). The special interest of this work is the result of cross-testing of A_{An} and A_{Ly} against gametophytes of *A. phyllitidis* and *L. japonicum*. While high concentrations of A_{An} promote antheridium induction on *L. japonicum* gametophytes, A_{Ly} is ineffective in the reciprocal test.

Factors related to the stability of the native antheridiogen were major impediments in the clear-cut demonstration of its presence in the gametophytes *Onoclea sensibilis*. In undisturbed cultures, only traces of the hormone (A_{On}) are found, but appreciable amounts are obtained by autoclaving or boiling the medium at low pH. A_{On} also appears to be uniquely different from other known antheridiogens. A measure of the difference between A_{Pt} and A_{On} is the fact that on chromatograms the latter travels nearly four times faster than A_{Pt}. The inability of A_{On} to induce antheridium formation on gametophytes of *A. phyllitidis* and *L. japonicum* seems to set it apart as a distinct chemical entity (Näf, 1965;

Näf *et al.*, 1969). In a similar way, cross reactivity tests between antheridiogens from *Microlepia speluncae* (Fellenberg-Kressel, 1969), *Ceratopteris thalictroides* (Schedlbauer, 1974) and *A. mexicana* (Nester and Schedlbauer, 1982) and gametophytes of ferns with known antheridiogen activity are always negative, indicating that each antheridiogen is a distinct chemical entity. That the *Ceratopteris* factor (A_{Ce}) is probably a distinct chemical substance is also suggested by its high Rf value on thin layer chromatograms, compared to A_{Pt} and A_{An} (Schedlbauer, 1976*a*). Somewhat differently, the antheridiogen from *Pityrogramma calomelanos* is effective in inducing antheridium formation on gametophytes of *Onychium siliculosum* (Adiantaceae; Dubey and Roy, 1985).

In summary, it appears that with one or two exceptions, there are differences in the nature of the hormonal stimulus that induces male sexuality in the gametophytes of different species of ferns. As the necessity for a hormone for antheridium induction has not been demonstrated in the large majority of ferns, it is unlikely that antheridiogen is a universal regulator of male sexuality in this group for plants.

Antheridiogen activity and gibberellins

Not long after the discovery of antheridiogens, their chemical identity to gibberellins became increasingly apparent. In Chapter 3 it was mentioned that GA induces dark-germination of spores of *Anemia phyllitidis*. At the same time as this observation was made, it was found that gametophytes of *A. phyllitidis* exposed to low levels of GA produce antheridia while those nurtured in the basal medium remain essentially vegetative (Schraudolf, 1962). The effect of GA is so dramatic that when freshly germinated spores with a single protonemal cell are exposed to the hormone, the next round of division of this cell yields an antheridium (Fig. 10.2). Later research established that antheridium induction by exogenous GA is mostly limited to four genera of Schizaeaceae namely, *Anemia, Lygodium, Mohria* and *Schizaea* (Döpp, 1962; Schraudolf, 1962; Voeller, 1964*b*; Voeller and Weinberg, 1969; Näf, 1966; Nester and Coolbaugh, 1986). The only known genus outside Schizaeaceae to form antheridia by GA application is *Vittaria*, although gemmae produced by gametophytes, rather than the latter themselves, respond to the hormone (Emigh and Farrar, 1977).

In the above studies, GA_3 was used as the source of gibberellin. Activity spectra of the different gibberellins on antheridium induction on gametophytes of *A. phyllitidis* and *L. japonicum* have shown that GA_7 and GA_9 are as effective as GA_3, while GA_1, GA_5 and GA_8, sponsor a delayed response or give a response only at high concentrations (Voeller, 1964*a*, *b*; Schraudolf, 1964, 1966*a*; Takeno and Furuya, 1975). In *A.*

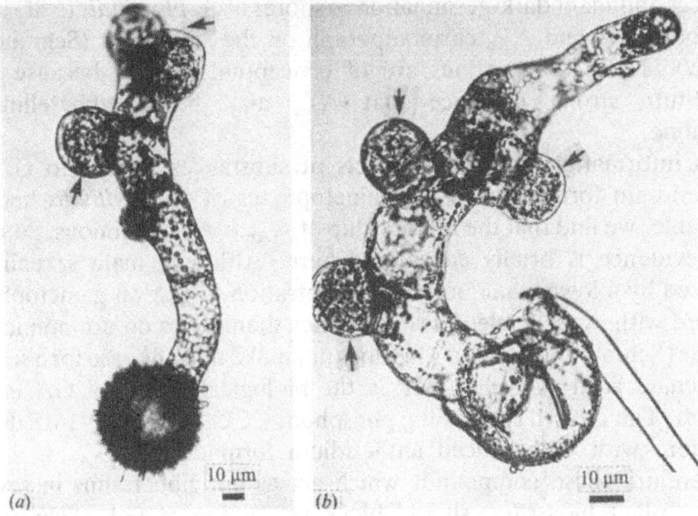

Figure 10.2. Antheridium (arrows) formation on the germ filaments of (*a*) *Anemia phyllitidis* and (*b*) *Lygodium japonicum* by gibberellic acid (10.0 mg/l).

phyllitidis antheridium assay, a C-2- substituted GA_4 derivative, 2,2,-dimethyl-GA_4 is slightly more active than authentic GA_4 (Schraudolf, 1982*b*), while a substituted phthalimide (AC-94,377) which acts like GA_3 is just as good as GA (Schraudolf, 1983*b*). Similarly, methyl esters of GA are more effective than GA in inducing antheridium formation on *L. japonicum* gametophytes (Yamane, Takahashi, Takeno and Furuya, 1979; Sugai *et al.*, 1987). The affinity of gametophytes to a modest range of gibberellins and related compounds implies that during antheridium formation, pheromone binding sites in the cell also recognize GA-like molecules.

Relationship between antheridiogens and GA

A matter of considerable interest is the extent of similarity in the effects of GA and A_{An} on antheridium induction on gametophytes of *A. phyllitidis*. At the morphological level, GA-induced antheridium exhibits the same developmental repertoire as that formed in response to A_{An} (Schraudolf, 1963); at the physiological level, sensitivity of gametophytes to GA and A_{An} is found to increase with their age (Schraudolf, 1966*b*; Näf, 1959, 1967*a, b*). Similarity in the effects of the two substances also transcends to dose-response curves, the time interval between hormone application and response, and the number of sperm formed in each antheridium (Voeller and Weinberg, 1967). Additionally, like GA, A_{An}

induces abundant dark-germination of spores of *A. phyllitidis* (Näf, 1966) and both GA and A_{An} chromatograph on the same spot (Schraudolf, 1966*d*). These observations are of conceptual interest because they constitute strong evidence that A_{An} may be a gibberellin-like hormone.

As information about the effects of substances related to GA on antheridium formation on the gametophytes of *A. phyllitidis* became available, we find that the relationship of A_{An} to GA is tenuous. Some of this evidence is briefly considered here. Although male sexuality is induced by a lower than normal concentration of GA on gametophytes treated with cyclic nucleotides, the latter themselves do not mimic GA effects (Schraudolf, 1977*a*). This does not make a strong case for a second messenger such as cyclic-AMP in the biological effects of GA in this system. The growth retardants, phosphon, CCC, and AMO-1618 do not interfere with GA-induced antheridium formation; if A_{An} is a true gibberellin, these compounds which act as antigibberellins in several higher plant bioassays, should block the action of GA (Schraudolf, 1967*c*). Compounds such as helminthosporol, helminthosporic acid, and dihydrohelminthosporic acid which show GA-like activity in plant growth but which differ markedly from GA in chemical structure, are less effective than GA in antheridium induction. Similarly, steviol, a terpenoid which shows some GA-like activity in bioassays is completely ineffective as an inducer of male sexuality (Schraudolf, 1967*d*). *Ent*-kaurenol, a putative precursor of GA is only barely active in inducing antheridium formation on the gametophytes of *L. japonicum*; moreover, compared to antheridia induced by GA, those induced by *ent*-kaurenol are underdeveloped (Takeno and Furuya, 1975). According to Schraudolf (1977b, 1980*a*), low concentrations of an inhibitor of DNA synthesis, 5-bromo-2-deoxyuridine (BUDR), and to a lesser extent 5-iodo-2-deoxyuridine and 5-bromo-2-deoxycytidine substitute for GA in antheridium induction on the gametophytes of *A. phyllitidis*. Due to disturbances in cell division, antheridia formed in the presence of these analogs appear as intermediates between the fully expressed phenotype and some covertly differentiated state. Although ^{14}C-BUDR is incorporated into nuclear and cytoplasmic DNA and causes aberrations in the chloroplast ultrastructure of the cells of *A. phyllitidis* gametophytes (Koop, 1973; Schraudolf and Šonka, 1979; Šonka and Schraudolf, 1979), the molecular target of the analog in the cell causing precocious antheridium differentiation is not determined. Another observation (Schraudolf, 1982*a*) that metronidazole (1,β-hydroxyethyl-2-methyl-5-nitroimidazole), an inhibitor of ferrodoxin-linked reactions enhances GA-induced male sexuality on *A. phyllitidis* gametophytes defies explanation in the light of the results presented so far. Obviously, use of some of the compounds in this list has no logical basis and it is mentioned here to illustrate how chemicals of different structure share at least in part with GA the propensity to induce antheridium formation on gametophytes. What common function could there be between antheridiogen and GA? Perhaps they give similar instructions to cells, telling them to make antheridia.

So, from this survey we are left with a key question. How close chemically are antheridiogens to GA and how distant are they from GA? As seen in the next section, chemical characterization of antheridiogens from gametophytes of *Anemia* and *Lygodium* has provided a partial answer to this question.

Chemical nature of antheridiogens

Knowledge of the chemical nature of antheridiogens is a prerequisite for an understanding of their physiological mode of action and for identifying the part of their chemical structure that accounts for hormonal activity. Since evidence from bioassays reviewed earlier points to the existence of more than one antheridiogen, understanding of the specific molecular configuration of the different antheridiogens is important to determine the structural basis for diversity. However, due to their low titer in the cells, isolation of antheridiogens in a pure form from gametophytes has been fraught with major difficulties.

Much of what we know about the chemistry of A_{Pt} is based on the analysis of a partially purified substance obtained from culture filtrates of *Pteridium aquilinum* gametophytes (Pringle *et al.*, 1960). Results of routine chemical analyses have shown that A_{Pt} is a complex carboxylic acid. Biological activity of the compound is tightly coupled to the carboxyl group, as it is destroyed by esterification and regenerated by hydrolysis of the ester (Pringle, 1961). The first antheridiogen isolated in a pure form is A_{An} from culture filtrates of *Anemia phyllitidis* gametophytes (Nakanishi *et al.*, 1971; Endo *et al.*, 1972). It has been characterized as a GA-like compound with two hydroxyl groups and a ring double bond. Although the antheridiogen of *A. hirsuta* is structurally identical to A_{An} (Zanno *et al.*, 1972), that of *A. mexicana* is different and is a C_{19}-GA-like compound (Nester, Veysey and Coolbaugh, 1987). Apparently, structural similarity of antheridiogens does not prevail at the species level.

On a general level, the chemical structure of A_{An} corresponds to that of a diterpene, a class of compounds composed of four isoprene units usually arranged to form three rings. Since gibberellins are four-ringed diterpenoids, a close structural relationship between A_{An} and gibberellins is apparent and has been subsequently confirmed by gas chromatography (Bürcky, 1977*b*). Like GA, pure A_{An} induces dark-germination of spores of *A. phyllitidis* and antheridia formation on the gametophytes of this species (Nakanishi *et al.*, 1971; Endo *et al.*, 1972). A_{An} acts as a weak gibberellin in bioassays for GA using higher plants (Näf, 1968; Sharp, Keitt, Clum and Näf, 1975; Bürcky, 1977*d*). From these observations it seems highly probable that A_{An} functions by chemical conversion within the cell into a gibberellin-type of compound. Recently, Corey and Myers (1985; see also Corey *et al.*, 1986) have chemically synthesized A_{An} and have designated it as antheridic acid. This compound is as active as the naturally derived hormone in inducing antheridium formation on the gametophytes of *A. phyllitidis* (Takeno *et al.*, 1987). The identity of the antheridiogens of *A. flexuosa* and *A. rotundifolia* to antheridic acid has also been established by chemical analyses and by bioassays (Yamane *et al.*, 1987).

Gas chromatography of a preparation of A_{Ly} from the culture filtrate of *Lygodium japonicum* gametophytes has also indicated its relationship to diterpenes (Bürcky, Gemmrich and Schraudolf, 1978). The substance causing antheridium induction in *L. japonicum* has been identified as GA_9-methyl ester (Yamane *et al.*, 1979). Since GA-like substances have been identified from sporophytes of ferns as well (Kato, Purves and Phinney, 1962; Muromtsev *et al.*, 1964; Yamane *et al.*, 1985), it seems that this group of hormones is important in the evolution of the reproductive biology of ferns.

The chemical nature of other antheridiogens has been much less investigated. Although work done thus far on the chemistry of antheridiogens represents a major advance, we still do not know the structural differences between A_{Pt} and A_{An}. As A_{Pt} exhibits a wide activity spectrum, knowledge of its chemical structure is essential for furthering our understanding of the way the hormone induces antheridium formation on gametophytes of unrelated ferns.

Physiological aspects of antheridiogen action

Antheridiogen is a chemical signal which communicates with the cells of the gametophyte fated to differentiate into antheridia. For antheridium differentiation to be adaptive to the needs of the gametophyte, the potential antheridium-bearing cells should be chemically linked with the rest of the gametophyte so that they do not produce antheridia uncontrollably. In other words, at various stages of development of the gametophyte there should be mechanisms to modulate antheridiogen synthesis or to make target cells insensitive to the hormone. Since the discovery of antheridiogens, a series of interesting observations have been made to show how this is achieved and in the process, what appeared to be an elegantly simple story has become highly complex.

Age of the gametophyte and sensitivity to antheridiogen

In many ferns the antheridiogen-sensitive phase is reduced to a few days in the life of the gametophyte. Since gametophytes of *Onoclea sensibilis* do not form antheridia spontaneously in agar culture, they will serve as our principal illustrative material to relate antheridiogen activity to the age of the gametophyte. By transferring gametophytes of different ages to a medium supplemented with A_{Pt}, it was shown that they completely become insensitive to the hormone at the heart-shaped stage (Näf, 1958, 1961*a*). The failure of older gametophytes to respond to A_{Pt} is not due to a low titer of the hormone in the medium, as once the gametophytes become insensitive, they fail to respond to even extremely high concentrations of A_{Pt}. It also appears that gametophytes continuously exposed to A_{Pt} right from the stage of germinated spores produce antheridia until they reach about nine times the size at which non-induced gametophytes become insensitive to the hormone (Näf, 1962*a*). This observation defies any logical explanation based on the relationship between the age of the gametophyte and antheridiogen action; something fundamental in the regulation of cellular metabolism in the gametophyte is involved here.

Physiological age is also an important factor that determines the sensitivity of gametophytes of *Pteridium aquilinum, Anemia phyllitidis* and *Lygodium japonicum* to antheridiogen. Gametophytes of *P. aquilinum* halt antheridium production shortly after they attain the heart-shaped stage and as they enter the archegonial phase. This occurs despite the fact that the maturing gametophyte continues to generate substantial amounts of antheridiogen in its cells (Döpp, 1950; Näf, 1958). Loss of sensitivity to antheridiogen with age is manifest in a different way in the gametophytes of *A. phyllitidis*. Young gametophytes potentiated by A_{An} form antheridia throughout their body, while with increasing age the responding area is restricted to the region posterior to the lateral meristem (Näf, 1959). In *L. japonicum* antheridia appear in response to A_{Ly} application in the posterior part of young gametophytes and in the anterior region of older gametophytes. As shown in Fig. 10.3, a critical period of only two days marks the transition of the hormone-sensitive phase to the hormone-insensitive phase of the gametophyte; generally, 12-day old or younger gametophytes are maximally sensitive to antheridiogen, whereas by 14 days their sensitivity decreases to 50% (Näf, 1960; Takeno and Furuya, 1980).

Role of the meristem. How can we account for the loss of sensitivity of gametophytes to antheridiogen with age? A concept which has received some attention in the literature treats antheridiogen sensitivity of gametophytes as a function of its meristem. It is generally observed that if

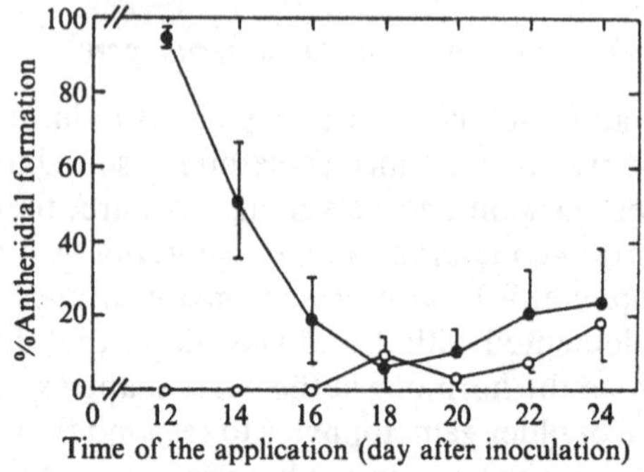

Figure 10.3. Effect of A_{Ly} (methanol extract of *Lygodium japonicum* prothalli) on antheridium formation on the gametophytes of *L. japonicum* of different ages; (●) with A_{Ly}; (○) without antheridiogen. Counts of antheridia formed were made seven days after incubation. Vertical lines represent ×2 standard error. (From Takeno and Furuya, 1980.)

meristemless parts of gametophytes which have long ceased to be sexually active are isolated surgically, they regain the competence to form antheridia (Czaja, 1921). In some cases viable cells spatially separated from the main body of the gametophyte by dead cells have been shown to differentiate antheridia, implicating physiological isolation by injury as an important alternative to surgical operation in sequestering the influence of the meristem (Parès, 1958). Other observations have led to the view that gametophyte maturation is associated with the synthesis of an inhibitor of antheridium differentiation. For example, a water extract of archegonium-bearing gametophytes of *P. aquilinum* suppresses the action of A_{Pt} on the gametophytes of *Dryopteris filix-mas* and causes a dimunition of vegetative growth (Döpp, 1959). A more direct approach and possibly one which will help in the identification of the inhibitor, is to monitor the effect of exogenous additives on antheridiogen action on gametophytes. Thus Döpp (1962) showed that the addition of IAA, NAA or indolebutyric acid (IBA) inhibits antheridium differentiation on *P. aquilinum* gametophytes, suggesting the identity of the inhibitor to an auxin-like substance. However, an earlier observation (Döpp, 1955*a*) that 2,3,5-triiodobenzoic acid, a well-known antiauxin, blocked antheridium formation in *P. aquilinum* gametophytes mitigates against this possibility. An important extension of this line of research is the demonstration that ABA acts as an antagonist of the antheridiogen system in *Ceratopteris pteridoides* (Chiang, 1976) and *C. richardii* (Fig. 10.4; Hickok, 1983). The low concentrations at which ABA suppresses antheridia formation in the latter without any detectable side effects on the gametophyte and the reversibility of inhibition by withdrawing ABA from the medium suggest that the latter acts as a specific block to

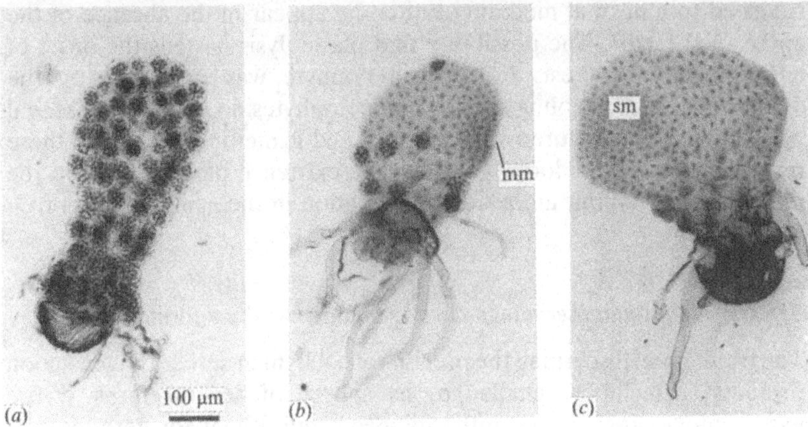

Figure 10.4. Inhibition of antheridiogen action on the gametophytes of *Ceratopteris richardii* by ABA. (*a*) A gametophyte grown for 25 days in a medium containing antheridiogen showing numerous antheridia. (*b*) A gametophyte with few antheridia and numerous rhizoids grown in a medium containing antheridiogen and 10^{-5}M ABA. (*c*) A gametophyte with no antheridia and few rhizoids grown in a medium containing antheridiogen and 10^{-4}M ABA. mm, marginal meristem; sm, submarginal meristem. Scale bar in (*a*) applies to all. (From Hickok, 1983; photographs supplied by Dr L. G. Hickok.)

antheridium differentiation. Given the long-standing preoccupation with the meristem, one could assume that an investigation of ABA metabolism in this tissue will go a long way in defining in hormonal terms its role in the development of male sexuality in the gametophyte.

Other experiments on antheridium induction on surgically treated gametophytes have led to a set of perplexing results. Schraudolf (1966*c*) found that surgically removed pieces of gametophytes or plasmolyzed whole gametophytes of *A. phyllitidis* form antheridia only in the presence of GA. This may be related to the rapidity of meristem regeneration following surgical or chemical treatments, but invites further study. According to Näf (1961*b*), meristemless segments of the gametophyte of *Onoclea sensibilis* regenerate antheridia in an agar medium spontaneously even in the absence of A_{Pt}; this observation is striking because whole gametophytes do not spontaneously form antheridia under this condition in the absence of a specific antheridium-inducing factor. By way of explanation it has been proposed that extirpation of the meristem eliminates a chemical block to antheridium formation, a block which in the normal life of the gametophyte would have been removed by antheridiogen. In a key experiment supporting this hypothesis, cellular continuity between the meristem and wings of the gametophyte is disrupted by plasmolysis. When the plasmolyzed gametophyte is returned to a normal medium, antheridia appear in the absence of the antheridial factor. The possibility that plasmolysis hastens the onset of antheridiogen synthesis in the gametophyte was ruled out by the demonstration that young sensitive gametophytes do not become sexual when they are cocultured with plasmolyzed gametophytes. From these results the prognosis looks good for the existence of substances in the meristem that inhibit antheridium formation in the gametophyte of *O. sensibilis*.

Effect of light on antherodiogen action and antheridium formation

The problem of discerning the precise role of light in antheridiogen action presents a technical challenge, as spores of several ferns whose gametophytes respond to antheridiogen require light for germination. Conversely, spores of ferns that germinate in complete darkness do so in the presence of GA which also induces antheridium formation on gametophytes. Because of the failure to appreciate these limitations, conclusions on the effect of light on antheridiogen-induced male sexuality in gametophytes have been mired in controversy.

The principal experiments have been done with gametophytes of *O. sensibilis*. The first report relevant to this discussion was by Miller and Miller (1970) who found that an atypical strain of gametophytes grown in a medium containing sucrose spontaneously form antheridia in the dark. Later Näf *et al.* (1974) showed that when seven-day old light-grown gametophytes are transferred to near darkness on a sucrose-agar medium, antheridia appear in 15 to 22 days; in the presence of A_{Pt} in the medium the first antherldum appears in four days. There is a further decrease in the lag period between A_{Pt} application and appearance of antheridium to two or three days with passing time in darkness before the hormone is applied. The conclusion from this work that light blocks the expression of maleness and that prolonged darkness or application of antheridiogen favors a decay of this block has however been questioned by Rubin, Robson and Paolillo (1984). Taking advantage of a low percentage of dark germination of *O. sensibilis* spores, these investigators found that the timing of spontaneous antheridium formation on gametophytes nurtured on ashed soil in complete darkness is not affected by light. It is likely that the response to A_{Pt} of gametophytes grown in light and transferred to near dark conditions is a secondary effect of the meristematic activity that occurs in the dark. The unfortunate consequence of these conflicting data is that we are still in the dark about any possible involvement of light even in a less pivotal role than that envisaged by Näf *et al.* (1974) in modulating antheridiogen action in *O. sensibilis* gametophytes.

Experiments on the gametophytes of *Pteris vittata* have provided us with a different set of data on the effects of light on antherodiogen action. When spores germinated in light to the stage of rhizoid initiation are transferred to darkness, they show great proclivity to generate antheridia spontaneously. This precocious antheridium formation is under phytochrome control. Interestingly enough, antheridia appear on gametophytes grown in light only in the presence of antheridiogen. These results have led to the suggestion that induction of male sexuality in the gametophyte of *P. vittata* is under dual control, one in the dark which is independent of antheridiogen action and the other in light which is stimulated by the hormone (Gemmrich, 1986*a, b*). Does dark incubation of germinated spores block the synthesis of antheridiogen? Investigation of this important question is needed to determine whether the hormone is absent or whether it is prevented from exercising its action in the dark.

Light quality rather than an endogenous hormone apparently controls antheridium formation on the gametophytes of *Polypodium* (= *Niphidium) crassifolium*. Here, antheridia are induced on gametophytes exposed to a period of darkness or to far-red light. Although phytochrome control of antheridium formation has been convincingly established in this system, any thought that the pigment might act by regulating antheridiogen synthesis was dispelled when it was shown that the culture medium from dark-grown antheridium-bearing gametophytes does not promote antheridium formation in light-grown gametophytes (Schraudolf, 1967*b*).

Gene activity during antheridium formation

Transformation of a gametophytic cell into an antheridium involves a cascade of regulatory events related to a temporal expression of gene activity, but the need for the synthesis of mRNA or new proteins for antheridium formation has not been clearly established. The use of metabolic inhibitors, which has formed the basis of most of the work, has yielded variable results. For example, although inhibitors of RNA synthesis (FLU) and protein synthesis (chloramphenicol) cause a noticeable lag in the frequency of cell division and GA-induced antheridium formation in the gametophytes of *Anemia phyllitidis* grown in light, gametophytes are revitalized to form antheridia when they complete a certain amount of vegetative growth in the presence of the inhibitor. This suggests that the block to antheridium formation is a secondary effect of the inhibition of cell division (Schraudolf, 1965, 1967*a*). However, in dark cultures actinomycin D inhibits GA-induced antheridium formation in the gametophytes of this same species without reducing cell number (Voeller and Weinberg, 1969). Comparable results have been reported on the effects of streptomycin, chloramphenicol and 5-bromouracil on red light-induced or antheridiogen-induced antheridium formation on the gametophytes of *Microlepia speluncae* (Kressel, 1965; Fellenberg-Kressel, 1969).

As the pattern of protein synthesis changes significantly when a cell undergoes differentiation, it is not surprising that quantitative differences in the activity of certain enzymes, along with enhanced synthesis of proteins, have been demonstrated during antheridium formation (Iqbal and Schraudolf, 1977, 1984*a*). The finer biochemical details of regulation of enzyme activity and protein synthesis have not yet been investigated, and so the significance of these changes in the context of gene activation remains uncertain.

Archegonium development and differentiation

The archegonium of fern gametophytes is a more complex and uniform structure than the antheridium. Basically it is a flask-shaped object and consists of an enlarged basal region buried in the gametophyte with a neck protruding from the surface. Morphological changes associated with the maturation of the archegonium include some remodeling of the basal region to form a large egg cell and a small ventral canal cell, situated just above the egg. The most stream-lined part of the mature archegonium is the neck which has a cellular jacket enveloping a single neck canal cell. Although the neck canal cell is generally binucleate, abnormal archegonia of certain primitive families of homosporous ferns have multinucleate or divided neck canal cells (Nishida and Sakuma, 1961).

Differentiation of the archegonium is the culmination of a series of mitotic divisions of a superficial initial cell (archegonium initial) located close to the apical notch of the gametophyte. As described in *Pteridium aquilinum*, this cell undergoes two transverse divisions to form a tier of three cells (Bell, 1960*b*). The shaping of the archegonium is mainly a matter of forming the neck region; this entails two successive perpendicular divisions of the upper cell of the tier followed by repeated transverse divisions. At the same time the inner cell divides to form the basal cells of the jacket or venter. Finally, the middle cell of the row (primary cell) is partitioned in a decidedly unequal way by a transverse wall. The small cell cut off towards the neck of the archegonium subsequently divides unaccompanied cytokinesis, giving rise to the neck canal cell. The lower cell (central cell) again divides to form a large basal egg cell and a small ventral canal cell. Generally, the formation of the anterior part of the organ is proceeding well before remodeling is completed posteriorly.

Control of archegonium differentiation

A widely accepted view concerning archegonium differentiation derives from the work of Döpp (1959) who thought that the usual site of archegonia on the gametophyte in the vicinity of the meristem is largely dictated by the relatively high metabolic state of the cells in this region. In support of this view, it was found that extirpation of the meristem from the gametophyte of *P. aquilinum* already in the archegonial phase halts further archegonium production, but leads to a surge in antheridium formation. The critical metabolic factor of the meristem which maintains continued expression of female sexuality on the gametophyte has not been identified.

In recent years, some attention has been focused on the role of naturally occurring inhibitors of archegonium initiation on fern gametophytes. Although gametophytes of *Lygodium japonicum* spontaneously form archegonia, female sexuality is completely suppressed in gametophytes originating from spores allowed to germinate and grow in a medium containing GA_3. Withdrawal of the hormone from the medium promptly reinstates archegonium-forming ability in the gametophytes (Takeno and Furuya, 1977). Subsequent work (Takeno, Furuya, Yamane and Takahashi, 1979) showed that cells of the gametophyte synthesize inhibitors of archegonium differentiation and secrete them into the medium. The inhibitor has been retrieved from the medium and identified as GA_9-methyl ester, the paradox being that this same compound also promotes antheridium differentiation. Perhaps this has some physiological significance in regulating the timing of sexual differentiation on individual gametophytes of a population to hinder self-fertilization (intragametophytic selfing). This could be achieved by the secretion of inhibitors of archegonium differentiation into the medium by the fast-growing archegonium-bearing gametophytes of a population which will turn off archegonium formation and promote antheridium differentiation on the slow-growing gametophytes. The result will be the preponderance of unisexual female and unisexual male gametophytes although eventually bisexuality will also be rampant by either male or female gametophytes forming organs of the opposite sex. The practical importance of this system in the reproductive biology of *L. japonicum* is underscored by the demonstration that self-fertilization hardly occurs in bisexual gametophytes in which sexuality is induced by GA (Takeno and Furuya, 1980, 1987).

General comments

Over the past 30 years experimental work has provided a respectable foundation for a better understanding of the role of endogenous hormones in the induction of sexuality in fern gametophytes. Although we know more about the control of antheridium differentiation than about archegonium differentiation, we are obviously a great distance from understanding the mechanism by which cells differentiate into antheridia. As developmental control is exerted by regulation of gene expression, much could be learned by examining the interaction of the genome with antheridiogens and other regulatory agents in a molecular framework.

On a general level, one of the perplexing facts seen in this survey is the relatively small number of ferns in which hormonal control of sexuality has been unequivocally established. This raises the question as to whether alternate control mechanisms exist to promote sexuality in the gametophytes of ferns. Further analysis of differentiation of sex organs on the gametophytes of a wide representation of ferns appears clearly in order and the results of such a study will have a significant impact on our understanding of the reproductive biology of ferns.

11

Sexuality and genetics of gametophytes

There has been increasing interest in recent years in the reproductive biology of ferns and much attention has been paid to the gametophytes. As is well-known, one of the traits that fern gametophytes have evolved is the ability for sexual reproduction. In sexually reproducing plants, the evolutionary importance of a mechanism that ensures outbreeding to preserve genetic diversity needs no emphasis. However, most homosporous ferns produce bisexual gametophytes that promote inbreeding and attainment of homozygosity in the progeny. Adaptations that prevent these deleterious features in the gametophyte facilitate genetic variability in the population and successful colonization of the species. In this chapter we will review some of the strategies in the reproductive biology of fern gametophytes that have contributed to genetic diversity and evolutionary potential of the species. A word of caution is necessary before we proceed further. Most of what we know in this area concerns homosporous ferns, those represented in nature by their green, thalloid gametophytes. Heterosporous ferns, as well as homosporous ferns with subterranean gametophytes have barely been studied.

The blight of homozygosity

To begin with, some comments about the breeding system of ferns are in order. Insofar as functional antheridia and archegonia are born on different gametophytes, heterosporous ferns are considered to be obligatorily heterothallic (self-incompatible, thus requiring two compatible gametophytes for sexual reproduction). On the other hand, members of the homosporous group are homothallic (self-compatible), since they form antheridia and archegonia on the same gametophyte. For complete obligatory homozygosity in the offspring, it is also crucial that the gametophyte of a homothallic fern produces both types of sex organs simultaneously. The breeding system of ferns must therefore be viewed from the perspective of the timing and occurrence of antheridia and archegonia on the same or different gametophytes. When this is considered in terms of the parentage of the sporophyte from which gametophytes originate, four categories of sexual reproduction are possible in ferns (Klekowski, 1969*a*). One is *intragametophytic selfing* involving the union of sperm and egg from the same gametophyte. The second is *intergametophytic selfing* by the fusion of sperm and egg from different gametophytes originating from the same parental sporophyte. Thirdly, fusion between gametes from different gametophytes, each originating from a different parental sporophyte might occur; this is designated as *intergametophytic crossing*. Finally, there is *intergametophytic mating* when fertilization occurs between gametes from different gametophytes of unspecified origin. Intragametophytic selfing and intergametophytic crossing are genetically equivalent to self-fertilization and cross-fertilization, respectively, in flowering plants. A diversity of studies of mating systems, carried out in unusually great detail, has established that intragametophytic selfing is widespread among homosporous ferns. Although evidence obtained from some allozyme studies of fern sporophytes (Werth, Guttman and Eshbaugh, 1985; McCauley, Whittier and Reilly, 1985; Soltis and Soltis, 1986) have supported this conclusion, other allozyme studies have provided evidence for the operation of outcrossing mechanisms as well as for mixed matings (Haufler and Soltis, 1984; Gastony and Gottlieb, 1985; Soltis and Soltis, 1987*a, b*). As we shall see later, breeding patterns alone, however, do not provide sufficient evidence to assess the genetic diversity in the population.

There has been little serious analysis of the genetics of ferns until Klekowski and Baker (1966) noted that homosporous, homothallic pteridophytes in general, and ferns in particular, have high chromosome numbers, whereas heterosporous, heterothallic pteridophytes have low chromosome numbers. For example, the mean gametic chromosome number of homosporous ferns is 54, *versus* 13 for the heterosporous group. As the majority of homosporous ferns that resort to intragametophytic selfing have had a polyploid origin, we have every reason to believe that this group of ferns has evolved a genetic system that is the product of polyploidy within the context of their mating habit. A caveat to this statement is that recent genetic evidence has suggested that not all ferns with high chromosome numbers are necessarily polyploid (Haufler and Soltis, 1986).

The critical issue in approaching the consequences of intragametophytic selfing lies with the extent of homozygosity displayed by the zygote. Since gametes are generated by simple mitotic divisions of the antheridium and archegonium, the sperm and egg formed on a single gametophyte have the same genotype. Intragametophytic selfing in this system therefore results in the formation of a diploid zygote that is completely homologous and bears every chromosome in the homozygous condition. The zygote cleaves repeatedly by mitotic divisions to form the embryo which is eventually transformed into a mature sporophyte. The dominant cytological event in the life of the sporophyte is sporogenesis involving meiosis in the sporocytes. Based on a genetic system that allows pairing of homologous chromosomes at meiosis, the sporophyte will yield a population of genetically uniform spores.

The main disadvantage of homozygosity thus acquired is that it gnaws away the evolutionary values of sexual reproduction and affords very little opportunity to store or release genetic variability. It has been proposed however that the polyploid condition in homosporous ferns bestows selective advantage over diploidy for storage and expression of genetic variability, thus compensating for their inbreeding potential. How this is accomplished is very much tied to the distinctive and elusive problem of chromosome pairing during meiosis. A variant form of meiosis that does not restrict pairing to homologous chromosomes but allows with equal fidelity occasional pairing within homoeologous chromosomes[1] has been envisioned as the hallmark of this system. In an operational sense, heterozygosity is maintained within the duplicated loci that are located on homoeologous chromosomes and is released through pairing and segregation of these chromosomes at meiosis (Klekowski, 1973*b*). In short, the blight of homozygosity imposed by the homothallic nature of the gametophytes is partially alleviated by homoeologous pairing at meiosis. In this context, the presumed polyploid nature of homosporous ferns acquires an adaptive significance because the duplication of gene loci by polyploidy is a mechanism for preserving recessive alleles in the population and for maintaining genetic variability.

Both cytological and genetic evidence has been presented for the occurrence of homoeologous pairing during meiosis in homosporous ferns. At the cytological level, the expression of spontaneous and induced chromosome aberrations such as paracentric inversions, translocations, and bridges in the homozygous sporocytes of sexual and apomictic ferns is considered as strong evidence for the homoeologous pairing hypothesis (Klekowski and Hickok, 1974; Bierhorst, 1975*b*; Hickok, 1977*a*). Genetic evidence is based on the documentation of duplicated loci and the segregation of marker genes in ratios that would be compatible with homoeologous recombination (Klekowski, 1971, 1976; Hickok, 1978*a*, *b*). Additional support for the occurrence of homoeologous heterozy-

[1] The following paragraph from Bierhorst (1975*b*) explains the difference between homologous and homoeologous pairing. 'The differences between homologous and homoeologous chromosomes are relative, not absolute. If a species is known to carry two diploid sets of chromosomes, ie, if it is an allopolyploid, the pairs within either of the two sets are considered homologs and pairs composed of one chromosome from each of the two sets are considered homoeologs. Homoeologous chromosomes in this sense should show various degrees of homology (and subsequent degrees of pairing) in the allopolyploid, depending upon the degree to which the chromosomes of a particular pair phylogenetically diverged in the two ancestors of the allopolyploid.'

gosity based on recombination between unlinked, duplicated loci is the presence of multiple bands in the electrophoretic profile of enzyme markers extracted from sporophytes of *Pteridium aquilinum* (Chapman, Klekowski and Selander, 1979). However, some of this evidence has been criticized (Lovis, 1977; Gastony and Darrow, 1983) and data directly contradicting these assertions have also been presented. For example, a comparison of electrophoretic patterns of isozymes in individual gameotophytes of *Pellaea andromedaefolia (= P. andromedifolia)* has indicated that genetic heterozygosity is encoded by alleles at a single locus than by recombination between paired homoeologous chromosomes (Gastony and Gottlieb, 1982). A reexamination of isozyme variability in the gametophytes of *P. aquilinum* using improved methods has also failed to uncover evidence of extensive gene duplication to imply a large reliance on homoeologous recombination in the parental sporophytes (Wolf, Haufler and Sheffield, 1987).

It should be apparent from this summary that despite their homothallic nature, homosporous ferns are fine-tuned to produce heterozygotes; whether this is accomplished by subtle alterations in the meiotic process or by some other means, remains to be settled. With this background, we will briefly examine the various gametophytic adaptations that favor heterozygosity.

Breeding systems

The breeding system of a plant formally describes the morphology and ontogeny of gametophytes, the duration of the gametophytic generation, development and display of sex organs and adaptations to thwart or favor a particular mating system. In the previous chapters the development of the gametophyte and differentiation of sex organs have been described, so the emphasis here will be on adaptations; these encompass the sequence of gametangial ontogeny, positioning of the gametangia on the gametophyte and the frequency of unisexual and bisexual gametophytes.

In respect to other plants, especially the angiosperms with their pronounced bisexuality, the free-living bisexual gametophytes of homosporous ferns lend themselves to the thought that a genetic self-incompatibility system is in operation. This seems a reasonable consequence of the principle that an interchange of genes contributing to genetic diversity usually underlies cross-fertilization. But such is not the case in homosporous ferns where no substantiated examples of self-incompatibility have been described to date. A case of genetic incompatibility reported in *P. aquilinum* (Wilkie, 1956) was subsequently explained under the genetic load hypothesis (Klekowski, 1972). An observation that in isolate cultures (where bisexuality is required and intragametophytic selfing and self-compatibility are obligatory) of gametophytes of *Cheilanthes farinosa*, no sporophytes ever appeared, can also be attributed to either genetic incompatibility or genetic load of the species (Khare, 1980). Thus, the present evidence suggests that intragametophytic selfing prevails in the majority of homosporous ferns which also possess adaptations that decrease the probability of this phenomenon.

Sequence of formation of sex organs and their spatial relations

The sequence of formation of antheridia and archegonia and their arrangement on the gametophyte influence the mating system of the species. There are several studies on the morphology of gametophytes in which occurrence of antheridia and archegonia has often been described, but detailed investigations on the timing of their formation from the perspective of reproductive biology have been undertaken only in a few cases. A generalization suggested by morphological studies is that a pattern in which antheridia form very early in the ontogeny of the gametophyte, followed shortly by archegonia is most common among homosporous ferns; this is designated as the 'A' type (Klekowski, 1969*a*; Klekowski and Lloyd, 1968). As shown in *Adiantum capillus-veneris*, a sufficient number of gametophytes in a population change from an initial male phase to a prolonged bisexual phase, so that a quantitative image of this process emerges (Fig. 11.1; Masuyama, 1972). As the simultaneous appearance of both sex organs on the same gametophyte is a definitive indication of intragametophytic selfing, we can conclude that true homozygosity prevails here. However, uncertainties surface when antheridia have already matured by the time archegonia appear. This change which diminishes the chances of intragametophytic selfing has been observed in *Pseudodrynaria coronans* (Polypodiaceae; Singh and Roy, 1977).

A sequence of antheridia and archegonia formation ('B' type) first described in the gametophytes of *Scolopendrium vulgare* (=*Asplenium scolopendrium;* Aspleniaceae; Andersson-Kottö, 1929) has been recently documented in a number of other ferns (Masuyama, 1975*a, b*). The gametophyte is at first male or asexual but becomes archegoniate after a period of growth. Although the situation might not seem different from the A type, what sets this apart as a category by itself is that the antheridium is nonfunctional or empty when the egg is most receptive. The mechanism responsible for the failure of antheridium function remains unclear. As the gametophyte matures, there is a new wave of antheridium production making it functionally bisexual. Gametophytes of this type are considered to be adapted for intergametophytic unions.

In several species of *Blechnum, Doodia, Woodwardia, Lorinseria, Stenochlaena* (Klekowski, 1969*b*, 1970*b*), and *Sadleria* (Holbrook-

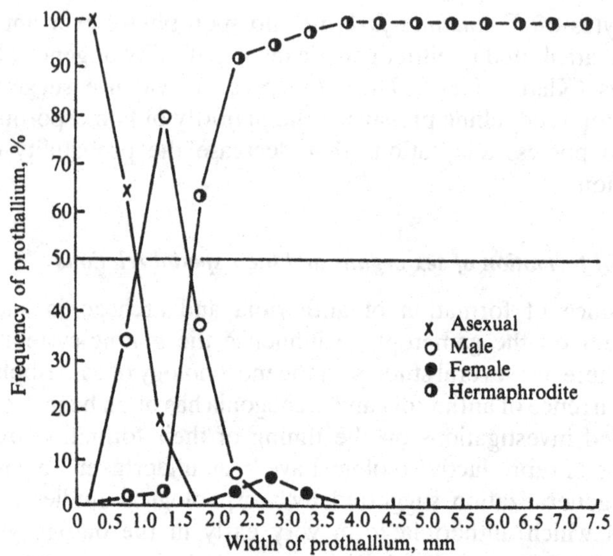

Figure 11.1. Changes in the sex expression of gametophytes of *Adiantum capillus-veneris* as related to their growth (expressed as width of gametophytes). (From Masuyama, 1972.)

Walker and Lloyd, 1973) (all Blechnaceae), *Ceratopteris thalictroides* (Klekowski, 1970*a*), *Cibotium glaucum* (Thrysopteridaceae; Lloyd, 1974), *Onoclea sensibilis* (Saus and Lloyd, 1976), and *Acrostichum aureum* (Lloyd, 1980), gametophytes form archegonia first followed by attainment of bisexuality. The significance of this sequence, designated as the 'C' type, is that it is most susceptible to intergametophytic selfing. This has been established by comparing the rate of sporophyte production between gametophytes grown in isolate cultures and composite cultures (where intergametophytic or intragametophytic selfing is possible). Demonstration of intragametophytic selfing as the normal mode of breeding in the species would require sporophyte production at similar rates in both cultures. Occurrence of intergametophytic selfing is inferred when the rate of sporophyte production in composite cultures outstrips that in isolate cultures. Experimental analyses of the reproductive biology of gametophytes by this technique have illuminated both the genetic mechanisms of mating and the mechanism underlying colonization of the species.

The type 'D' sequence is set in motion by the development of antheridia on the gametophytes. As the antheridial number decreases, there is an increase in the number of bisexual gametophytes. The latter phase is followed by an archegoniate phase. This sequence has been described in *Phlebodium aureum* (Ward, 1954) and *Adiantum pedatum*

(Masuyama, 1972) and is interesting in that unlike the previous types, there is a developmental block to continued antheridium formation in the ontogeny of the gametophyte. The significance of this sequence as an adaptation to promote intragametophytic or intergametophytic selfing is unclear.

In the examples cited above, the sequence of the formation of sex organs is fixed for a given species. This is not to be considered as universal as intraspecific differences occur in some species. A case in point is *Phegopteris decursive-pinnata* (Thelypteridacea) whose gametophytes indulge in two different types of matings. One is displayed by gametophytes of the diploid cell lineage in which the ontogenetic sequence of formation of antheridium and archegonium favors intergametophytic selfing; the other sequence occurs in tetraploids which increases the probability of intragametophytic selfing (Masuyama, 1979).

The success of the mating systems described above greatly depends upon the spatial distribution of sex organs. A common arrangement finds both antheridia and archegonia on the ventral side of the gametophyte with archegonial necks pointing toward the antheridia, an adaptation obviously inviting intragametophytic selfing. On the other hand, if archegonial necks are turned away from the antheridia, intragametophytic selfing is stymied (Klekowski, 1969*b*).

In summary, the appearance and disappearance of antheridia and archegonia that predict a particular type of breeding evidently requires institution of hitherto unknown regulatory controls. Nonetheless, the great complexity of the breeding programs is evident from this comparative survey, even given our present lack of understanding of how the programs operate.

Population effects

The long history of field observations of homosporous ferns has led to the identification of adaptations that affect the breeding systems of gametophytes. One of the most visible adaptations is the changing proportion of male, female and bisexual gametophytes in a population. It is the norm that spores derived from a single sporophyte give rise to only one kind of prothalli that may be either male, female or bisexual; more frequently, they start off as males and end up as bisexuals. This kind of monogametophytic system found in the majority of homosporous ferns does not lend itself to intergametophytic selfing. A different type of monogametophytic system where members of a population are entirely archegoniate for a period of time and later become bisexual is found on the gametophytes of *Woodwardia fimbriata* and *W. virginica*. From an adaptive point of view, the unavailability of antheridia during the initial archegoniate phase results in intergametophytic selfing as the primary reproductive strategy in these species (Klekowski, 1969*b*).

Though encountered rarely, populations of gametophytes encompassing two distinct sexual types such as males and females (*Platyzoma microphyllum*), males and bisexuals (*Ceratopteris thalictroides*) or females and bisexuals (*Doodia caudata*) also exist. This is the bigametophytic system which, in varying degrees, increases the dominance of intergametophytic selfing (Klekowski, 1969*a*). It is also instructive to consider the trigametophytic system described in *Gymnogramme (=Pityrogramma) calomelanos* where three kinds of gametophytes (males, females and bisexuals) are found at one time in the population (Tourte, 1967; Cousens, 1975). In terms of the dynamics of reproductive biology, the trigametophytic system is the least investigated, but overall it is geared to both intragametophytic and intergametophytic unions.

We will now shift the perspective of this account to the various other factors that control sex expression in populations of gametophytes from time to time. In *Onoclea sensibilis* reautoclaving the agar medium (Klekowski and Lloyd, 1968) changes the system from monogametophytic to bigametophytic; however, unlike soil, agar as a substrate suppresses maleness and promotes femaleness (Rubin and Paolillo, 1983). Density of gametophyte population also affects the timing of sexuality and the proportion of sex expression types in *O. sensibilis* (Rubin *et al.*, 1985) and other ferns (Klekowski and Lloyd, 1968; Rashid, 1970; Cousens, 1979; Cousens and Horner, 1970; Lloyd and Gregg, 1975; Warne and Lloyd, 1987). As we have seen in the previous chapter, sex expression is known to depend on interactions between neighboring gametophytes by the production of antheridiogens and inhibitors of archegonium differentiation (Döpp, 1950; Pringle, 1961; Schedlbauer, 1976*b*; Schedlbauer and Klekowski, 1972; Haufler and Gastony, 1978; Haufler and Ranker, 1985; Takeno *et al.*, 1979). However, in spite of the complex developmental control of antheridium formation exerted by the hormone, it has been questioned whether outcrossing is necessarily achieved by the antheridiogen system (Willson, 1981). Based on sex ratio similarities in isolate and composite cultures of *Pteridium aquilinum* gametophytes, Wilkie (1963) has implicated nuclear or cytoplasmic genetic factors in the control of sex expression. On the other hand, a recent work with two homozygous strains of *C. richardii* which exhibit strikingly different antheridiogen sensitivities, has provided clear evidence for the inheritance of sex expression by nuclear genes (Scott and Hickok, 1987). Other factors that control the proportion of different kinds of gametophytes in a population include ionic concentration of the medium (Tourte, 1967), extremes of pH (Guervin and Laroche, 1967; von Aderkas and Cutter, 1983*b*), and assorted compounds like amino acids (Sossountzov, 1955), fungicides (Schedlbauer, 1978), and auxins (Hickok and Kiriluk, 1984). From this it appears that under natural conditions, availability of the right mix of gametophytes to promote a particular breeding pattern may be an enormously complex process requiring the interplay of a variety of extracellular and intracellular factors.

Genetic effects

Aspects of the reproductive biology of fern gametophytes just described are very informative, but by themselves they do not lead one to conclude whether a particular type of mating is actually accomplished in nature or not. Consequently, it becomes difficult to rule out other roadblocks to sexual reproduction in the gametophyte unless predictions on breeding systems based on the appearance and spatial relations of sex organs are tested independently. Genetic studies of sporophytes have proved to be extremely useful to test some of the predictions. As first shown by Klekowski and Lloyd (1968), the formation of a completely homozygous zygote following intragametophytic selfing is admirably suited for a population analysis to determine the genetic load, that is, the presence of accumulated recessive genes which are homozygous in the diploid zygote or sporophyte; the rationale is that since gametes of a single gametophyte are genetically identical, selfing will simultaneously screen all the chromosomes for recessive deleterious genes which should show up in the resultant zygote or sporophyte. In an operational sense, any bisexual gametophyte that does not form a sporophyte in an isolate culture after a prolonged period can thus be assumed to be charged with sporophytic lethals. Accordingly, the complete absence or occurrence at low frequency of intragametophytic selfing in a population is consistent with the presence of genetic load while high frequency intragametophytic selfing is correlated with its absence. On the other hand, sporophytes formed in composite cultures of gametophytes are presumed to be heterozygous owing to the probability of intergametophytic mating and thus show genetic load. In this latter system, occasional absence of sporophytes in functional females can be ascribed to the presence of dominant sporophytic lethals.

Besides the pervading genetic influences like mutation rate and polyploidy, ecological and physiological factors like breeding system, population size, and habitat diversity of gametophytes determine the frequency of genetic load in a population. The relationship between the mating system of gametophytes and genetic load was first established by Klekowski (1970*c*) in *Osmunda regalis*. In this investigation involving over 100 sporophyte genotypes from several different populations, it was found that the frequency of gametophytes forming sporophytes in isolate cultures varied from 0 to 95% with an average of 41%. How can one explain the failure of intragametophytic selfing in more than 50% of the combined population? One explanation, based on the genotype of the gametophytes, is that they are lethal in the sporophytic generation. This however would require at least the demonstration that sporophytes are formed in random interpopulational pairs of gametophyte cultures. Such results have been obtained (Klekowski, 1973*a*) and have led to the conclusion that recessive lethal gene combinations are present heterozygously in the sporophytes. At the phenotypic level, genetic load is expressed in a variety of ways, the most common of which is the termination of embryogenic development before the emergence of sporophytic organs such as the leaf, root, and shoot apex ('early sporophytic lethals'). Occasionally, gametophytes produce viable sporophytes only after repeated attempts at fertilization ('leaky lethality'); this has been attributed to unknown cytoplasmic differences in the egg population of a gametophyte, with only a few possessing the ability to fuse with a sperm.

The experiments on *O. regalis* are among the most impressive demonstrations of genetic load in fern populations. As shown in Table 11.1, this work spawned a whole new series of investigations into the

Table 11.1 *Genetic load in ferns, expressed as percentage of gametophytes bearing recessive sporophytic lethals per sporophyte sampled*[a]

Species	Genetic load % (mean)	Reference[b]
Acrostichum aureum	5.0–45.0 (3.7)	Lloyd, 1980
A. danaeifolium	4.0–32.0 (10.3)	Lloyd and Gregg, 1975
Asplenium platyneuron	8.5–28.6 (19.0)	Crist and Farrar, 1983
Athyrium filix-femina	73.0	Schneller, 1979
Blechnum spicant	2.0–90.0 (28.9)	Cousens, 1979
Ceratopteris pteridoides	0–0.75 (0.42)	Warne and Lloyd, 1981
C. thalictroides from Old World and Hawaii	0–6.0	Klekowski, 1970*a*
C thalictroides from New World	0–25.0 (2.0)	Lloyd and Warne, 1978
Cheilanthes farinosa	100.0	Khare, 1980
C. tenuifolia	95.0–100.0 (98.5)	Khare and Roy, 1977
Cibotium chamissoi	26.6–90.0 (59.0)	Lloyd, 1974
Cyclosorus dentatus (=*Christella dentata*)	60.0	Khare, 1980

Table 11.1 (*cont.*)

Species	Genetic load % (mean)	Reference[b]
C. parasiticus (=*Christella parasitica)*	30.0	Khare and Kaur, 1979
Dicranopteris linearis (Gleicheniaceae)	20.0–65.0 (36.6)	Lloyd, 1974
Lygodium flexuosum	18.2–19.2 (18.6)	Lal and Roy, 1983
Microsorium scolopendria	0–10.0 (7.2)	Lloyd, 1974
Nephrolepis exaltata	0	Lloyd, 1974
Onoclea sensibilis	6.0–41.0 (26.0)	Ganders, 1972
Osmunda regalis	5.0–100.0 (59.5)	Klekowski, 1970*c*
Phegopteris decursive-pinnata		
diploid	28.0–59.0 (46.4)	Masuyama, 1979
tetraploid	0	
Pityrogramma calomelanos	0	Singh and Roy, 1977
Pseudodrynaria coronans	86.7	Singh and Roy, 1983
Pteridium aquilinum subsp. *aquilinum* (3 varieties)	6.0–100.0 (38.8)	
P. aquilinum subsp. *caudatum* (2 varieties)	0–87.8 (41.3)	Klekowski, 1972
Sadleria cyatheoides	44.0	Holbrook-Walker and Lloyd, 1973
S. souleyetiana (=*S. souleytiana)*	44.0	Holbrook-Walker and Lloyd, 1973
Stenochlaena tenuifolia	48.0	Klekowski, 1970*b*
Tectaria macrodonta	77.5	Khare, 1980
Thelypteris palustris	30.0–92.4 (40.0)	Ganders, 1972
Woodwardia fimbriata	0	Klekowski, 1969*b*

[a] Modified from Lloyd, R.M. (1974). *Ann. Mo. Bot. Gard.* **61**, 318–31.
[b] Except for the following, references to data in the Table are given in the reference list at the end of the book.
1 Khare. P.B., and Kaur, S. (1979). *Proc. Indian Acad. Sci.* **88**B, 243–7.
2 Khare, P.B., and Roy, S.K. (1977). *Genetica* **47**, 183–5.
3 Lal, M., and Roy, S.K. (1983). *Curr. Sci.* **52**, 560–2.
4 Singh, S.P., and Roy, S.K. (1983). *Curr. Sci.* **52**, 924–6.
5 Warne, T.R. and Lloyd, R.M. (1981). *Bot. J. Linn. Soc.* **83**, 1–13.

genetics of breeding systems of gametophytes to determine the frequency of genetic load and causes for sporophyte lethality in a number of ferns. These studies have convincingly demonstrated that gametophytes which show an ontogenetic sequence of sex organ formation favorable for intragametophytic selfing have a high propensity for self-fertilization; this is in turn correlated with a low genetic load. Similarly, gametophytes which show a sexual ontogeny favorable for intergametophytic selfing have a high frequency of cross-fertilization episodes and the resultant sporophytes have a high genetic load. The deleterious recessive sporophytic genes are usually expressed as abortive zygotes, abortive embryos which appear as swollen tissue within the archegonium, small abnormal sporophytic tissue, or as abnormalities in the production of the first leaf or root (Ganders, 1972; Saus and Lloyd, 1976; Masuyama, 1979). Although deleterious genes are identified at the morphological level, more specific genetic tests have also been devised to analyze the factors inhibiting sporophyte formation in self-sterile gametophytes (Masuyama, 1986).

The high degree of homozygosity attained by fern sporophytes following intragametophytic selfing is to be regarded essentially as an adaptation to stabilize the gene pool of the population. This might be envisaged to occur by two different routes. For example, lethal or deleterious recessive alleles can be eliminated by natural selection in homozygous sporophytes over a long period of time. Alternately, frequent occurrence of intragametophytic selfing will eliminate recessive lethals very rapidly. The advantages posed by these mechanisms might appear to contradict the classical view that homozygosity is a bottleneck to genetic variability in a population.

Sex expression in heterosporous ferns and other pteridophytes

As stated at the outset of this chapter, not much work has been done on the reproductive biology of heterosporous ferns. In contrast to homosporous ferns, where sex determination takes place after spore germination, in the heterosporous group it is accomplished at certain specific stages in the life cycle of the sporophyte. This makes it difficult to follow and manipulate particular sexual cell types within the complex body of the sporophyte.

Ontogenetic studies of sporangium development in certain genera of heterosporous ferns seem to indicate the existence of a series illustrating the divergence of the two sporangial types. For example, in *Marsilea*, differences between the megasporangium and microsporangium appear early during archesporial development. In *Salvinia* (Salviniaceae), the two types of sporangia appear seemingly different at the conclusion of archesporial development. Most striking is the situation in *Azolla* where both types of sporangia begin their development in the same way. However, preparatory to meiosis in the megasporocytes, microsporangial initials arise on the stalk of the megasporangium. From this point on, the question as to whether a sporangium eventually becomes the male or the female type, depends on the abortion of initials of the opposite sex type. Thus, the two kinds of sporangia diverge rather late in ontogeny (Bierhorst, 1971; Konar and Kapoor, 1974). Some early experiments on *Marsilea* (Shattuck, 1910) suggested that differentiation of spore types may be related to the exposure of the sporangium at the time of meiosis to unfavorable conditions such as cold stress, drought or low light intensity, but these studies were not pursued further.

Although *Platyzoma microphyllum* is not a heterosporous fern in the strict sense of the term, its incipient heterospory offers two unique features to study sex expression. One is that gametophytes of both kinds of spores are photosynthetic and exosporic, similar to those produced by homosporous ferns. Secondly, heterosporous condition is not associated with strict dioecism. This is evident from the observation that although filamentous gametophytes born out of small spores bear only antheridia and spatulate gametophytes formed from large spores produce only archegonia, in the absence of fertilization, a wave of antheridium formation is initiated on the female gametophytes (Tryon, 1964). Recently, Duckett and Pang (1984) have shown that the underlying genomic program of sexuality determined during sporogenesis can be modified by subculturing fragments of gametophytes. Regardless of their origin from a male or a female type, as the fragments reconstitute the basic organization of the gametophyte from which they are isolated, they bear secondary antheridia and archegonia (Fig. 11.2). The import of these results is that experimental manipulations cause the transfer of sex determination from the sporophyte to the gametophyte. It is not too far-fetched to hope that future explorations might uncover some treatments which may even induce the formation of spatulate females from small spores and filamentous males from large spores.

Among other pteridophytes, investigations on sex determination on the gametophytes of *Equisetum* are germane to the present discussion. The protracted and at times controversial history of research on the sexuality of *Equisetum* gametophytes has been reviewed by Duckett (1970) and so only the recent work is considered here. The point of contention in these studies has centered on the occurrence and timing of bisexuality. It now appears that gametophytes of most species are potentially bisexual, despite the fact that they begin their life as males or females. During a period of growth extending in some cases even up to a year, the females become bisexual, but the males never produce archegonia (Duckett, 1970, 1972). This does not mean that maleness is irrevocably fixed, as subculture of regenerates from male or female gametophytes of some species induces production of a new line of both antheridia and archegonia on them (Duckett, 1977, 1979*a*; Duckett and Pang, 1984). This example suggests that dioecism in *Equisetum* gametophytes is to be regarded rather as a useful way of distributing the

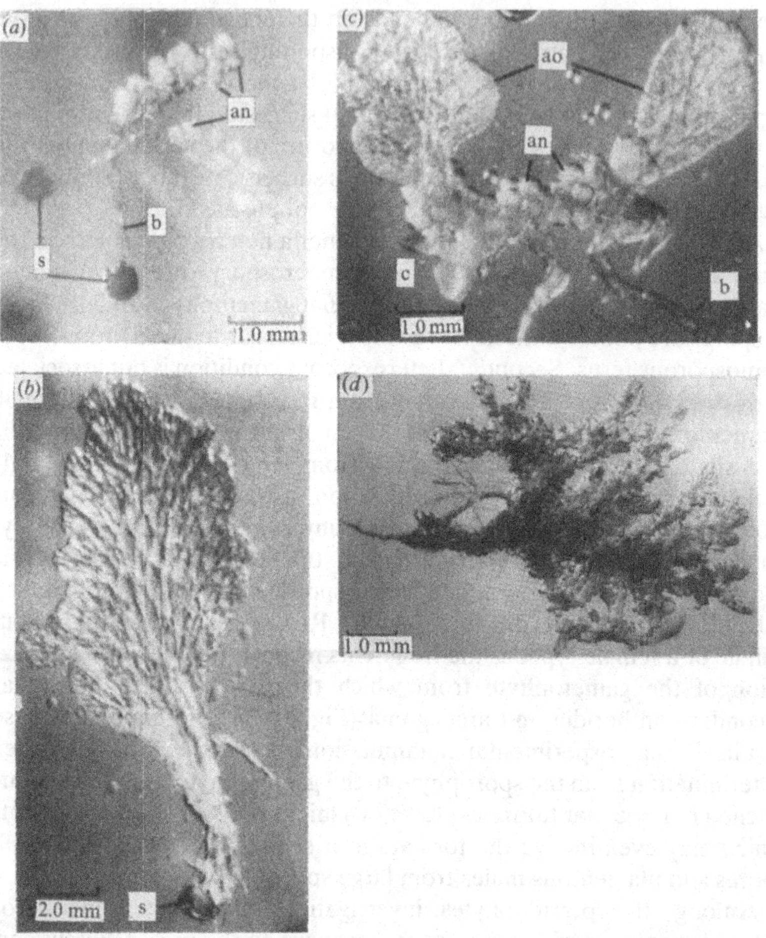

Figure 11.2. Differentiation of sex organs on the gametophytes of *Platyzoma microphyllum*. (*a*) Two filamentous male gametophytes grown from small spores showing antheridia. (*b*) A female gametophyte from a large spore. (*c*) Basal end of a male gametophyte showing two adventitious outgrowths and antheridia. (*d*) Antheridial branches, with numerous antheridia, regenerated from an originally female gametophyte. an, antheridium; ao, adventitious outgrowth; b, elongated basal cell; c, cut surface of the gametophyte; s, spore. (From Duckett and Pang, 1984.)

male and female reproductive organs, than as a causal element in the process of sex determination.

An important question raised by these results appears to be why some *Equisetum* spores regenerate male and others female gametophytes. The answer is not known, but it has been found, for example, that conditions for vigorous vegetative growth favor females whereas less propitious regimens foster males (Duckett, 1970; Mohan Ram and Chatterjee, 1970). Among other factors, light quality, light intensity, density of sowing, pH of the medium, hormones, and genotype (Mohan Ram and Chatterjee, 1970; Hauke, 1971, 1977; Srinivasan and Kaufman, 1978; Laroche, Guervin and Le Coq, 1978; Guervin, Laroche and Le Coq, 1979; Le Coq, Laroche and Guervin, 1980) are known to favor one sexual type over the other. Close to the heart of the question posed above, there is evidence to indicate that some spore populations are dimorphic and are thus functionally heterosporous; this could serve as a basis for the pattern of sex determination (Srinivasan and Kaufman, 1978). However, at the present time there are few compelling reasons to choose among these various possibilities.

The spatial and temporal aspects of distribution of sex organs on *Equisetum* gametophytes suggest that there is little chance for intragametophytic selfing. At the same time, the sparsity of bisexual gametophytes and the frequency of dioecism in nature attest to the prevalence of intergametophytic selfing. The predictive value of these observations in determining the breeding system of *Equisetum* gametophytes has been demonstrated by both laboratory experiments and field observations (Duckett, 1979*b*; Duckett and Duckett, 1980).

Natural reproductive biology of ferns

The significance of the diversity of breeding systems in the reproductive biology of ferns in nature is very poorly understood. Yet, it holds the key to so much that it is characteristic of the genetic structure of fern sporophyte populations. Moreover, a proper breeding system ensures the long-range dispersal of gametophytes as colonizers and thus extends the geographical range of the species. To explain the high incidence of genetic load observed in populations of *Osmunda regalis* in an ecological framework, Klekowski (1973*a*) has hypothesized that occasional colonizers in an open habitat initially resort to intragametophytic selfing. This results in the production of a range of homozygous sporophytes which do not show any measurable amount of genetic load. There is little doubt that members of a colony and those of neighboring colonies recognize each other with the passage of time. Consequently, at later stages of colonization, there is an increasing tendency toward intergametophytic selfing and crossing and production of heterozygotes carrying a high genetic load. This hypothesis has been tested in a comparative study of the genetic load among populations of *Ceratopteris thalictroides* (Lloyd and Warne, 1978) and *Blechnum spicant* (Cousens, 1979). The presence of low levels of genetic load in disjunct or small populations of these ferns lends support to the contention that they are of recent origin and are thus capable of intragametophytic selfing. At the other extreme is

Pseudodrynaria coronans which has a high level of genetic load and recessive sporophyte lethality. These preclude the chances of a single spore functioning as a propagule to raise a new population of sporophytes which consequently exhibit a sporadic distribution in nature (Singh and Roy, 1977). The problem of maintaining genetic variability in these populations remains acute, so does that of breaking the natural geographic isolation barriers. Based on a study of disjunct populations of *Asplenium platyneuron* which exhibit low genetic loads, Crist and Farrar (1983) have suggested that the plants successfully colonize distant habitats through a regular cycle of sporophyte regeneration from isolated gametophytes and sporogenesis by homoeologous recombination and pairing.

The importance of antheridiogens as hormonal cues in influencing gametophytic mating systems in natural populations of ferns has been underscored in several studies. In a pioneer work, Tryon and Vitale (1977) showed that field populations of heavily antheridiate gametophytes of *Asplenium pimpinellifolium (=A. auriculatum)* and *Lygodium heterodoxum* are always close to archegoniate ones, while those with a few antheridia occur far away from the females. Although no determinations of antheridiogen activity of the gametophytes were made, this observation has led to the view that the hormonal system does operate in nature, facilitating outcrossing in the species. Additional insight into the role of antheridiogens in the outcrossing of ferns has been derived from studies on *Bommeria* (Haufler and Gastony, 1978; Haufler and Soltis, 1984). Gametophytes of *B. ehrenbergiana* and *B. subpaleacea* are well disposed toward obligate outbreeding which requires the interaction of two different gametophytes to produce a sporophyte and the action of a native antheridiogen to induce male sexuality. The limited distribution of both species of *Bommeria* seems to be a consequence of the breeding system requiring the chance proximity of the developing gametophytes and antheridiogen action. Induction of male sexuality mediated by antheridiogens is also important in other genera like *Athyrium* (Schneller, 1979) and *Cyclosorus* (=*Christella*; Haufler and Ranker, 1985) in establishing gametophytic populations in nature that promote outcrossing.

Unfortunately, very little information has come to light concerning the impact of other aspects of the breeding system of ferns on their natural reproductive biology. We are thus obviously a great distance from gaining a complete understanding of how reproductive adaptations have actually affected the distribution and the genetic and evolutionary characteristics of ferns in the field.

General comments

One of the traits strongly affected by natural selection among ferns is the ability of a population to preserve variability. The image that emerges from this survey is that despite the frequent chance afforded for intragametophytic selfing, many homosporous ferns maintain heterozygous, lethal-shielding genomes. The mechanism that makes this possible has been attributed to genetic factors. Whether heterozygosity is attained by restriction of intragametophytic selfing or by specialized recombination effects during sporogenesis is not established. What is significant is that a reproductive system which allows outbreeding as well as self-fertilization has contributed to the success of ferns as colonizers of newly formed habitats and as components of mature vegetation.

12

Gametogenesis and fertilization

In this chapter we will take a close look at the manner in which the antheridium and archegonium give rise to gametes and at the act of sexual recombination. Formation of sex organs and the various physiological and genetic interactions between gametophytes represent one phase of the reproductive biology that terminates in sexual fusion and gene exchange. The final acts in this drama are gametogenesis and fertilization. In ferns, the function of gametogenesis is to construct two specialized reproductive cells, the spermatozoid (sperm), produced by the antheridium and the egg, produced by the archegonium. Fertilization involves the union of the sperm with the egg which is generally housed in a privileged location in the archegonium. In the first part of this chapter we will examine spermatogenesis, the origin of the sperm and follow it up with oogenesis, the formation of the egg and conclude our traverse of the gametophytic landscape with an account of fertilization.

Topics covered in this chapter have been reviewed by Duckett (1975), Bell and Duckett (1976), DeMaggio (1977), and Bell (1979*b*) and the reader is referred to these sources for additional information.

Spermatogenesis

Our knowledge of spermatogenesis in ferns is of recent vintage, derived largely from some careful work on *Pteridium aquilinum, Marsilea vestita,* and *Ceratopteris thalictroides*, which has provided a rich heritage of new information and some excellent electron micrographs. Details of the structure of spermatozoids of these ferns are sufficiently distinctive from other plant cells that a direct immersion in their study can be confusing. For this reason, we will first describe the salient features of the organization of these cells and follow up with a discussion of the evolution of their ultrastructure in relation to function.

The relatively simple architectural design of the antheridium in homosporous ferns has contributed to the accessibility of spermatogenesis in this group while the complexity of the process has restricted its study to just two species. The point of departure for our discussion of

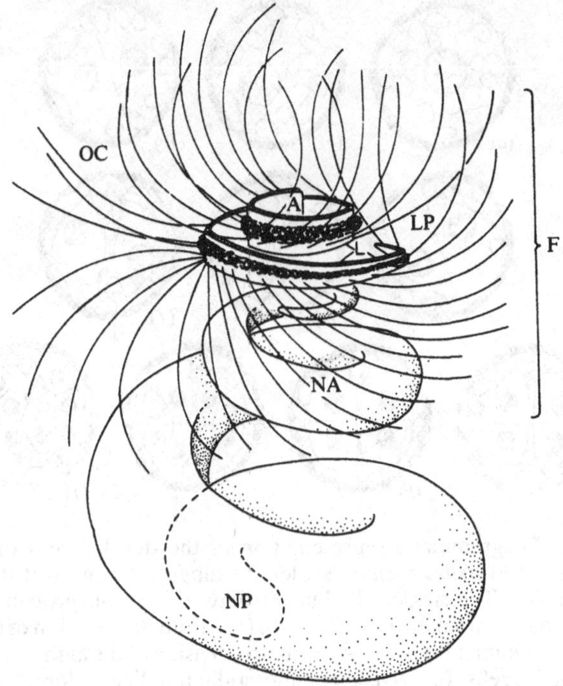

Figure 12.1. A side-view of the reconstruction of a mature spermatozoid of *Pteridium aquilinum*. A, anterior tip of the multilayered structure; F, flagella; L, lamellar strip; NA, anterior tip of the nucleus; NP, posterior tip of the nucleus; OC, osmiophilic crest. (From Duckett, 1975.)

spermatogenesis in *P. aquilinum* is the antheridium in which a centrally placed spermatogenous cell is surrounded by a jacket layer of cells. Soon after it is formed, the spermatogenous cell undergoes cleavage which distributes the cytoplasm among 32 small isodiametric cells known as spermatocytes. These are directly transformed into spermatozoids, which are of course programmed for terminal differentiation. The spermatozoid is a coiled helix which bears anteriorly a bunch of about 40 flagella, containing the standard 9 + 2 axoneme (Manton, 1959; Bell and Duckett, 1976; Elmore and Adams, 1976). A side-view of the spermatozoid in Fig. 12.1 shows that the coil is a left-handed spiral of about five gyres. Except for minor differences, the structure of the mature spermatozoid of *C. thalictroides* is similar to that of *P. aquilinum* (Duckett, Klekowski and Hickok, 1979). A study of spermatocyte differentiation in *Lygodium japonicum* has revealed that transient cytoplasmic bridges connect the spermatocytes thus making the spermatogenous cell a coenocyte (Vaudois, 1983).

Research on spermatogenesis in *Marsilea* is facilitated by the rapid and

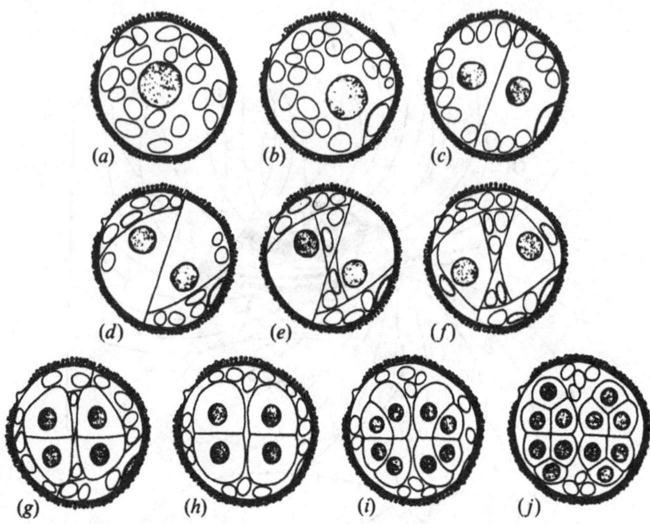

Figure 12.2. Diagrammatic representation of the development of the microgametophyte of *Marsilea vestita.* Nuclei are stippled; open, oval structures are starch grains. (*a*) The single-celled microspore. (*b*) Asymmetric division of the microspore to form a small prothallial cell. (*c*) Symmetric division of the large cell to form antheridium initials. (*d*) Asymmetric division of the antheridium initial to form two jacket cells. (*e*) Division of antheridium initials to form two triangular jacket cells. (*f*) Formation of additional jacket cells by the division of the antheridium initials; the other cell derived from each antheridium initial is now known as the spermatogenous cell. (*g*) Formation of four spermatogenous cells. (*h*) Formation of eight spermatogenous cells. (*i*) Formation of 16 spermatogenous cells. (*j*) Division of spermatogenous cells to yield a total of 32 spermatocytes. (From Hepler, 1976.)

synchronous development of the microspores; it takes no more than 12 hours for completion of spermatogenesis from the time microspores are incubated in water. In *M. quadrifolia* and *M. vestita*, typical development of the male gametophyte initially entails the division of the contents of the microspore into a large cell and a small cell (prothallial cell). The large cell divides to form two antheridium initials which are subsequently transformed into antheridia each containing eight spermatogenous cells surrounded by a sterile jacket layer of six cells. In the final round of divisions, each spermatogenous cell yields 2 spermatocytes (Fig. 12.2). In *Regnellidium*, the formation of two prothallial cells seems to be the norm, this being accomplished by a second division of the large cell. In *Azolla*, a total of four to eight spermatogenous cells are produced in each microspore by division sequences slightly different from those seen in *Marsilea* (Sharp, 1914; Higinbotham, 1941; Mizukami and Gall, 1966; Konar and Kapoor, 1974; Hepler, 1976). The spermatocyte metamorphoses into the spermatozoid which is a pear-shaped structure with a

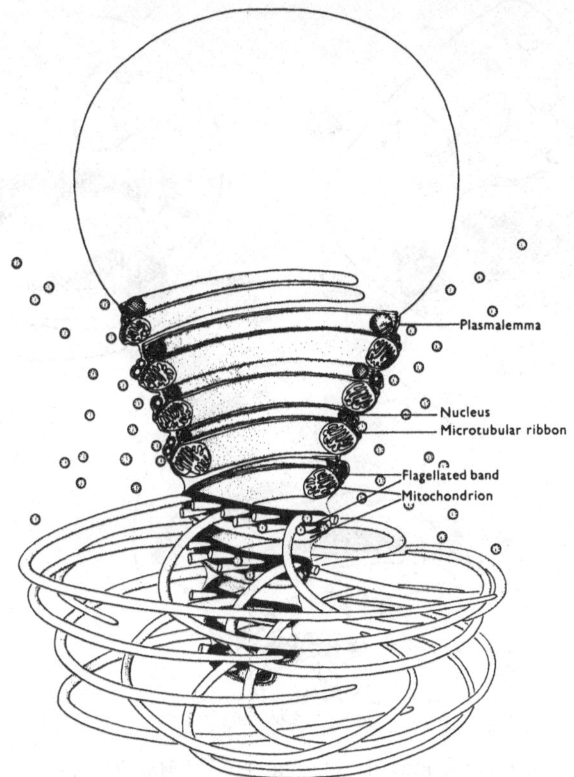

Figure 12.3. Reconstruction of a swimming spermatozoid of *Marsilea vestita*. Some flagella have been excised from the anterior region. From the posterior region, half of the spermatozoid is removed to reveal the cut ends of the organelles of the coil. The cytoplasmic vesicle has been shed and the plasmalemma assumes an expanded shape. (From Myles and Bell, 1975.)

tapered anterior end containing a left-handed spiral coil of about ten turns and 100 to 150 flagella gracing the entire edge of the coil (Fig. 12.3). In the mature spermatozoid, the cytoplasm is restricted to its bulbous portion as well as to the region within the coils of the spiral (Rice and Laetsch, 1967; Myles and Bell, 1975; Myles and Hepler, 1977).

At the ultrastructural level, the presence of an elongate nucleus with extremely condensed chromatin conforming to the shape of the coil, a single mitochondrion extending almost the entire length of the coil and a cytoskeletal framework provided by a ribbon of microtubules are characteristic features of the coil encircling the spermatozoid. At the anterior tip of the coil are found a few layers of plate-like repeating units associated with the band of microtubules; for want of a better term, this composite outfit is known as the multilayered structure (MLS). As described below,

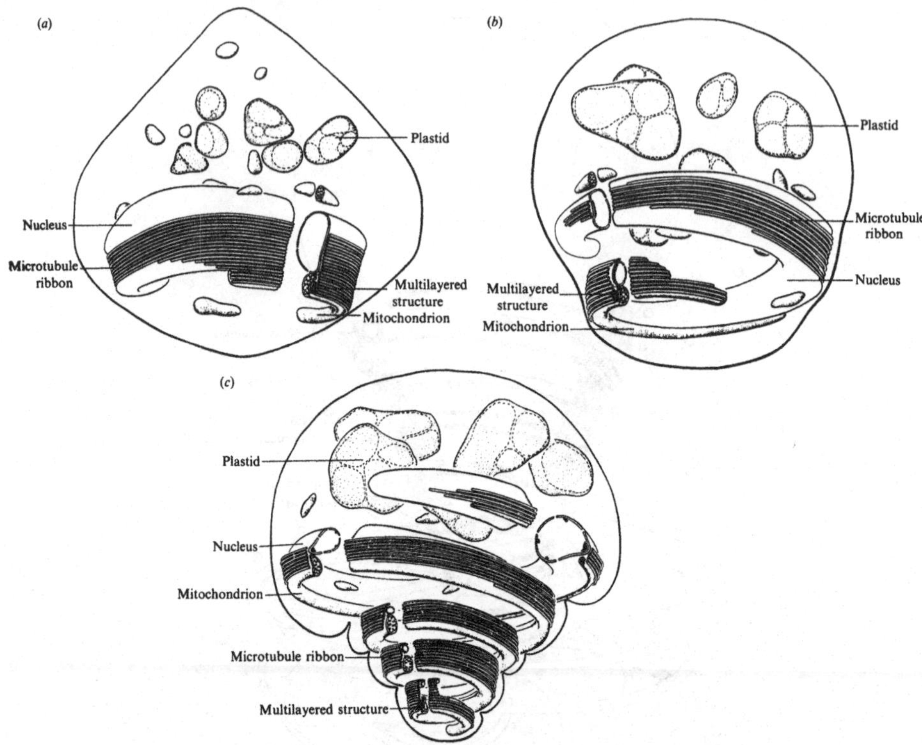

Figure 12.4. Reconstructions of a spermatozoid of *Marsilea vestita* at three stages of development. (*a*) Early stage of development of the spermatozoid. The nucleus is beginning to coil; a ribbon of microtubules runs along the outer edge of the nucleus and the coil mitochondrion; the multilayered structure is found between the microtubule ribbon and the mitochondrion; the other mitochondria are found near the nuclear envelope; plastids are restricted towards the posterior region of the cell; basal bodies have been omitted from the diagram. (*b*) Mid-stage development of the spermatozoid. The nucleus, coil mitochondrion and microtubule ribbon have extended considerably; the multilayered structure makes about one complete gyre; positions of other mitochondria and plastids are the same as in the early stage; the flagellated band and flagella have been omitted from the diagram. (*c*) Late stage of development of the spermatozoid. The nucleus has made about ten gyres and is attenuated at the anterior end; elongation of the coil mitochondrion continues; the microtubule ribbon increases to its maximum number of microtubules; the flagellated band and flagella are omitted from the diagram. (From Myles and Hepler, 1977.)

evolution of the final form and structure of the spermatozoid from an initially angular or spherical spermatocyte involves problems which are both geometrical and mechanical as well as those concerned with cellular reorganization and organelle morphogenesis. For orientation purposes, Fig. 12.4 illustrates diagrammatically three stages in the development of the coil and associated structures of the spermatozoid of *M. vestita*. An

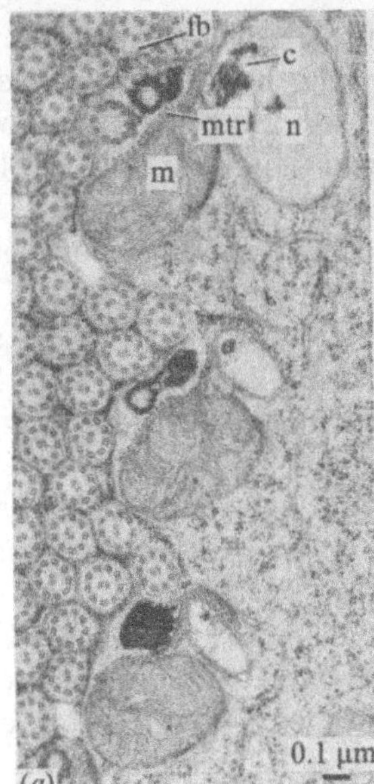

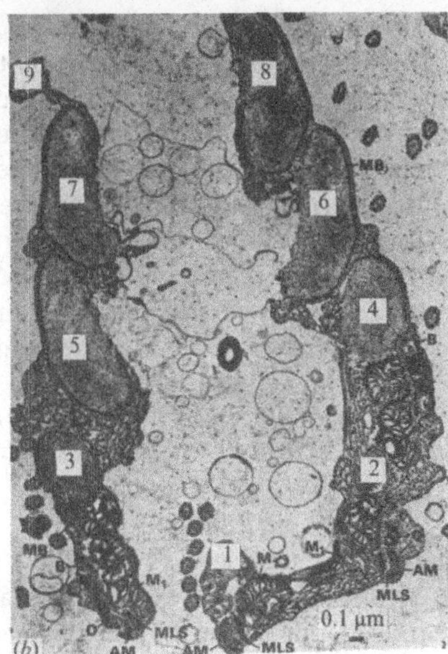

Figure 12.5. Fine structural details of spermatozoids. (*a*) Cross section through three gyres of the organelle coil of the spermatozoid of *Marsilea vestita*. The nucleus (n) has almost reached its final shape and the chromatin (c) has condensed considerably. The microtubule ribbon (mtr) is external to the nucleus and coil mitochondrion (m) and separates them from the flagellated band (fb). (From Myles and Hepler, 1977; photograph supplied by Dr D. G. Myles). (*b*) A median section of the spermatozoid of *Pteridium aquilinum* in the venter of the archegonium. There are nine sections of the nucleus (numbered from 1 to 9, beginning at the anterior end). AM, auxiliary band of microtubules at the anterior end of the osmiophilic material; B, basal body; M_1, M_2 mitochondria; MB, microtubular band (confined to the nucleus in the posterior part of the spermatozoid); MLS, multilayered structure; O, osmiophilic material. (From Bell *et al.*, 1971).

overview of organelle disposition in the spermatozoids of *M. vestita* and *P. aquilinum* is given in Fig. 12.5.

The blepharoplast and the multilayered structure

The blepharoplast is the earliest progenitor of the motile apparatus of the spermatozoid and its only known function is the elaboration of basal bodies of the flagella. Structural groundwork for the blepharoplast is laid down during the late division stage of the spermatogenous cell. In *M. vestita*, the genesis of this organelle is foreshadowed by the appearance of a flocculent material on one side of the telophase nucleus during the penultimate division of the spermatogenous cell. Within the sphere of this aggregate, two plaques appear, each of which develops into a blepharoplast. During the final division of the spermatogenous cell, the blepharoplasts separate from one another as they are propelled to opposite poles of the spindle. Microtubules seem to be associated with the blepharoplast at this stage of its development. Indicative of the role of the blepharoplast in the formation of the flagellar basal body is the fact that centrioles (microtubule organizing centers) destined to become basal bodies soon appear toward the periphery of the granular matrix of the blepharoplast (Mizukami and Gall, 1966; Hepler, 1976). There is also some suggestive evidence for the origin of centrioles from the blepharoplast during spermatogenesis in *Polypodium vulgare* (Vazart, 1964).

Details of blepharoplast development during spermatogenesis in *Pteridium aquilinum* (Tourte and Hurel-Py, 1967; Bell, 1974*c*; Vaudois and Tourte, 1979; Marc and Gunning, 1986; Marc *et al.*, 1988) and *Platyzoma microphyllum* (Doonan, Lloyd and Duckett, 1986) have been elucidated by electron microscopy and by immunofluorescence microscopy using antibodies to tubulin and other cytoskeletal proteins. The earliest appearance of the blepharoplast is in each of the two spermatogenous cells. In the young spermatocyte, the organelle is seen as radially arranged tubules imprisoned in the cytoplasm. At a later stage of development, the tubules appear as a central granular matrix enclosed by a shell of about 40 cylinders corresponding to early centrioles. Intensive assembly of tubulin at this location indicated by cross-reactivity with anti-tubulin antibodies is consistent with the idea that the incipient centriole is a focal point of the differentiating motile apparatus.

The MLS is thought to play a cardinal role in the assembly of microtubules involved in the shaping of the spermatozoid. It originates in the blepharoplast and is found in close proximity to a mitochondrion on one side. In the electron microscope, the MLS displays an intricate morphology, consisting of an outer stratum of microtubules, a layer of fine partitions and a dense plaque on the mitochondrial side. The microtubule ribbon does not remain as a passive appendage to the MLS, but it extends posteriorly to keep pace with the elongating coil (Bell, Duckett and Myles, 1971; Bell, 1974*b*, *c*; Myles and Hepler, 1977). Although this overview of the structure of MLS illuminates its complexity, the problem is to convince ourselves that its role in some phase of spermatogenesis is a demonstrable reality.

Shaping of the spermatozoid

There are clearly many diverse mechanisms that contribute to the shaping of the male gamete. These include, besides the formation of the coil itself, modification of the nucleus and mitochondria and the production of flagella. As the coil begins to appear, the nucleus elongates parallel to the longitudinal axis of the microtubule ribbon until it makes about five to ten gyres. At the same time, the chromatin is transformed from a well-dispersed to a condensed state. Since condensation renders the chromatin incapable of transcription, gene activation for sperm differentiation must occur before condensation sets in. That the ordering of chromatin condensation is somehow influenced by microtubules is indicated by two independent lines of evidence. Firstly, experimental intervention that affects microtubule assembly during spermatogenesis, such as colchicine application also disrupts chromatin condensation (Cave and Bell, 1979; Myles and Hepler, 1982). Secondly, in the nonmotile spermatozoid of a male sterile mutant of *Ceratopteris thalictroides* in which normal flagella are absent, nuclear deformations are associated with chaotic configurations of microtubules in the immediate vicinity of the nucleus (Duckett *et al.*, 1979). As to the role of the mitochondria, it is believed that they come to lie around the elongating nucleus and fuse with each other. The mitocondrion of the MLS also gets into the act with other mitochondria possibly fusing together to form a giant mitochondrion which later encircles the entire ten gyres of the coil. This forms an efficient energy-transducing system necessary to drive the flagellar apparatus. A patch of osmiophilic material found close to a band of microtubules is assumed to strengthen the anterior end of the spermatozoid of *P. aquilinum* (Myles and Bell, 1975; Myles and Hepler, 1977; Bell and Duckett, 1976).

The microtubule ribbon in association with the nuclear envelope appears to play a key role in the shaping of the nucleus, as demonstrated by colchicine treatment of spermatozoids of *M. vestita* (Myles and Hepler, 1982). In the presence of the drug, the nucleus elongates, but it does not attain the normal spiral shape. Therefore, the mechanical force for preservation of the elongate form of the nucleus appears to be independent of the presence of microtubules. Seen in this light, the shape-generating force is thought to include a molecular system associated with the nuclear envelope, perhaps in the form of a proteinaceous complex. A role for the nuclear envelope in nuclear shaping is reinforced by the observation that in the spermatozoids of both *P. aquilinum* (Bell, 1978) and *M. vestita* (Myles, Southworth and Hepler, 1978*b*), it is structurally modified in the region that is in contact with the microtubule ribbon. A crucial matter obviously in need of elucidation is how the mechanism for shape *determination* of the nucleus is put in place and becomes functional.

Flagellar morphogenesis. One of the signal achievements of the spermatozoid is the fabrication of the flagella, which provide for its motility. In *M. vestita,* flagella have their origin in rows of basal bodies embedded in a dense material known as the flagellated band. The microtubule ribbon separates the flagellated band from the other organelles of the coil. A freeze-fracture study has shown that changes during flagellar morphogenesis are accompanied by the appearance and rearrangement of ordered particle arrays on the plasma membrane (Myles and Bell, 1975; Myles *et al.*, 1978*a*). In *P. aquilinum*, the structure analogous to the flagellated band is an osmiophilic layer in which the flagella originate and which interconnects them (Duckett and Bell, 1971). Once the basal bodies are organized, they rapidly assume and stabilize the elongate form characteristic of fully formed flagella.

The importance of protein synthesis, especially of tubulin, the component of microtubules, for the assembly of the motile apparatus of the spermatozoid hardly needs emphasis. A study of ^{14}C-cysteine incorporation into developing antheridia of *C. thalictroides* has shown that although the label is readily incorporated into the early stage spermatocytes, differentiating spermatocytes and spermatozoids do not incorporate any label. This finding has come as a surprise, but it does not mean that tubulin polypeptides are not available when they are most needed. More likely, tubulin precursors are synthesized at an earlier stage of spermatogenesis and assembled at a later stage (Schedlbauer, Cave and Bell, 1973). The presence of callose and lipids in the wall of the spermatocyte which typically insulates the cell contents from the external milieu (Cave and Bell, 1973) has not simplified this interpretation as the wall may function as a barrier for the entry of metabolites. As far as their properties are concerned, flagellar tubulin isolated from fern spermatozoids has much in common with tubulin characterized from other plants and from animals (Ludueña, Myles and Pfeffer, 1980; Iqbal and Schraudolf, 1984*b*).

The work reviewed above provides an outline of the order and timing of the events during spermatogenesis. It defines spermatogenesis as a sequence of well-defined cellular episodes when the structural framework of the spermatozoid is established. The approach cannot however tell us much about the molecular biology of conformational changes in the chromatin. This question can be addressed using current techniques to isolate large amounts of spermatozoid DNA.

Release of the spermatozoid

The final stage of spermatogenesis begins with the liberation of spermatozoids from the antheridium. Only a few minutes of contact with water is all that is required to force antheridial dehiscence and spermatozoid discharge. The sperm is now ready for a motile life of a short duration. Fern spermatozoids begin their development with a lot more cytoplasm than is necessary for effective genetic recombination. Spermatozoids of both *P. aquilinum* and *M. vestita* shed some of this excess baggage after they are released from the antheridium. In the latter species where this has been studied in exquisite detail (Myles and Hepler, 1977), the anterior coil including the flagella becomes encircled in a ring of cytoplasm which migrates from the posterior region of the spermatozoid. It is this cytoplasm enriched with organelles such as ribosomes, mitochondria, and ER that is sloughed off from the spermatozoid after its release. The spermatozoid however harbors other organelles including plastids which form a small aggregate towards its posterior end. In *P. aquilinum*, shedding of the cytoplasm occurs after the spermatozoid has penetrated the archegonial canal (Bell and Duckett, 1976). At this stage, organelle fate is probably determined by events that occur in certain preferred regions of the spermatozoid. As spermatozoids quickly swim away, successful fertilization is contingent upon their being attracted to the vicinity of the archegonium.

Oogenesis

The processes of oogenesis in ferns bear little resemblance to those just described for spermatogenesis. To understand the basis for this, one must consider their respective goals in the context of post-fertilization development. As we have seen, the function of the spermatozoid is to deliver its payload of nucleus and cytoplasm to the egg. On the other hand, oogenesis is geared to the production of a cell that has the potential of becoming an embryo. This means that the egg has to contain within its single-celled domain all the reserve materials needed to build the embryo up to the point when it becomes autotrophic. This, as well as the fact that the egg acquires a number of specializations and unusual cytological features during oogenesis makes this process very complex indeed.

Light microscopic investigations of oogenesis in ferns abound. As described in a previous chapter, the general scenario is one in which the egg has its origin in the archegonium. In the final stage of differentiation of this structure, the lowermost cell in a row of three generated by the division of the primary cell is anointed as the egg (Fig. 12.6). However, ultrastructural studies have revealed that the primary cell engages in nucleo-cytoplasmic dialogue related to oogenesis even before the egg is demarcated. Its significance is to remind us that there is much more to oogenesis in ferns than the formation of the egg by a simple mitotic division of a progenitor cell.

Ultrastructural cytology of oogenesis

For no other ferns does there exist ultrastructural observations on oogenesis comparable to those published on *Pteridium aquilinum*. We begin our account here with the primary cell, which shows a five-fold

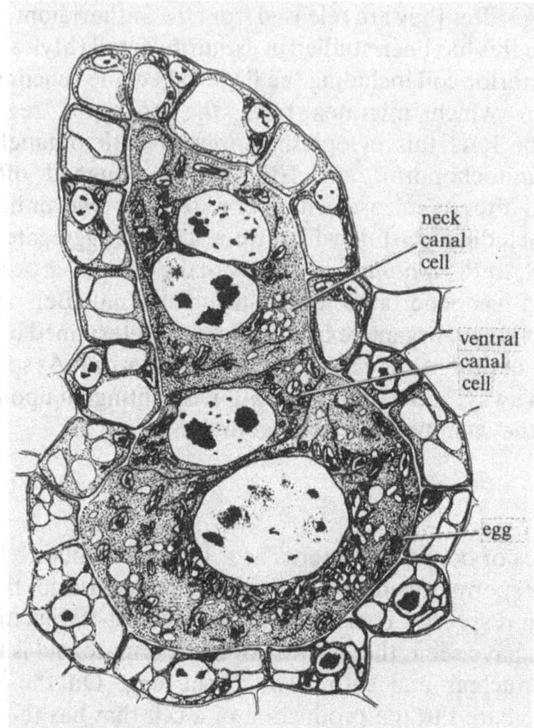

Figure 12.6. Diagrammatic representation of the young archegonium of *Pteridium aquilinum*. (From Tourte, 1975; print supplied by Dr Y. Tourte.)

increase in volume in preparation for the division. Major ultrastructural upheavals in this cell are the dedifferentiation of plastids resulting in the loss of grana and starch, the precipitous drop in the number of large vacuoles and the increasing frequency of small vesicles in the cytoplasm. Although the mitochondrial profile shows little change, the cytoplasm on the whole becomes very dense. Following the division of the primary cell into a small neck canal cell and a large central cell, the latter inherits much of the cytoplasm. Ultrastructural studies have shown that the cytological chores initiated in the primary cell such as plastid dedifferentiation continue unabated in the central cell. However, vesicles are the most prominent cytoplasmic components of the central cell and their aggregation into small masses is the most conspicuous change in this cell. Although a mitochondrial origin and an autophagic function have been suggested for the vesicles, there is no hard evidence to support either of these views (Bell, 1969; Bell and Mühlethaler, 1962*b*; Bell and Duckett, 1976; Tourte, 1970).

An unequal division of the central cell heralds the formation of a highly cytoplasmic egg, occupying the base of the archegonium and a ventral canal cell, lying in the axial row next to the egg. One of the most striking aspects of the evolution of the egg of *P. aquilinum* is the continuity of the cytoplasm from the primary cell to the central cell and eventually to the egg. Thus, the essential interactions of oogenesis begin when the primary cell is carved out of the archegonial initial.

Maturation of the egg

Given the fact that much of the egg cytoplasm is a holdover from the central cell, it is not surprising that the ultrastructural profile of the newly formed egg varies very little from that of the latter. However, within a few hours after it is formed, the egg displays some remarkable structural changes as it readies for fertilization. In the egg of *P. aquilinum* the first noticeable change is the elimination of the vesicles. These seem to migrate to the periphery of the egg from where they are probably discharged at the surface. As the vesicles disappear, a new electron-opaque membrane is deposited outside the plasmalemma of the egg (Bell and Mühlethaler, 1962*a*). This membrane is probably analogous to that present around eggs of *Dryopteris filix-mas* (Menke and Fricke, 1964) and *Histiopteris incisa* (Dennstaedtiaceae; Bell, 1980), and as revealed by staining reactions, consists predominantly of lipids (Cave and Bell, 1974*b*).

Another facet of egg maturation concerns the nucleus which assumes a highly irregular outline and sends out evaginations into the cytoplasm (Fig. 12.7). These outgrowths should not be mistaken for fixation artifacts, because histochemical evidence suggests that they are generated by localized enzyme activity of the nuclear envelope (Cave and Bell, 1975). Nuclear extensions have been described in the eggs of *P. aquilinum* (Bell and Mühlethaler, 1962*b*; Tourte, 1975), *D. filix-mas* (Bell, 1974*a*), *H. incisa* (Bell, 1980), and *Lygodium japonicum* (Bell and Pennell, 1987), but they are absent in the eggs of members of the primitive families of Osmundaceae and Marattiaceae (Bell, 1986). From a functional point of view, the evaginations might be thought to prepare the egg cytoplasm for sporophytic growth following fertilization. Nonetheless, nucleo-cytoplasmic interaction in the egg is probably a contributing, but not a complete, explanation for the complex way the nucleus behaves in the mature egg. In *Marsilea vestita,* the egg nucleus becomes lobed or cup-shaped, but this is different from the volatile changes found in *P. aquilinum* and other ferns referred to above (Myles, 1978). Nuclear differentiation in the egg of *M. vestita* also appears to depend upon the cytoskeletal framework, as drugs which interfere with microtubule assembly induce multinucleate condition (Kermarrec and Tourte, 1984).

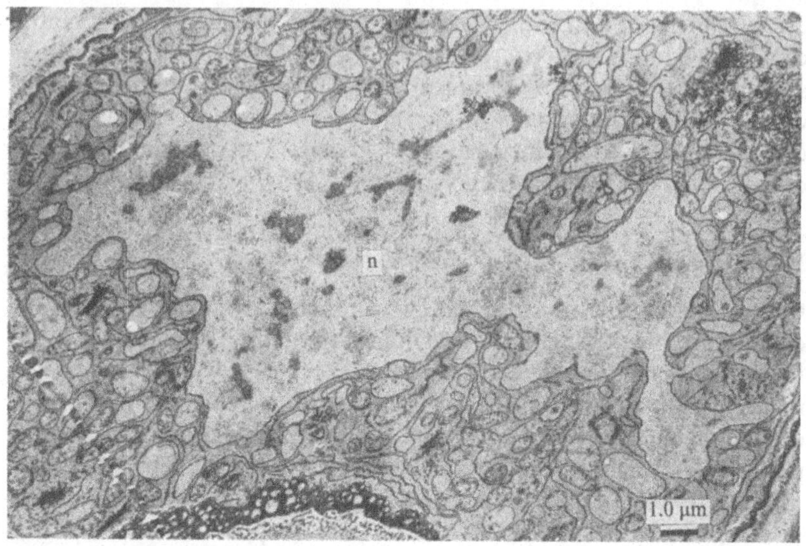

Figure 12.7. Electron micrograph of a mature egg of *Pteridium aquilinum* showing nuclear evaginations. n, nucleus. (From Tourte, 1975; photograph supplied by Dr Y. Tourte.)

In maturing eggs of *P. aquilinum* (Tourte, 1968; Bell, 1983) and *L. japonicum* (Bell and Pennell, 1987), concomitant with the appearance of nuclear protrusions, many nucleolus-like bodies move into the evaginations. Although these inclusions have been implicated in the transfer of transcribed information from the nucleus to the cytoplasm (Tourte, 1975), this view has been challenged (Bell, 1983; Bell and Pennell, 1987). The acidity of the nucleolar bodies and their failure to synthesize RNA make it hazardous to speculate on their function connected with information transfer in the maturing egg.

The maturation process of the fern egg is unusual in another fundamental way; the egg displays a cycle of appearance and disappearance of plastids and mitochondria. In the immature egg of *P. aquilinum*, the cytoplasm contains vacuolate bodies identified as plastids which disappear as the egg matures. However, the mature egg regenerates a new set of rather ill-defined plastids which have very few internal lamellae, but which possess novel features such as the presence of electron-opaque aggregates of ferritin, an iron-protein complex (Sheffield and Bell, 1978) and unusual arrays of tubules (Bell, 1982). In the immature egg, mitochondria appear at best as distended vacuolate bodies with fragments of cristae. A few days before maturity of the egg there is an abundance of large mitochondria of various forms freely scattered in the cytoplasm. Collectively, these studies have provoked a colorful, but controversial interpretation of the origin of plastids and mitochondria. It has been proposed that the second generation of plastids and mitochondria are formed *de novo* in the egg from nuclear evaginations that peel off into the cytoplasm, but there is minimal evidence to support this view (Bell, 1979*a*; Bell and Mühlethaler, 1962*b*; 1964*a*). In any event, what these profound changes mean is that the mature fern egg is in a high metabolic state. Viewed from a different perspective, they also mark the time when the sporophytic pattern of gene activity becomes partly established without being masked by gene activity for gametophytic growth. All that is required to activate the sporophytic program fully is the stimulus of fertilization.

As the egg approaches complete maturity, the ventral canal cell disintegrates, followed by the breakdown of the neck canal cell. The mature egg in readiness for fertilization appears to be isolated from the surrounding cells of the gametophyte by the presence of the special lipoidal membrane and by the absence of plasmodesmatal connections. It is apparent that by a series of predictable steps, the egg has become by now largely self-sufficient and independent of parental control.

Biochemical cytology of oogenesis

Due to the technical difficulty of isolating eggs at various stages of development, very little work worthy of modern biochemistry has been done on developing fern eggs. Hence, the focus of our attention here will be on the biochemical cytology of oogenesis as it relates to the pattern of nucleic acid and protein synthesis and accumulation.

Studies on DNA metabolism during oogenesis in *P. aquilinum* have sought to monitor the cytoplasmic control over nuclear activities in the egg. Ultraviolet microspectrography has shown that the cytoplasm of the primary cell of the axial row contains very little DNA, but substantial accumulation occurs in the cytoplasm of the central cell and the egg cell (Sigee and Bell, 1968). Consistent with the absorption data, autoradiography of ^{3}H-thymidine incorporation has revealed that appreciable DNA synthesis occurs in the cytoplasm of the mature egg (Bell, 1960*b*; Tourte, 1975). Since the number of mitochondria and proplastids of the egg multiply at this time, it is not surprising that radioactivity due to increased cytoplasmic DNA synthesis is mostly, if not entirely, associated with these organelles (Bell and Mühlethaler, 1964*b*; Sigee and Bell, 1971). Although the evidence is incomplete, it appears that the cytoplasmic DNA is derived largely by replication within the cytoplasm without direct nuclear participation (Sigee, 1972).

Major chores of oogenesis in the fern egg are the fabrication and accumulation of ribosomes. This obviates the need for the fertilized egg to produce its own ribosomes to support protein synthesis during the

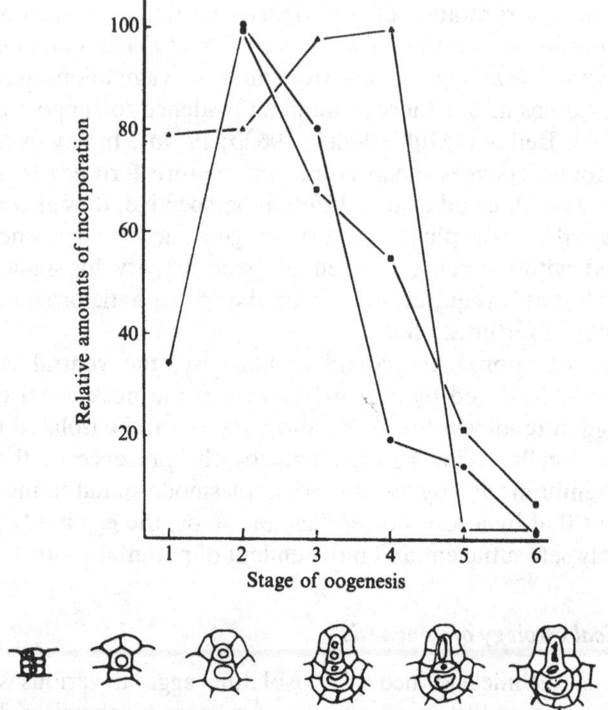

Figure 12.8. Incorporation of ^{3}H-uridine (▲), ^{3}H-leucine (●) and ^{3}H-arginine (■) in the archegonia of *Pteridium aquilinum* during oogenesis. The highest level of incorporation of each isotope is expressed as 100 to facilitate comparison. Stages of archegonium development indicated on the abscissa are also shown diagrammatically. (From Cave and Bell, 1974*a*.)

early stage of embryogenesis. The ribosome package includes 18S, 28S and 5S RNA as well as the ribosomal proteins. In *P. aquilinum*, the first wave of RNA synthesis seen in the primary cell signals the beginning of ribosome enrichment of the cytoplasm of the future egg. A pattern of intense RNA synthesis continues through the formation of the central cell and egg, attaining a peak in the young egg cell, but there is hardly any ^{3}H-uridine incorporation in the mature egg (Fig. 12.8). Protein synthesis, monitored by autoradiography of ^{3}H-leucine and ^{3}H-arginine incorporation, peaks during early oogenesis, but declines precipitously thereafter (Jayasekera and Bell, 1971; Cave and Bell, 1974*a*).

The changing pattern of RNA synthesis raises questions about the timing of information processing during oogenesis. For example, is continued RNA synthesis necessary for development of the egg? Does RNA, synthesized at a certain stage of oogenesis serve as a stockpile of information for initiation of sporophytic growth after fertilization? To obtain partial answers to these questions, archegonial differentiation and egg morphogenesis were followed in the gametophytes of *P. aquilinum* of different stages of development grown in a medium containing 2-thiouracil, an inhibitor of RNA synthesis. The results were revealing in that young gametophytes fail to differentiate new archegonia when reared in the presence of the inhibitor. Most interesting was the fate of eggs that matured in the presence of 2-thiouracil; they either abort or yield, after fertilization and in the absence of the inhibitor, sporophytes which revert to gametophytic growth. In view of the effect of 2-thiouracil in curtailing RNA synthesis in the gametophyte, these observations point to the need for continued transcription for archegonial differentiation and embryogenic development. The failure of eggs treated wth 2-thiouracil to form normal sporophytes upon fertilization indicates that RNA synthesized during oogenesis probably serves as informational RNA for initiating the sporophytic mode of growth (Jayasekera and Bell, 1972; Jayasekera, Cave and Bell, 1972). From this it is apparent that the early stages of embryogenesis in *P. aquilinum* must be the product of maternal gene expression rather than zygotic gene expression. Oogenesis in ferns therefore represents the closest analogy to the situation very clearly established in animal oocytes.

Fertilization

Fertilization is an event of fundamental importance as it triggers the developmental program of the diploid phase of the organism in a rapid burst of metabolic and mitotic activity. Ferns constitute the only major group of vascular plants in which fertilization takes place in an aqueous milieu. This element of the reproductive biology of ferns illustrates the highly adaptive nature of the genetic recombination system, as these plants grow where water is plentiful. Mechanisms that facilitate successful gametic union include those concerned with the motility of the spermatozoid, its attraction to the archegonium and the forces that waft it through the archegonial neck.

Locomotion of spermatozoids

A rapid rotary motion, initiated soon after their discharge from the antheridium, is characteristic of spermatozoids of ferns. Details of the locomotion of spermatozoids of *Lygodium japonicum* and *Marsilea vestita* deduced from high speed cinematography show that they swim with flagella oriented in a latero-posterior fashion. The flagella execute in a coordinated, nonsynchronous manner a sequence of continuous helical waves thrusting the spermatozoid forward (Bilderback, Bilderback, Jahn and Fonseca, 1974). The motion of spermatozoids in a field devoid of any orientation influences is random in a manner suggestive of Brownian movement (Wilkie, 1954; Brokaw, 1959). Results of recent research have indicated that a number of parameters of the medium such as anion enrichment and changes in pH (Evans and Bozzone, 1977, 1978; Evans and Conway, 1980) adversely affect the motility of spermatozoids of *Pteridium aquilinum*. A sidelight to these results is that as low pH and high concentrations of anions simulate acid rain conditions, they provide an explanation for the low sporophyte production in areas of acid rain precipitation.

Chemotaxis

In the course of a short time, spermatozoids crowd around the open neck of the archegonium in significant numbers. It is clear that this cannot be accounted for by random movements, but then how does it happen? It has been proposed that spermatozoids move chemotactically, in response to a gradient in the concentration of a chemical substance. In *P. aquilinum*, which is by far the best and the longest studied case of chemotaxis in a fern (beginning around 1884), malic acid or its salt works beautifully as the attractant. There still remains the question as to how the information available in a chemical gradient is translated into directional stimulus to the spermatozoid. Brokaw (1958) has proposed that as spermatozoids line up with respect to an electric field when a malate salt is present in the medium, a voltage gradient produces an effect that is easily picked up by their sensory elements. It must be emphasized that for all the complexities of the mechanism of chemotactic orientation of spermatozoids, the chemical agent produced in the archegonium has not been identified.

Spermatozoids attracted to the archegonia are trapped in a mucilaginous substance that is released during archegonial opening. Very likely this material contains some of the chemotactic factor. The contents of the disintegrating neck canal cell and ventral canal cell are now explosively released from the archegonium making an unhindered passageway for the spermatozoid (Ward, 1954). The entrapped spermatozoids orient themselves towards the open passageway of the archegonium as if following a directional influence and enter the archegonial canal.

Final phase of spermatozoid movement and gametic union

The movement of the spermatozoid through the archegonial channel to the vicinity of the egg is an active process that involves changes in its cytoskeletal properties. In *P. aquilinum*, the initial morphological change in the spermatozoid entering the archegonium is the unwinding of the coil and loss of its characteristic helical form; however, spermatozoids that manage to reach the venter of the archegonium display a remarkable developmental resiliency by the successful recovery of the helical form. Accompanying the change in shape, the spermatozoid appears contrite. The implication, consistent with the disruption of the nucleus and flagella and separation of the microtubule band from the nucleus, is that digestive enzymes are probably at work in the archegonial cavity. A closer look at the spermatozoid as it straightens during passage through the archegonial neck also reveals the loss of the vesicular cytoplasm with the associated amyloplasts and mitochondria. The spermatozoid now displays one of its most impressive capabilities, namely, the ability to survive as a functional cell in the form of a cytoskeleton consisting of a coiled rod-like nucleus and a mitochondrion coupled to a motile apparatus (Duckett and Bell, 1971).

It was mentioned earlier that in *M. vestita*, the spermatozoid divests itself of a part of the cytoplasm soon after its release from the antheridium. As the spermatozoid moves toward the archegonium, it is bereft of its remaining posterior vesicle of plastid-enriched cytoplasm and the nuclear envelope. The reduced size of the spermatozoid, coupled with its spiral symmetry and presence of flagella facilitates its movement through the archegonial canal to the vicinity of the egg (Myles, 1978). An open cavity just above the egg apparently serves as a landing stage for spermatozoids as several of them are observed in this cavity before fertilization.

Sufficient observations on the final stages in the life of spermatozoids and eggs as independent cells exist so that a general picture of the process of fertilization emerges, but the actual mechanism by which gametic union is accomplished has not been fully elucidated. A particularly arresting feature of the egg of *P. aquilinum* just before fertilization is the presence of several coiled spermatozoids in a clear oval area immediately beneath the extra egg membrane toward its upper region. This space, known as fertilization cavity, has nothing to do with the egg cytoplasm and is clearly separated from it by a unit membrane. How spermatozoids penetrate the egg membrane to reach the fertilization cavity is not known, although it is thought to be due to their helical movements which apparently rupture the membrane (Duckett and Bell, 1972). If this were true, it would bolster the view that the membrane is not a barrier to wide hybridization in ferns; indeed, a recent demonstration that spermatozoids of *Gymnogramme (=Pityrogramma) calomelanos* can break the egg membrane of *Pteridium aquilinum* and lodge in the fertilization cavity weighs against any type of species-specificity of the membrane (Vaudois, 1980).

Although details of the very beginning of conjugation in *P. aquilinum* are lacking, the topographic stability of the spermatozoid seems to be affected in the unfamiliar environment of the egg. An early change seen in the spermatozoid coming in contact with the female cytoplasm is the reconstitution of an envelope at the surface of the chromatin. Other changes include the unwinding of the coil, separation of the microtubular band from the nucleus and mitochondria and loss of shape of these organelles. Entrance of the spermatozoid also causes physical disruption of the egg cytoplasm and the remnants of the spermatozoid such as the flagellar profiles seen adrift in the cytoplasm adds to the complexity of its ultrastructure. In sea urchins, changes in the electrical properties of the egg surface and exocytosis of the cortical granules of the egg apparently prevent polyspermy, the entrance of more than one sperm into the egg. In other animals, a special membrane constituted immediately after fertilization serves the same function. Since none of these changes occur following fertilization in *P. aquilinum*, prevention of polyspermy is attributed to unknown physiological changes that take place in the egg cytoplasm after entry of one spermatozoid (Duckett and Bell, 1972; Bell, 1975).

Progress in our understanding of fertilization in *M. vestita* has resulted from the electron microscopic observations of Myles (1978); the sequence of events is shown in Fig. 12.9. Among important questions concerning the dynamics of fertilization in this fern, one is the mechanism by which the spermatozoid penetrates the thick wall around the egg. It now appears that the spermatozoid wiggles its way through a hole in the egg membrane that is only slightly wider than itself. Whether membrane lysis occurs before or after the spermatozoid arrives on the scene is not known. A highlight of this work is the description of the moment of syngamy which is initiated when the plasma membranes of the spermatozoid and egg fuse. Fusion of the gametic membranes creates a hollow bridge between the egg and the spermatozoid into which the egg cytoplasm flows. This leads to the establishment of a transient complex analogous to a fertilization cone seen in various animals. Although the entire organelle complement of the fusing spermatozoid is lodged in the egg cytoplasm, fertilization is intimately tied to changes in the structure of the nucleus. Nuclear transformation is initiated soon after the spermatozoid establishes contact with the egg cytoplasm and entails extensive decondensation of the chromatin and reconstitution of a new nuclear envelope with the aid of the egg ER. Another study of fertilization in *M. vestita* (Faivre, Kuligowski and Tourte, 1982) has shown that as soon as the spermatozoid comes in contact with the nuclear membrane of the egg, the nucleus of the former separates from the other organelles and enters the territory of the female nucleus. Presence of colchicine in the medium hinders nuclear fusion at fertilization (Kuligowski-Andres, Faivre and Tourte, 1982; Baron-Ferrand, Kuligowski, Chenou and Tourte, 1984).

The final change seen in the fertilized egg before it begins to divide is the formation of a new extracellular membrane over its outer surface

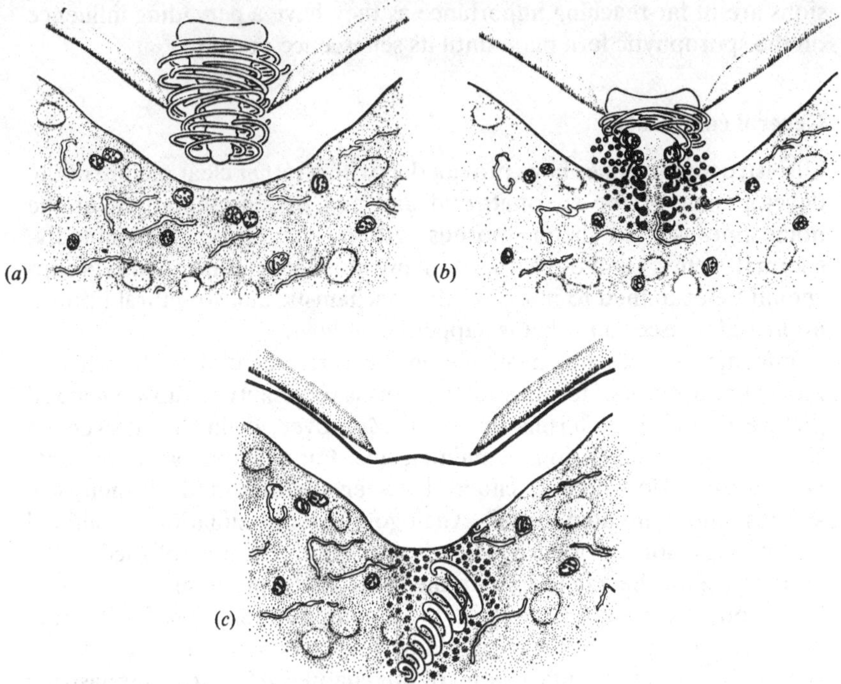

Figure 12.9. Diagrams of the top center part of the egg of *Marsilea vestita* showing penetration of the spermatozoid. (*a*) Passage of the spermatozoid through the hole in the wall of the egg. (*b*) Spermatozoid fusing with the egg leading to the formation of the fertilizaton cone. (*c*) Movement of the spermatozoid deeper into the egg cytoplasm (From Myles, 1978.)

beginning in the region of spermatozoid penetration and spreading circumferentially. It is believed that this membrane is important in preventing polyspermy (Myles, 1978).

Some fascinating work has been done to follow the fate of the paternal genome in the egg nucleus during fertilization in *M. vestita* by autoradiographic localization of ^{3}H-thymidine-labeled spermatozoids. The results show that rather than stirring up the egg nucleoplasm, the male chromatin mixes gradually with the latter after fertilization. The presence of labeled organelles other than the nucleus in the egg cytoplasm, although trifling in quantity, seems to indicate that paternal cytoplasm participates in the organization of the sporophyte cytoplasm (Kuligowski-Andres and Tourte, 1978; Tourte, Kuligowski-Andres and Barbier-Ramond, 1980; Faivre *et al.*, 1982).

Fertilization marks the end of the gametophytic phase. It also portends the beginning of the sporophytic phase, since the zygote nucleus almost immediately begins to divide in the embryogenic pathway. These divisions are of far-reaching importance as they have a pervading influence on the sporophytic fern plant until its senescence or death.

General comments

Ultrastructural studies have provided us with a rather clear picture of the development of the spermatozoid and egg in representatives of the homosporous and heterosporous groups of ferns. Undoubtedly, spermatogenesis and oogenesis of many more species from both groups should be examined to integrate the mechanistic and structural information and to ascertain what is happening *in vivo*.

Information presently available on the ultrastructural, metabolic and molecular aspects of fertilization in ferns is too scanty to draw a general picture of the mechanisms involved. Moreover, definitive answers to some interesting questions remain elusive. For example, what prevents polyspermy? How do eggs choose between spermatozoids of their own species and unrelated ones? Analogous to the situation in animal embryology, some evidence indicates that there is a large reliance on the metabolism of the egg to initiate the sporophytic program after fertilization, but it does not encourage the assumption that post-fertilization development of the egg is triggered by conserved maternal mRNA. On the whole, from the vantage point of this chapter, it is worth emphasizing that while there is no shortage of biochemical and molecular methods to study gametogenesis and fertilization in ferns, the internal location of the egg within the archegonium is a major obstacle to such studies.

Part IV

Developmental options

13

Apogamy – an alternate developmental program of the gametophyte

Fertilization which heralds the diploid or sporophytic phase, is a normal feature in the life cycle of ferns and partly accounts for the typical alternation of generations. A number of fern species also exhibit a way of life in which a sporophyte with the gametic number of chromosomes is born out of the gametophytic cell without fertilization. From both developmental and cytological perspectives, the parenchymatous cell of the gametophyte, dedifferentiating as it does during transformation into a sporophyte, is an obvious target for experimental analysis. With the accounts of sexual reproduction presented in previous chapters as a background, we will explore here the methods and mechanisms by which a sporophyte is regenerated from the gametophyte, while keeping innocent of sex.

Apogamy is the preferred term used to designate the developmental and reproductive adaptation of gametophytic cells that are deflected into the sporophytic pathway without sexual union. The term as used generally includes the regeneration of any organ of the sporophyte from the cells of the gametophyte. In some early accounts of apogamy, the term was extended to include production within the gametophyte of characteristic tissue elements of the sporophyte such as tracheids. However, as gametophytes that do not regenerate apogamous sporophytes differentiate tracheids, when supplied with sucrose or hormones in the medium (see chapter 9), it would seem logical to consider apogamous regeneration and tracheid differentiation as separate processes. For this reason, apogamy is used in a restricted sense in this chapter.

Homosporous ferns display two types of apogamous development. In the first type known as obligate apogamy, one or both sex organs may be absent or underdeveloped and dedifferentiation of any cell of the gametophyte into a sporophyte is the invariable rule. In most ferns that have functional sex organs, fertilization followed by sporophyte development occurs normally. In such cases, apogamy serves as an alternate pathway for the completion of life cycle. This second type of apogamy, which is most prone to experimental manipulations, is known as induced or facultative apogamy.

In this chapter we will first take a brief look at the occurrence of obligate apogamy in ferns. This is followed by an account of induced apogamy with emphasis on its developmental and cytological aspects. The early studies on apogamy have been reviewed at great length by Steil (1939, 1951).

Obligate apogamy

More than a century ago, Farlow (1874) observed that several anomalous gametophytes of *Pteris cretica*, that apparently had no archegonia, regularly produced sporophytic outgrowths by proliferation from the gametophyte rather than by fertilization. Comparison of the asexually formed sporophytes with those formed on sexual races of the same species provided a convenient framework to examine what appeared to be a remarkable developmental feat. It was found, for instance, that in the asexual buds which do not form a foot or an equivalent organ, the first appendage to arise is the leaf, followed by the root and still later by the stem. In contrast, a foot is always present in a sexually produced sporophyte serving to separate the latter from the gametophyte. Less obvious is the fact that the formation of asexual buds is preceded by the appearance of scalariform elements in the gametophyte and eventually an intimate connection is established between the vascular tissues of the gametophyte and those of the sporophyte. On the other hand, vascular tissues are absent in gametophytes programmed to produce sporophytes by syngamy.

DeBary (1878) confirmed the occurrence of anomalous sporophytes on the gametophytes of *P. cretica,* and reported their presence in *Aspidium (=Dryopteris) filix-mas* (Aspleniaceae) and *A. (=Cyrtomium) falcatum* as well. He also proposed the term apogamy to describe the phenomenon. Since these early reports, examples of apogamy have become legion in ferns. A list of members of filicalean ferns showing obligate apogamy, based on embryo development and/or sporophyte outgrowth is given in Table 13.1. The development of apogamous sporophytes in a representative genus, *Cheilanthes*, is shown in Fig. 13.1. Despite conflicting accounts of gametangial development, obligate apogamy is attributed to the lack of functional sex organs on the gametophyte. Unfortunately, no one has analyzed the causes for the dysfunction of sex organs in the apogamous gametophyte and in consequence, we know precious little about the extent to which the egg and sperm perform in the apogamous gametophyte. A recent study has shown that in the apogamous gametophyte of *P. cretica*, antheridia are fully functional in terms of sperm motility and chemotactic response of sperm to an open archegonium. However, cells of the archegonial wall collapse during development preventing opening of the neck to provide a passageway for the

Table 13.1 *Members of filicalean ferns showing obligate apogamy*

Family	Subfamily	Genera and species	References[a]
Adiantaceae	Adiantoideae	*Actinopteris australis*	Stokey, 1948*a*
		Adiantum lunulatum	Mehra, 1938*a*
		Bommeria pedata	Gastony and Haufler, 1976
		Cheilanthes alabamensis	Whittier, 1965
		C. castanea	
		C. feei	Steil, 1933
		Gymnogramme (=Pityrogramma) calomelanos	Woronin, 1908
		Notholaena incana	Nayar and Bajpai, 1964
		N. eckloniana (=Cheilanthes eckloniana)	Woronin, 1908
		N. marantae	
		N. sinuata	Woronin, 1908; Nayar and Bajpai, 1964
		Pellaea adiantoides (=P. viridis)	Steil, 1915*a*
		P. atropurpurea	Steil, 1911; Hayes, 1924; Nayar and Bajpai, 1964
		P. cordata (=P. cordifolia); P. flabellata	Steil, 1933
		P. (=Notholaena) flavens	Woronin, 1908
		P. flexuosa (=P. ovata)	Nayar and Bajpai, 1964
		P. glabella	Pickett and Manuel, 1925; Whittier, 1968
		Pellaea (=Notholaena) nivea	Woronin, 1908
		P. ovata	Steil, 1951
		P. sulcata (=P. striata)	Steil, 1918
		P. viridis (=Cheilanthes viridis)	Steil, 1918; Nayar and Bajpai, 1964
	Pteridoideae	*Pteris argyraea*	Steil, 1918
		P. biaurita	Mehra, 1938*b*
		P. cretica	Farlow, 1874; DeBary, 1878; Steil, 1918; Kanamori, 1972

Table 13.1 (*cont.*)

Family	Subfamily	Genera and species	References[a]
		P. fauriei, P. kiuschiuensis, P. natiensis, P. nipponica, P. oshimensis, P. setulosocostulata	Kanamori, 1972
		P.sulcata (= *P. biaurita*)	Steil, 1918
Aspleniaceae	Asplenioideae	*Asplenium aethiopicum*	Braithwaite, 1964
	Athyrioideae	*Athyrium felix–femina* (=*A. filix-femina*)	Farmer and Digby, 1907
		A. okuboanum	Kanamori, 1972
		Diplazium (=*Athyrium*) *hachijoense*	Kanamori, 1972
	Dryop-teridoideae	*Aspidium angulare* (=*Polystichum aculeatum*)	Heilbronn, 1910
		A. (=*P.*) *auriculatum*	Steil, 1918
		A. (=*Cyrtomium*) *falcatum*	DeBary, 1878; Allen, 1914 Steil, 1918; Whittier, 1964*a*
		A. (=*Polystichum*) *tsus-simense*	Steil, 1915*a*; Patterson, 1942; Kanamori, 1972
		A. (=*Dryopteris*) *filix-mas*	DeBary, 1878; Heilbronn, 1910;
		A. varium (=*D. varia*)	Steil, 1918
		Cyrtomium fortunei	Steil, 1918; Kanamori, 1972
		Dryopteris borreri (=*D. affinis*) *D. paseudo-mas* (=*D. affinis*) *Lastrea pseudo-mas* (=*D. affinis*)	Farmer and Digby, 1907; Döpp, 1939; Duncan, 1943; Bell, 1959
		D. chinensis	Kanamori, 1972; Kurita, 1981
		D. fuscipes *D. hondoensis*	Kanamori, 1972
		Dryopteris × *hakonecola*	Kanamori, 1972
		D. paleacea (=*D. wallichiana*)	Loyal, 1960
		D. remota	Fischer, 1919; Döpp, 1932
		Nephrodium hirtipes (=*Dryopteris atrata*)	Steil, 1915*b*, 1919*a*; Duncan, 1943; Kanamori, 1972
	Tectarioideae	*Lastrea chrysoloba* (=*Ctenitis falciculata*)	Steil, 1915*a*

Table 13.1 (*cont.*)

Family	Subfamily	Genera and species	References[a]
		Tectaria trifoliata	Steil, 1944
Blechnaceae		*Doodia caudata*	Heim, 1896
Grammitidaceae		*Xiphopteris serrulata*	Stokey and Atkinson, 1958
Hymenophyllaceae		*Crepidomanes latemarginale*	Yoroi, 1976
		Trichomanes alatum	Bower, 1888
		T. auriculatum	Stokey, 1948
		T. kaulfussii (= *T. trigonum*)	Georgevitch, 1910
		T. kraussii	Woronin, 1908
		T. pinnatum	Bierhorst, 1975*b*
Osmundaceae		*Osmunda cinnamomea*	Wuist, 1917
		O. claytoniana	
		O. javanica	Sarbadhikari, 1939
Parkeriaceae		*Ceratopteris* hybrid	Hickok, 1979
Polypodiaceae	Polypodioideae	*Polypodium dispersum*	Evans, 1964
		Phlebodium aureum	Ward, 1963
Schizaeaceae		*Anemia tomentosa*	Atkinson, 1962
Thelypteridaceae		*Phegopteris polypodioides* (= *P. connectilis*)	Wuist, 1917

[a] Except for the following, references to data in the Table are given in the reference list at the end of the book:

1 Bower, F.O. (1888). *Ann. Bot.* **1**, 269–305.
2 Fischer, H. (1919). *Ber. Deut. Bot. Ges.* **37**, 286–92.
3 Georgevitch, P. (1910). *Jahrb. Wiss. Bot.* **48**, 155–70.
4 Hayes, D.W. (1924). *Trans. Amer. Micros. Soc.* **53**, 119–35.
5 Heilbronn, A. (1910). *Flora* **101**, 1–42.
6 Heim, C. (1896). *Flora* **82**, 329–73.
7 Kurita, S. (1981). *J. Jap. Bot.* **56**, 50–8.
8 Loyal, D.S. (1960). *J. Indian Bot. Soc.* **39**, 608–13.
9 Pickett, F.L., and Manuel, M.E. (1925). *Bull. Torrey Bot. Cl.* **52**, 507–14.
10 Steil, W.N. (1911). *Bot. Gaz.* **52**, 400–1.
11 Steil, W.N. (1915*a*). *Science* **41**, 293–4.
12 Steil, W.N. (1915*b*). *Bot. Gaz.* **59**, 254–5.
13 Steil, W.N. (1918). *Bull. Torrey Bot. Cl.* **45**, 93–108.
14 Steil, W.N. (1933). *Bot. Gaz.* **95**, 164–7.
15 Stokey, A.G. (1948*a*). *J. Indian Bot. Soc.* **27**, 40–9.
16 Whittier, D.P. (1968). *Am. Fern J.* **58**, 12–9.
17 Woronin, H.W. (1908). *Flora* **98**, 101–62.
18 Wuist, E.D. (1917). *Bot. Gaz.* **64**, 435–7.
19 Yoroi, R. (1976). *J. Jap. Bot.* **51**, 257–67.

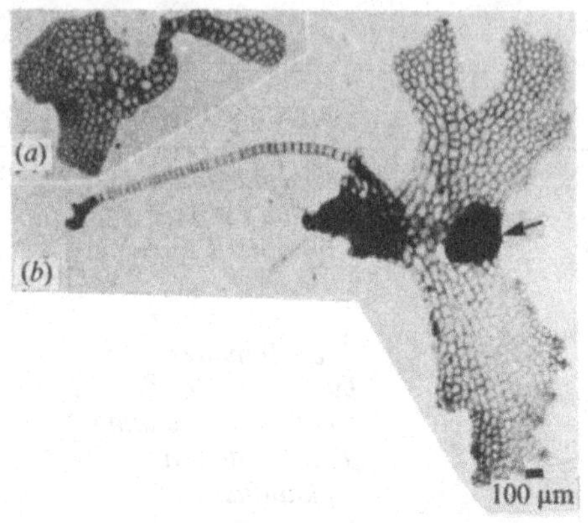

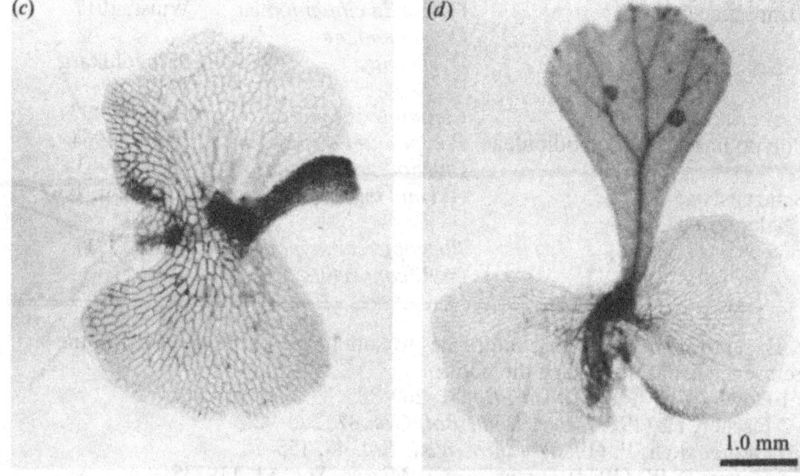

Figure 13.1. Development of the sporophyte by obligate apogamy in *Cheilanthes*. (*a*) A cordate gametophyte of *C. alabamensis* with a secondary gametophyte from the basal cell. (*b*) Gametophyte with an early stage of the apogamous sporophyte (arrow). (*c*) Apogamous sporophyte of *C. tomentosa* with a young leaf. (*d*) Apogamous sporophyte with mature leaf and root. Scale bar in (*b*) applies to (*a*) and (*c*) also. (From Whittier, 1965; photographs supplied by Dr D. P. Whittier.)

sperm (Laird and Sheffield, 1986). Thus, there is good circumstantial evidence for the impairment of archegonial function during obligate apogamy and now that we are alerted to its possibility, hard evidence may soon turn up.

On the basis of cytotaxonomical and cytological studies, it has been established that several wild and cultivated fern species, besides those listed in Table 13.1, have evolved apogamously (Manton, 1950; Manton and Sledge, 1954; Tryon and Britton, 1958; Bell, 1960*a*; Kurita, 1961; Mehra, 1961; Mehra and Singh, 1957; Mehra and Verma, 1960; Panigrahi, 1962; Walker, 1962). Although the number of spores in the sporangium might also give clues to the occurrence of apogamy (Knobloch, 1969; Knobloch, Tai and Ninan, 1973), in the absence of cytological data, detection of apogamy from spore counts in the sporangium can be misleading (Walker, 1979).

There is considerable evidence for the view that many ferns exhibiting obligate apogamy have a hybrid origin (Manton, 1950; Tryon and Britton, 1958; Mehra and Bir, 1960). Such hybridization might involve either mating between gametes of different species or between gametes of different degrees of polyploidy. When gametophytes destined to produce apogamous sporophytes form viable sperm, they may function as male parents and hybridize with the sexual race. According to Walker (1962), in crosses involving sperm of apogamous strains and eggs of sexual strains, the apogamous trait appears dominant, yielding hybrids that are self-fertile and true-breeding. It has also been noted that apogamous races of certain ferns are generally more wide-ranging in distribution than their sexual counterparts (Tryon, 1968; Evans, 1968; Gastony and Haufler, 1976). This is not surprising as apogamous plants have a relatively rapid rate of gametophyte maturation; they are also not constrained by the need for a film of free water for sperm movement and fertilization.

Apart from certain isolated reports that apogamy is initiated on the gametophytes sooner in a medium containing sucrose than without it (Whittier, 1964*a*, 1965), the causal factors that promote obligate apogamy have not been subject to experimental investigations. Spontaneous regeneration of sporophytic buds appears to be a cryptic property of the gametophytic cells, one of the cues for its expression being the apparent absence of functional sex organs. This bestows on the cells of the gametophyte an ability to manifest totipotency and retrieve the full range of the developmental program encoded in their genome. An analogous situation is seen in certain angiosperms in which pollen grains and unfertilized egg cells regenerate embryo-like structures and plants with the gametic chromosome number.

Induced apogamy

As the term induced apogamy conveys a broad meaning, it is not surprising that conditions favoring apogamy in sexually reproducing species of ferns exist in nature. Most importantly, these include a lack of sufficient moisture in the habitat, a reduced vitality of the gametophytes due to insufficient mineral nutrition, and changes in the intensity of illumination. Thus, by growing gametophytes under water stress, in soil deficient in mineral salts, or in light regimens of varying fluences, several investigators have successfully induced growth of apogamous sporophytes on gametophytes of numerous ferns which reproduce sexually in a perfectly normal way. It is doubtful whether any of these factors promote apogamy directly. Therefore, we will not consider this work here again; rather, in the following discussion, focus will be on recent investigations on induced apogamy which have produced new insights into our understanding of the phenomenon.

One important factor that has accounted for unexpected progress in our knowledge of induced apogamy in fern gametophytes has been the use of tissue culture techniques. Although apogamy has been induced in gametophytes of diverse ferns by *in vitro* techniques, it has been thoroughly studied in *Pteridium aquilinum*. An early experiment by Whittier and Steeves (1960) revealed that gametophytes of *P. aquilinum* var. *latisculum*, grown under a 12-hour photoperiod in a medium supplemented with glucose, give rise to a range of sporophytic structures such as entire plantlets, single leaves, and unbranched cylindrical processes. Chromosome counts of these outgrowths established their haploid or apogamous origin. Apparently, the presence of a carbon energy source in the medium enables the same genome to express a different developmental program resulting in a radically new phenotype.

Thus far, apogamy has been successfully induced by tissue culture methods in homosporous filicalean ferns. Some reports of induced apogamy in heterosporous ferns appear to be cases of sporophyte regeneration from the egg in the absence of fertilization. In *Regnellidium diphyllum*, sporophyte formation ensues when megagametophytes are isolated and cultured in a medium containing a high concentration of sucrose and a moderate level of 2,4-D (Loyal and Chopra, 1976). Varying degrees of sporophyte regeneration also occur from megagametophytes of *R. diphyllum* and *Marsilea vestita* cultured in distilled water (Mahlberg and Baldwin, 1975). A recent study (Bordonneau and Tourte, 1987) has shown that when unfertilized megagametophytes of *Pilularia globulifera* are cultured, the egg disintegrates, but the cells outside the egg regenerate a callus from which embryo-like structures are formed. This comes closest to a case of induced apogamy seen in heterosporous ferns.

Apogamy has also been induced in *Botrychium*, a non-filicalean fern. Gametophytes of *B. dissectum* are capable of growth in the dark or in light when they are cultured in a mineral salt medium supplemented with sucrose. Apogamous development that occurs on light-grown gametophytes ranges from buds with a single immature leaf to sporophytes with both leaves and adventitious roots, with a primary root being absent (Whittier, 1976). In *B. multifidum* apogamy takes a more restricted nature that does not reveal true leaf-like structures (Gifford and Brandon, 1978).

Tissue culture techniques have shown that sporophytic structures regenerate not only from the original cells of the gametophyte, but also from undifferentiated cells of gametophytic origin. This was first noted when gametophytes of *P. aquilinum* were grown in a mineral salt medium supplemented with yeast extract and glucose. During subculture of a type of proliferative growth formed on the gametophyte, single leaves and entire plantlets are found to regenerate apogamously (Steeves *et al.*, 1955). According to Kato (1963), germination of spores of *Pteris vittata* in a complex medium leads to the formation of a callus; subculture of the callus in the dark followed by its transfer to light results in the regeneration of sporophytes. Apogamy is triggered on the callus formed on spores of *Ampelopteris prolifera* (Thelypteridaceae) germinated in a medium containing 2,4-D by the simple expedient of removing the auxin source (Mehra and Sulklyan, 1969). Similar results have been reported with calluses induced by auxin and coconut milk on gametophytes of *Adiantum trapeziforme* (Padhya and Mehta, 1981).

It would appear from this brief survey that the original cells of the gametophyte, as well as cells dedifferentiated from it, display a propensity to form sporophytic structures by simple treatments in culture. Although no generalizations can be made, it is clear that the cultural milieu in which the gametophyte is trapped is responsible for diverting it to a different phenotypic expression. The importance of some ingredients of the culture medium in inducing apogamy is discussed below.

Role of sugar

In an early study on the induction of apogamy on gametophytes of *P. aquilinum* var. *latisculum* (Whittier and Steeves, 1960), optimum response in terms of the number of sporophytes per culture was obtained in a medium containing 2.5% glucose. In later investigations sucrose was shown to produce the best apogamous growth when gametophytes are reared in white light, blue light or far-red light (Whittier, 1964*b*; Whittier and Pratt, 1971). Other sugars such as fructose, maltose, and ribose are equally as effective as glucose or sucrose, while mannose, lactose, galactose, and xylose are poor substrates. Gametophytes that have their photosynthates as the sole energy source produce no apogamous structures. The widespread application of exogenous sugar to promote apogamy has been established in gametophytes of other isolates of *P. aquilinum* var. *latisculum* and in tissues of gametophytic origin in other ferns (Whittier and Steeves, 1962; Mehra and Sulklyan, 1969; Kato, 1970*a*; Padhya and Mehta, 1981).

In the gametophytes of *P. aquilinum* cultured in an optimum concentration of sucrose (4%), the formation of apogamous sporophytes is accomplished in three arbitrary phases. During an experimental period of 56 days, the first two periods each of 14 days duration are termed the *gametophytic* and *initiative* phases, respectively, and the third lasting 28 days is termed the *developmental* phase. Not all three phases are equally dependent on the presence of sucrose in the medium; production of the largest number of apogamous structures occurs when gametophytes are initially exposed to 4% sucrose during the first 28 days of growth and then transferred to a medium containing 0.1% sucrose during the last 28 days of growth (Whittier, 1964*b*). Interpretation of the role of sugar in apogamy is complicated by the fact that dark-grown gametophytes do not form apogamous structures even in the presence of an adequate supply of sugar in the medium.

One of the problems apparent from an analysis of the role of sugar in apogamy was the need to be certain that exogenous sugar in the medium is not acting as an osmoticum. Towards this end, gametophytes of *P. aquilinum* were grown in media containing suboptimal levels of glucose supplemented with concentrations of mannitol or carbowax sufficient to bring the osmolality to that of media containing 1.0 to 2.5% glucose. Absence of apogamous regeneration on gametophytes grown in media of high osmolality suggested that sugar does not induce apogamy by acting as an osmoticum (Whittier and Steeves, 1960). However, in a later work (Whittier, 1975) it was found that increasing the osmolality of the medium containing 3% sucrose by the addition of 0.5% mannitol promotes apogamy in the gametophytes to the same extent as that obtained with 4% sucrose. It would thus appear that a concentration of sucrose in the medium above that required as a combustible source of energy might function in an osmotic role in inducing apogamous growth on the gametophytes. A partial replacement of sucrose by an osmoticum has also proved optimal for the promotion of apogamy on gametophytes of *Matteuccia struthiopteris* (von Aderkas, 1984).

Role of ethylene and other hormones

A chance observation by Whittier and Pratt (1971) that gametophytes of *P. aquilinum* grown in sealed containers produce a large number of apogamous buds has led to a refreshingly new line of inquiry into the factors controlling apogamy. This has, in turn, given rise to the view that ethylene is involved in a casual role in apogamy. It was found that when various numbers of gametophytes are grown in glass containers of constant volume, the apogamous response is high in containers with large numbers of gametophytes (Fig. 13.2). On the other hand, when a constant number of gametophytes are grown in containers of various

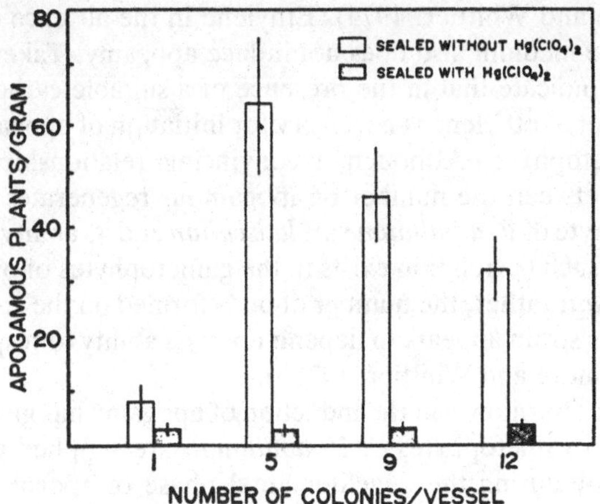

Figure 13.2. Effect of varying number of gametophytic colonies per culture vessel on the induction of apogamy in *Pteridium aquilinum*. Vertical lines indicate standard error of the mean. (From Elmore and Whittier, 1973.)

volumes, the maximum number of apogamous sporophytes are produced on gametophytes cultured in the smallest container. A correlation between accumulation of ethylene in the culture vessel and apogamous bud induction was established by gas chromatographic demonstration of ethylene production by the gametophyte. Administration of exogenous ethylene to the gametophytes in a continuous flow system elicits an increase in the number of apogamous buds formed, while inclusion of the ethylene absorbant, mercuric perchlorate, in the culture vessel completely eliminates apogamy (Elmore and Whittier, 1973). As the effect of ethylene is not foreshadowed by any distinct morphological or anatomical changes in the gametophyte, it has been concluded that rather than acting as a triggering hormone, ethylene establishes a physiological state in the cells of the gametophyte that enables them to differentiate sporophytes, even in the absence of the inducing agent (Elmore and Whittier, 1975*a*). From these results one could separate an ethylene-dependent induction phase of apogamy from an ethylene-independent developmental phase when actual differentiation of the apogamous bud sets in.

In view of the established role of a carbohydrate in the medium for the promotion of apogamy in cultured gametophytes of *P. aquilinum*, more conclusive evidence appeared necessary to prove that apogamy is under the influence of ethylene. The question is this: if ethylene is required for apogamy, will the complete evacuation of the gas from the external and internal milieu of the gametophyte eliminate apogamy even though a carbohydrate is present in the medium? This was indeed found to be the case (Elmore and Whittier, 1974). Ethylene in the absence of a carbohydrate in the medium also does not induce apogamy. Taken together, these results indicate that in the presence of a suitable carbohydrate, a threshold level of ethylene is necessary for initiation of apogamous buds on the gametophyte. Although a convincing relationship has been established between the number of apogamous regenerates formed on the gametophyte of *P. aquilinum* var. *latisculum* and its ability to produce ethylene, no such correlation exists in the gametophytes of other strains of *P. aquilinum*; rather, the number of buds formed on the gametophyte of a particular strain appears to depend upon its ability to respond to the hormone (Elmore and Whittier, 1975*b*).

Use of other hormones in the induction of apogamy has given variable results. When gametophytes of *P. aquilinum* are supplied with NAA, IAA, and GA during the developmental phase of apogamy, the formation of apogamous buds is not affected, while the same hormones supplied during the initiation phase promotes apogamy. The hormones are not effective in the absence of sucrose in the medium suggesting that like ethylene they act through a pathway involving a byproduct of carbohydrate metabolism rather than by altering the basic biochemistry of the cell (Whittier, 1966*a*). In the gametophytes of *Pteris vittata* cultured in light in a medium containing high concentrations of sucrose, apogamous buds are initiated when the medium is supplemented with IAA, GA or tryptophan (Kato, 1970*a*). An increase in the number of apogamous buds also occurs on gametophytes grown in a medium containing 2-thiouracil, although the basis for this effect is mysterious (Kato, 1965*c*). Morphactin is another inhibitor with ill-defined functions that promotes apogamy (Sharma and Singh, 1983).

Gene dosage and apogamy

Cells of the haploid gametophyte consist of only a single copy of the genome. A question relevant to our understanding of the normal fern life cycle is the influence of the number of gene copies in the cell on the extent of apogamous growth. A prerequisite to study this question is the availability of gametophytes of different ploidy levels. Polyploid sequences of fern gametophytes can be produced and maintained in culture by simple experimental manipulations, such as, repeated induction of gametophytes with sporophytic chromosome number by apospory (see Chapter 14), and embryo formation and sporophyte growth by fertilization from the aposporous units (Wetmore *et al.*, 1963). In an early study using a polyploid series of gametophytes of *Polypodium* (= *Phlebodium*) *aureum*, Heilbronn (1932) found that opportunities for apogamy increase with an increase in the DNA content of the cells. Generally, haploid and tetraploid gametophytes regenerate sporophytes by fertilization and apogamy, respectively, while gametophytes with a diploid genome form sporophytes by fertilization or by apogamy. According to Whittier (1966*b*), diploid gametophytes of *Pteridium aquilinum* produced from haploid ones by self-fertilization and apospory, generate more apogamous sporophytes than their haploid counterparts. These limited data suggest that an important provision for apogamy is an increase in gene dosage, while normal sexual reproduction is favored by a lower genomic DNA content of the cells.

Origin and development of the apogamous sporophyte

As revealed by anatomical studies, formation of the apogamous sporophyte involves several distinct histological changes that are integrated into an overall morphogenetic process in which cells behind the apical notch of the gametophyte participate. Moreover, the process of sporophyte regeneration in the gametophytes is roughly the same by obligate apogamy and induced apogamy, although differences in the type of cells that initially participate do exist.

In the first description of obligate apogamy in *Pteris cretica*, Farlow (1874) identified a swelling, presumably caused by repeated divisions of a group of cells on the lower surface of the gametophyte, as the progenitor of the apogamous embryo. Later investigators (Steil, 1919*a*; Patterson, 1942; Kanamori, 1967, 1972) traced the apogamous bud to a group of meristematic cells of the thickened cushion in the anterior portion of the gametophyte. By repeated anticlinal and periclinal divisions, these cells give rise to a prominent structure on which an apical cell is formed. Shortly after this, tracheid differentiation occurs in the cellular mass; the tracheids extend into the gametophyte or establish a connection with those present therein. In *Pellaea* (Nayar and Bajpai, 1964) and *Cheilanthes castanea* (Whittier, 1970), formation of the meristematic cluster of cells has been attributed to the activity of one to three initial cells in the apical region of the cordate gametophyte.

In an analysis of induced apogamy in the gametophytes of *Doodia caudata,* Duncan (1941) found that preparatory to the formation of the sporophytic outgrowth, division and elongation of cells in the cushion lead to the formation of a cylindrical peg. A subsequent phase of internal divisions followed by organization of leaf and stem initials at the tip of the peg completes the formation of the apogamous sporophyte. Observations on induced apogamy in *P. aquilinum* have also focused on the role of cells in the apical cushion whose activities carry it toward an increasingly complex morphology (Whittier, 1962). Although the cushion is initially homogeneous, it soon forms nests of small nonvacuolate cells whose active proliferation assuredly aids in the appearance of a three-dimensional structure on which primordia of leaf and stem arise. An

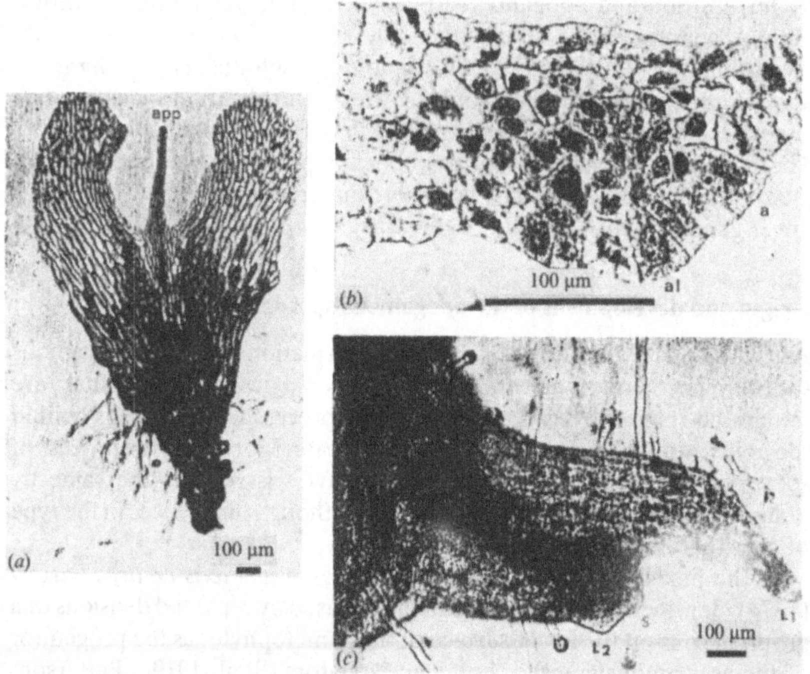

Figure 13.3. Early stages of development of the apogamous embryo. (*a*) A heart-shaped gametophyte of *Pteris fauriei* bearing a cylindrical structure (app). (*b*) Longitudinal section of the apogamous embryo of *Cyrtomium fortunei* showing the stem apical (a) and leaf apical cells (al). (*c*) External view of the apogamous embryo of *C. fortunei* bearing two leaves (l_1, l_2) and stem (s). (From Kanamori, 1972; photographs supplied by Dr K. Yasuda.)

important question from these investigations concerns the nature of the physiological gradients set up in the tissue preparatory to the initiation of localized meristematic activity. Such organized growth could be signaled by diffusible factors coming from neighboring cells, but we do not have even a glimpse of the nature of these cytoplasmic determinants.

Descriptive accounts of the origin of apogamous sporophytes from gametophytes of other ferns are also well represented in the literature (Allen, 1914; Mehra, 1938*a*, *b*; Duncan, 1943; Nayar and Bajpai, 1964; Kanamori, 1967, 1972); a few landmark stages in the development of the apogamous embryo are shown in Fig. 13.3. Although some details of the process differ in the different genera, the differences are minor and eventually an embryo-like structure emerges from the gametophyte. In the order of organ formation in the apogamous embryo, the root and stem follow the leaf; occasionally the stem appears first, followed later by the leaf and root. Once the primary organs are delimited, the presumptive sporophyte functions as a self-organizing system, comparable to the sporophyte produced by the sexual process.

Cytology of apogamy

The final outcome of apogamy is a fully developed sporophyte poised on the threshold of sporogenesis. Spore formation leads to the development of the gametophytic phase and completion of the life cycle. As the apogamous sporophyte has the gametic number of chromosomes, this means that unless a mechanism to double the chromosome number is incorporated at a discrete stage in the life of the sporophyte, a progressive dimunition in chromosome number will result in a succession of life cycles. The cytological analysis of apogamy in ferns has centred around the search for mechanisms to insure that chromosome number remains the same in both generations when apogamy is the preferred method of sporophyte formation. This topic is discussed in considerable detail in a book by Manton (1950).

For a number of years, the field was dominated by the observations of Farmer and Digby (1907) on the behavior of nuclei in the cells of the gametophyte of *Lastrea pseudomas* (= *Dryopteris affinis*). It was claimed that preparatory to the initiation of apogamous growth, migration of nuclei from adjacent cells in the wings or cushion of the gametophyte, and their fusion occurred. This process with all the hallmarks of a kind of information transfer was generally interpreted as 'pseudo-fertilization' and was thought to provide the cytological framework to explain the origin of a diploid apogamous sporophyte. However, the overwhelming failure of subsequent workers (Allen, 1914; Steil, 1919*a*; Döpp, 1932, 1933, 1939; Heun, 1939; Patterson, 1942; Duncan, 1943; Mehra, 1944) to confirm these observations has relegated them to a historical status and for this reason the work is no longer quoted in the literature. Although now discredited, this work had a period of great usefulness when it stimulated much research in different laboratories in the world that eventually unraveled the mystery surrounding the cytology of apogamy.

The clue to an understanding of the chromosomal changes during apogamy leading to the stability in chromosome number appears to lie in the sporangium. We saw in Chapter 2 that sporangia of the vast majority of homosporous ferns contain a group of 16 sporocytes each of which yields four spores by meiosis. As was first recognized by Allen (1914), the numerical aspect of sporogenesis is an important pointer in the analysis of the cytology of apogamy. The thrust of this work was the observation that 16 sporocytes that are initially produced in the sporangium of the apogamous sporophyte of *Aspidium* (= *Cyrtomium*) *falcatum* do not directly undergo meiosis. Rather, they fuse in pairs to form a set of eight sporocytes with a diploid number of chromosomes. These new sporocytes undergo meiosis to yield 32 haploid spores which appear regular in every respect. Somewhat similar deviations from the normal pattern of sporogenesis have been described in apogamous sporophytes of *Nephrodium hirtipes (=Dryopteris atrata)* (Steil, 1919*a*) and *Polystichum tsus-simense* (Aspleniaceae; Patterson, 1942). The changes are seen during the division of the eight-celled archesporium, resulting in a restitution nucleus with a diploid number of chromosomes in each of the eight sporocytes. The latter subsequently yield haploid spores by meiosis. These studies underscore the fundamental importance of cell and nuclear fusions in maintaining the alternation of generations in the life cycle of apogamous ferns.

Döpp's (1932, 1933, 1939) accounts of sporogenesis in *Dryopteris remota* and *Aspidium (=Dryopteris) filix-mas* have illuminated the cytological aspects of apogamy in ferns in a truly satisfactory way. While confirming the observations of Steil (1919*a*) on the incomplete nuclear and cell divisions preceding meiosis as the basic cytological modification in the apogamous sporophyte, Döpp drew attention to the presence of spores of different sizes and of sterile spores in the sporangium and suggested that irregularities in the division of sporocytes with restitution nucleus or in the failure of sporocytes to undergo the usual incomplete nuclear and cell divisions accounted for their presence. Independent investigations by Manton (1950) have confirmed Döpp's observations and from a modest beginning with *Cyrtomium falcatum*, Manton (1950) has extended the cytological description of sporogenesis to apogamous sporophytes of a number of ferns. This has brought to an end the long debate and has led to the view that something close to the variations in sporocyte ontogeny described by these investigators, designated as Döpp-Manton scheme is operating in the majority of continually apogamous ferns to maintain a constant chromosome number in both generations. According to Manton (1950), four different types of cytological abnormalities that may even occur in the different sporangia of a single plant, accompany apogamy in ferns. In one case, a full complement of 16 sporocytes are formed as in the sexual species. Although these cells prepare for meiosis, due to irregular chromosome pairing, the process is halted prematurely and aborted spores result. The most frequent situation is one in which mitotic divisions are initiated in a complement of eight sporocytes. A metaphase plate is formed in the usual way and the split chromosomes move to the equatorial region of the spindle. A subsequent meiosis yields 32 spores each having the same number of chromosomes as the original sporophyte. In a third type of sporangium, the restitution nucleus formed in one or more of the eight sporocytes is found to cleave imperfectly. Meiosis in these nuclear segments is highly irregular and yields aborted spores. Perhaps the most interesting type of sporogenesis is the one in which four homozygous, autotetraploid giant sporocytes are formed as a result of two abnormal premeiotic divisions. These cells undergo a perfectly normal meiosis to yield 16 spores, each having twice the number of chromosomes of the sporophyte from which it is derived. This probably accounts for the occurrence of many tetraploids among apogamous ferns which are predominantly diploid in the wild.

Other cytological aberrations might also account for the unreduced chromosome complement of spores produced on apogamous plants. For example, apogamous plants of *Asplenium aethiopicum* produce 16 sporocytes like their sexual counterparts. However, instead of the characteristic rearrangement of chromosomes that occurs during the heterotypic division, the bivalents fall apart as orphaned univalents. The latter slowly congregate at the equatorial region of the mitotic spindle without pairing. Consequently, the heterotypic division ends in a restitution nucleus in each of the sporocytes, which go through the homotypic division to produce 32 viable diplospores (Braithwaite, 1964). The upshot of this is that both the gametophytic and sporophytic generations have the same ploidy level which may be haploid or diploid. Fig. 13.4 illustrates the essential features of the Döpp-Manton and Braithwaite schemes as they affect sporogenesis and the life cycle in apogamous ferns. Details of sporogenesis in *Trichomanes insigne (=T. bipunctatum)* (Mehra and Singh, 1957), *T. proliferum (=Gonocormus prolifer*; Bell, 1960*a*; Braithwaite, 1969, 1975), *Asplenium macilentum (=A. auritum* var. *obtusum)* (Walker, 1966) and *A. flabellifolium* (Lovis, 1973) collected from various geographical locations in the world are remarkably similar to those occurring in *A. aethiopicum*. Although the cytological features of sporogenesis suggest the existence of an apogamous life cycle in these ferns, embryogenic studies have not been carried out to confirm this. In a related condition described in *Polystichum acrostichoides* × *P. braunii*, irregularities during sporogenesis range from abnormal meiosis of 16 sporocytes yielding aborted spores to the formation of giant spores due to the absence of divisions (Morzenti, 1962). According to Evans (1964), the inherent requirement of an apogamous form of *Polypodium dispersum* for an unreduced chromosome complement in the sporophyte is satisfied by the complete absence of meiosis in the sporocyte. In order for this scheme to be operative, sporogenesis is presumed to be accomplished by a single mitotic division of 16 sporocytes to yield 32 spores. A nuclear history similar to that observed in *P. dispersum* has been described in *Pteris cretica* var. *albo-lineata* (Heun, 1939) and in certain African Hymenophyllaceae such as *Microgonium (=Trichomanes) erosum* and *Selenodesmium guineense* (Tilquin, 1981). Walker (1979, 1985) has however questioned the claim of a complete somatic alternation of generations and suspects that the same cytological irregularity that occurs in *A. aethiopicum* is the mode in *P. dispersum*.

In a meiotic mutant of *Ceratopteris*, meiosis in the sporocytes is

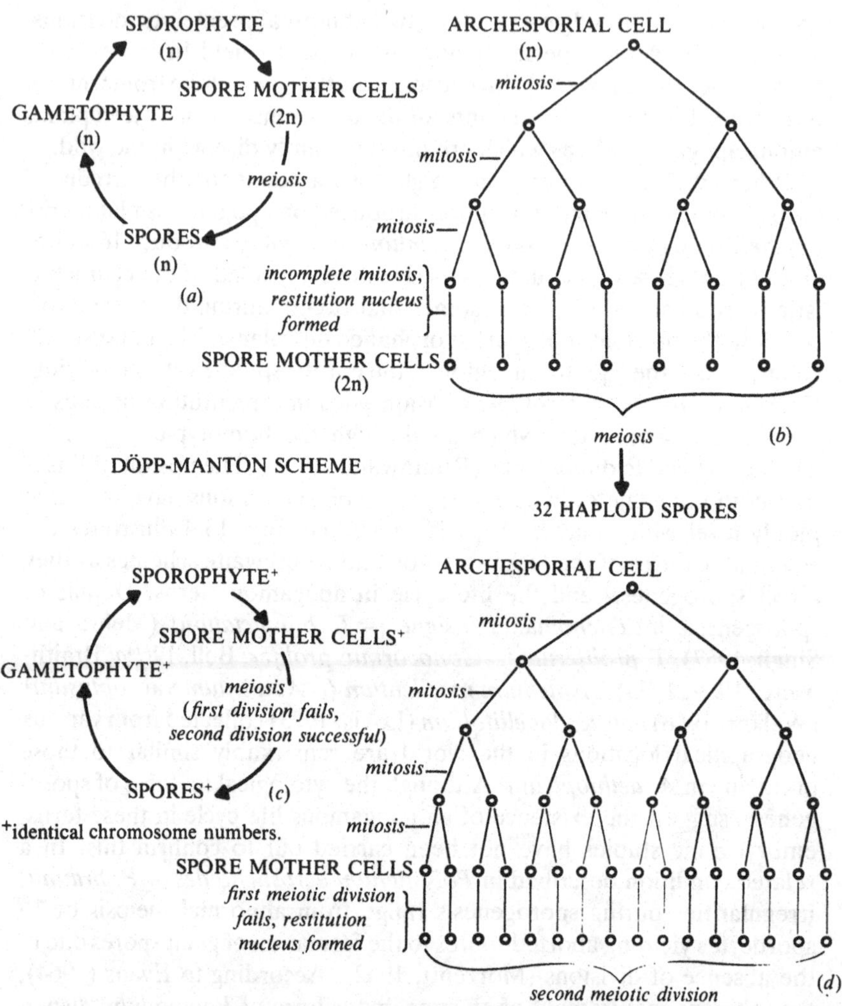

Figure 13.4. Diagrammatic representations of sporogenesis in apogamous ferns and its effect on the alternation of generations. (*a*) Life cycle according to Döpp-Manton scheme, in which the only diploid cell is the spore mother cell. (*b*) Division of the archesporial cell and sporocytes according to Döpp-Manton scheme. (*c*) Life cycle according to Braithwaite scheme, which shows the same ploidy level in both gametophytic and sporophytic generations. (*d*) Division of the archesporial cell and sporocytes according to Braithwaite scheme. (Modified from Klekowski, 1973*b*).

hindered by chromosome stickiness, leading to the formation of restitution nuclei during the heterotypic phase. This is followed by a homotypic division to yield dyads of unreduced spores. Because sporophytes develop regularly from gametophytes by apogamy, the chromosome number is stabilized throughout the gametophytic and sporophytic phases (Hickok, 1977*b*). Similar meiotic adaptations have also been noted during sporogenesis in hybrid sporophytes of other collections of *Ceratopteris* that allow production of unreduced spores (Hickok and Klekowski, 1973). In another mutant series of this fern, mitotic restitution is presumed to occur in the somatic tissues of the apogamous sporophyte at intervals during development to assure diploidy at the time of meiosis (Hickok, 1979).

The remarkable cytological peculiarities that accompany apogamy illustrate the marvelous plasticity of ferns, in which the basic mechanism of sporogenesis is modified in various ways to compensate for the absence of sexual recombination. It also underscores the importance of the sporocytes as cytological irregularities are invariably related to their number and their subsequent division.

General comments

Apogamy in ferns may be regarded as an evolutionary invention enabling sporophyte formation on the gametophyte in a nonaqueous external environment. In this context it is reasonable to assume that apogamous ferns have evolved genetic and physiological control mechanisms at crucial stages in their life cycle. From both the developmental and evolutionary perspectives, induced apogamy is of great interest, because its occurrence can be experimentally manipulated to provide insight into the conditions that control the periodic expression of gametophytic and sporophytic phases in the typical fern life cycle.

Apart from the potential to produce sex organs, the ability to regenerate apogamous sporophytes is probably the most important justification for the existence of the fern gametophyte as a free-living entity. Like other developmental phenomena, apogamy is modulated by gene activation. To explore how genetic information is used to direct sporophytic changes in the cells of the gametophyte, it is necessary to examine the individual cells where the first signs of change are detected. This is easier said than done. In any event, the implications of apogamy in the life cycle of ferns are immense and will become clear in the next chapter, when apospory is discussed.

14

Apospory – formation of gametophytes without meiosis

While the gametophyte featured in our previous discussions is born out of a reduction division of the sporocyte, in this final chapter we will pay some attention to the development of gametophytes directly from sporophytic tissues bypassing meiosis.

The reproductive strategy which results in the generation of gametophytes without meiosis or sporulation is a facultative property of sporophytic tissues of certain homosporous ferns and is known as apospory. Aposporously formed gametophytes grow without any restrictions and behave in every respect like those evolved from germinated spores except that their germ cells display the same chromosome number as the sporophyte. For this reason, apospory is considered to account for the natural polyploidization in ferns, although other factors might also be involved. Unlike apogamy, apospory is of sporadic ocçurrence in nature and only a few ferns have been shown to display this phenomenon consistently in their life cycle. However, aposporous regeneration of gametophytes is readily induced when different parts of the sporophyte are challenged by simple manipulative or cultural techniques. The early work on apospory has been reviewed by Steil (1939, 1951).

Although Steil (1939) considers apospory in a broad sense to include the formation, without meiosis, of rhizoids, gametophytic cells, sex organs or sperm on the sporophyte, the term will be used in this book in a restricted sense to include regeneration from the sporophytic tissue of a more or less complete gametophyte that perpetuates a persistent diploid genome in its cells. As a morphogenetic event, apospory is of wide scope, since virtually any organ of the sporophyte can regenerate a facsimile of the gametophyte under appropriate conditions.

Natural apospory

Abnormal outgrowths of fern sporophytes have periodically attracted the attention of early investigators and the reversion of sporangia to structures which were prothallial in appearance was one which aroused some curiosity as well. A chance observation by Druery in the 1880s that the free ends of aborted sporangia of *Athyrium filix-foemina (= A. filix-femina)* dilated into gametophyte-like outgrowths is said to mark the discovery of apospory in ferns. The subsequent appearance of archegonia and antheridia on the outgrowths gave unmistakable evidence of their gametophytic nature. Druery's observations were confirmed by no less a personage than Bower who designated the phenomenon as apospory. Following these reports, in the remaining years of the nineteenth century, Druery and Bower described fresh cases of natural or obligate apospory in *Polystichum angulare, Trichomanes alatum, T. kaulfussii (=T. trigonum), T. pyxidiferum, Dryopteris pseudo-mas (=D. affinis)*, and *Scolopendrium vulgare (=Asplenium scolopendrium)* (see Steil, 1939 for references). In later years, natural apospory has been described only in *Asplenium dimorphum* (Goebel, 1905), *Pteris sulcata (=P. biaurita)* (Steil, 1919*b*), *Osmunda javanica* (Sarbadhikari, 1936), and *Tectaria trifoliata* (Steil, 1944).

A rather special aspect of obligate apospory in *S. vulgare (= A. scolopendrium)* is the striking appearance of sporophytes that initiate apospory. They attain only a fraction of the height of normal sporophytes and bear fronds which have no sporangia. Moreover, except for the vascular system, the fronds also lack tissue differentiation. These phenotypic differences between normal and apospory-prone sporophytes have led to pioneering genetic studies on the inheritance of apospory in this species (Andersson-Kottö, 1932; Andersson-Kottö and Gairdner, 1936). The main outcome of these studies is the demonstration that aposporic trait is recessive to normal and is determined by a single Mendelian gene; a simple segregation of three normal to one aposporous phenotype occurs in the F_2 generation from heterozygotes between the two.

It should be emphasized that with one or two exceptions, obligate apospory was discovered on plants maintained in cultivation for decorative purposes. Unfortunately, no attempt was made in these studies to determine whether the phenomenon is due to genetical or physiological imbalances resulting from selection for some characters over a long period of time.

Induced apospory

Experimental approaches have been important in inducing apospory in ferns that are not obligatorily aposporous. Compared to their unpredictable occurrence in nature, consistent production of aposporous regenerates in a wide range of ferns by experimental means has shown how changes in environmental and nutritional conditions can generate replicas of the gametophyte directly from the sporophyte.

The first experimental induction of aposporous gametophytes was achieved in *Athyrium filix-foemina* (= *A. filix-femina*) by the simple expedient of keeping pieces of a mature frond in damp soil under conditions of high humidity (Stansfield, 1899). Subsequent researchers were able to induce apospory far more easily on the primary or juvenile leaves than on adult leaves. In these latter investigations, the specific conditions conducive to apospory were found to vary somewhat but they essentially involved a wounding reaction of the leaf by detaching it from the parent plant and/or subjecting it to a starvation diet. When such leaves are buried in moist loam, sand, peat, or vermiculite or floated in nutrient solutions, aposporous outgrowths appear regularly from the leaf blades or margins. Occasionally, aposporous regenerates are obtained as unexpected bonus from experiments designed for other purposes. Often, sporophytic parts of species which are naturally aposporous are found to regenerate gametophytes more readily by specific treatments than in nature. The success of these experiments implies that under unfavorable conditions, the genetic blueprint of the cells of the sporophyte is activated to produce structures that are nutritionally less demanding than the sporophyte itself to prolong the life of the plant to the genetically permissible extent.

In Table 14.1 are listed filicalean ferns in which apospory has been induced on leaves; two typical cases are illustrated in Fig. 14.1.

Table 14.1 *Members of filicalean ferns in which apospory has been induced on leaves*

Family	Subfamily	Genera and species	References[a]
Adiantaceae	Adiantoideae	*Adiantum fulvum*	Beyerle, 1932
		A. pedatum	Morel, 1963
		Anogramma chrysophylla (=*A. chaerophylla*)	Beyerle, 1932
		A. leptophylla	
		Gymnogramme (=*Pityrogramma*) *farinifera*	Woronin, 1908
		G. hookeri (=*P. calomelanos* var. *ochracea*	Beyerle, 1932
		G. (=*P.*) *calomelanos*	Hurel-Py, 1950*a*, *b*
		Hemionitis palmata	Beyerle, 1932
		Nothochlaena (=*Notholaena*) *sinuata*	Köhler, 1920
		Pellaea (=*Notholaena*) *nivea*	Goebel, 1905; Woronin, 1908
	Pteridoideae	*Pteris cretica*	Lawton, 1932; Bristow, 1962

Table 14.1 (*cont.*)

Family	Subfamily	Genera and species	References[a]
		P. tremula	Beyerle, 1932
		P. vittata	Takahashi, 1969
Aspleniaceae	Aspenioideae	*Asplenium ebenoides*	Morlang, 1967
		A. rhizophyllum (=*Camptosorus rhizophyllus*)	
		A. nidus; A. serratum	Beyerle, 1932
		A. platyneuron	Lawton, 1932
		A. ruta-muraria	Meyer, 1953
		Scolopendrium vulgare	Köhler, 1920
		(=*Asplenium scolopendrium*)	Farmer and Digby, 1907; Lawton, 1932
	Athyrioideae	*Athyrium filix-femina*	Stansfield, 1899 Farmer and Digby, 1907; Beyerle, 1932; Lawton, 1932
		Cystopteris fragilis	Lawton, 1932
		Matteuccia struthiopteris	von Aderkas, 1986
		Woodsia obtusa	Lawton, 1932
	Dryopteridoideae	*Aspidium marginale* (=*Dryopteris marginalis*)	Lawton, 1932
		Dryopteris erythrosora	Takahashi, 1962
		Lastrea paleacea (=*Dryopteris wallichiana*)	Stansfield, 1899
		L. pseudomas (=*D. affinis*)	Farmer and Digby, 1907
		Nephrodium hirtipes (=*Dryopteris atrata*)	Steil, 1919*a*
		Aspidium capense (=*Polystichum capense*)	Beyerle, 1932
		Polystichum acrostichoides	Lawton, 1932
		P. aculeatum	Stanfield, 1899
	Tectrioideae	*Tectaria maingayi*	Beyerle, 1932
Blechnaceae		*Doodia caudata*	Duncan, 1941
		Lomaria (= *Blechnum*) *capensis*	Beyerle, 1932
		Woodwardia virginica	Lawton, 1932
Cyatheaceae		*Alsophila tomentosa*	Beyerle, 1932
		Dicksonia fibrosa	Beyerle, 1932
Davalliaceae	Davallioideae	*Davallia canariensis*	Beyerle, 1932
	Oleandroideae	*Nephrolepis biserrata*	Beyerle, 1932
		N. cordifolia	Sulklyan and Mehra, 1977

Table 14.1 (*cont.*)

Family	Subfamily	Genera and species	References[a]
Dennstaedtiaceae	Dennstaedtioideae	*Dennstaedtia punctilobula*	Lawton, 1932
		Pteridium aquilinum	Farlow, 1889; Steil, 1949; Bell and Richards, 1958; Takahashi, 1962; Sheffield and Bell, 1981*a*
Hymenophyllaceae		*Trichomanes kaulfussii* (= *T. trigonum*)	Georgevitch, 1910
		T. kraussii	Woronin, 1908
Osmundaceae		*Osmunda cinnamomea*	Partanen, 1965
		O. regalis	Manton, 1932; Lawton, 1932
Parkeriaceae		*Ceratopteris pteridoides*	Loyal and Chopra, 1977
		C. thalictroides	Köhler, 1920; Beyerle,1932; Gottlieb, 1972
Polypodiaceae	Drynarioideae	*Drynaria heraclea*	Beyerle, 1932
		D. rigidula	Beyerle, 1932
	Microsorioideae	*Polypodium iridoides* (= *Microsorium punctatum*)	Steil, 1921
	Platycerioideae	*Platycerium alcicorne* (= *P. bifurcatum*)	Köhler, 1920
	Pleopeltoideae	*Microgramma vacciniifolia*	Hirsch, 1975
		Polypodium lycopodioides (= *Microgramma polypodioides*)	Bally, 1909
	Polypodioideae	*Polypodium* (= *Phlebodium*) *aureum*	Bally, 1909; Beyerle, 1932
		P. vulgare	Köhler, 1920
Schizaeaceae		*Lygodium japonicum*	Takahashi, 1969
Thelpteridaceae		*Ampelopteris prolifera*	Mehra and Sulklyan, 1969
		Phegopteris polypodioides (= *P. connectilis*)	Brown, 1918
		Thelypteris palustris	Beyerle, 1932 Bell and Richards, 1958; Sheffield and Bell, 1981*a*
Thrysopteridaceae		*Cibotium schiedei*	Beyerle, 1932

Table 14.1 (*cont.*)

[a] Except for the following, references to data in the Table are given in the reference list at the end of the book.
1 Bally, W. (1909). *Flora* **99**, 301–10.
2 Brown, E.W. (1918). *Bull. Torrey Bot. Cl.* **45**, 391–7.
3 Farlow, W.G. (1889). *Ann. Bot.* **2,** 383–5.
4 Georgevitch, P. (1910). *Jahrb. Wiss. Bot.* **48**, 155–70.
5 Gottlieb, J.E. (1972). *Bot. Gaz.* **133**, ,299–304.
6 Köhler, E. (1920). *Flora* **113**, 311–36.
7 Steil, W.N. (1921). *Bull. Torrey Bot. Cl.* **48**, 203–5.
8 Steil, W.N. (1949). *Am. Fern J.* **39**, 19–22.
9 Woronin, H.W. (1908). *Flora* **98**, 101–62.

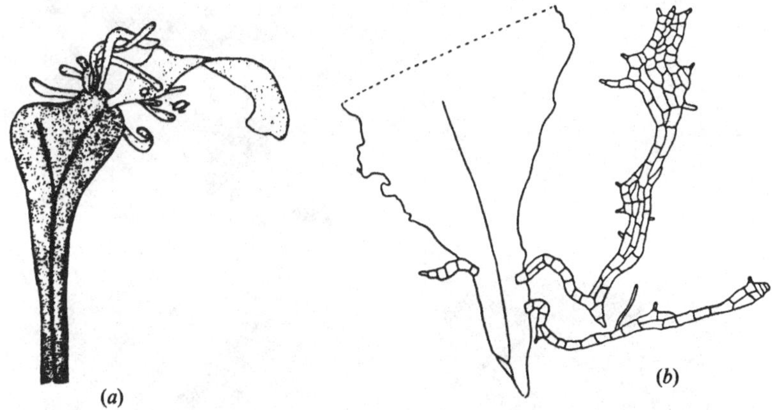

Figure 14.1. Examples of induced apospory in ferns. (*a*) A frond of *Pteris sulcata* (=*P. biaurita*) with gametophytes regenerating on both surfaces; a, antheridium. (From Steil, 1919*b*). (*b*) Part of a frond of *Woodsia obtusa* with filamentous and planar gametophytes. (From Lawton, 1932.)

The ability of detached leaves to form gametophytic outgrowths is a remarkable phenomenon. Other parts of the sporophyte are also capable of this morphological feat (Fig. 14.2), but none have been found to match the leaf blade in its versatility. Although sporadic origin of gametophytes has been recorded on the roots of certain ferns (Lawton, 1932; Hurel-Py, 1950*a, b*; Durand-Rivières, 1960; Takahashi, 1962; Mehra and Palta, 1971; Loyal and Chopra, 1977; Sulklyan and Mehra, 1977; Cheema, 1979), the conditions under which they frequently occur have been carefully followed in *Pteridium aquilinum* (Partanen and Partanen, 1963; Munroe and Sussex, 1969). Generally, aposporous outgrowths appear only in excised roots grown under conditions that are not growth-supporting, such as an exhausted or stale medium with a low carbohydrate status. Intact roots of young sporophytes transplanted to a fresh medium lacking carbohydrates also regenerate gametophytes profusely.

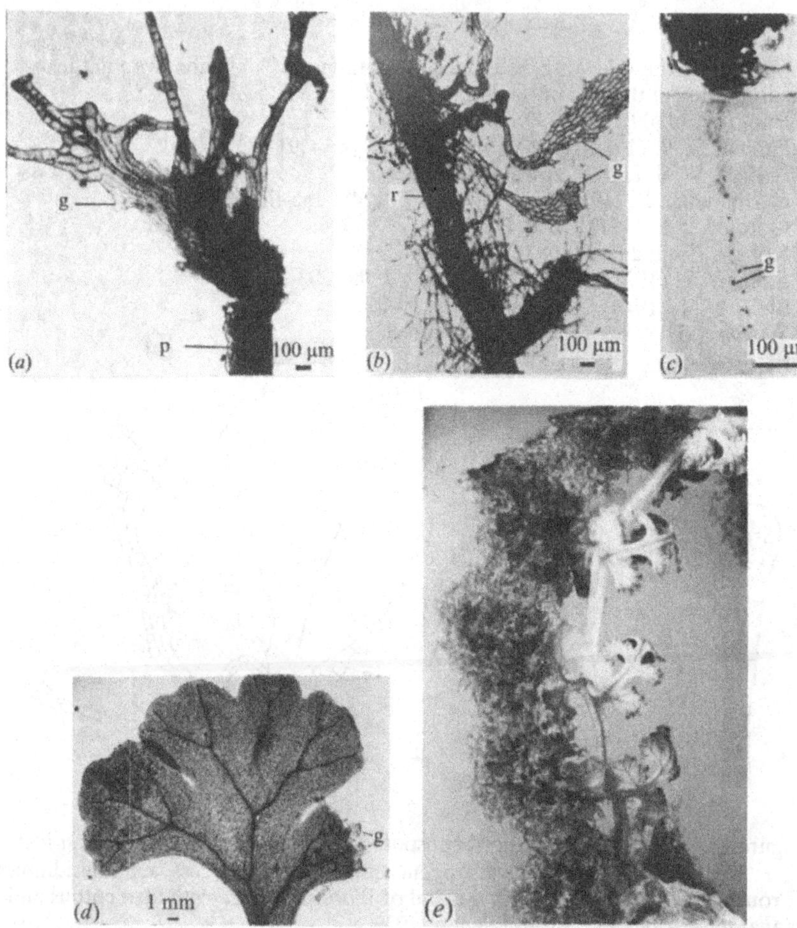

Figure 14.2. Aposporous origin of gametophytes from different parts of the sporophyte. (*a*) An excised petiole segment of *Nephrolepis cordifolia* showing formation of gametophytes. (*b*) An excised root of *N. cordifolia* showing formation of gametophytes. (From Sulklyan and Mehra, 1977). (*c*) A young sporophyte of *Pteridium aquilinum* growing on a solidified medium. Roots extending into agar have produced numerous gametophytes. (From Munroe and Sussex, 1969). (*d*) Early stage in the formation of aposporous gametophytes from a frond of *P. aquilinum* (Photograph supplied by Dr E. Sheffield). (*e*) A frond of *P. aquilinum* attached to a plant showing aposporous regenerates on pinnules on the side which was in contact with the medium. (From Sheffield and Bell 1981*b*; photograph supplied by Dr E. Sheffield). g, aposporous gametophyte; p, petiole; r, root. Scale bar in (*d*) applies to (*e*) also.

Gametophyte initiation on roots is apparently a light-dependent process and is promoted by the presence of kinetin in the medium. Aposporous outgrowths have also been induced on sporophytic hairs (Steil, 1919*b*, 1944; Takahashi, 1962), leaf stipes (Beyerle, 1932; Takahashi, 1962, 1969; Ward, 1963; Mehra and Sulklyan, 1969), runner segments (Sulklyan and Mehra, 1977), and shoot apical meristems (von Aderkas, 1986) of diverse ferns.

Apospory in tissue culture

Tissue culture has been an invaluable tool in the study of apospory especially in plants that do not directly or readily regenerate gametophytes by treatments which are successful in others. In these studies, by supplying selected nutrients to sporophytic organs or tissues grown in aseptic culture, attempts have been made to identify the effects of specific substances on the recrudescence of gametophytic growth. Bristow (1962) set the stage for these investigations by the demonstration that leaves of *Pteris cretica* cultured in a medium containing 1.0 mg/1 2,4-D and 2% sucrose form a compact callus. Under these conditions of culture, the callus is not committed to any particular pathway of differentiation but freely forms sporophytes when it is transferred to an auxin-free medium supplemented with 1.0% sucrose or regenerates gametophytes when sucrose is omitted from the medium (Fig. 14.3). The evidence from this work is consistent with the view that nutrient-limiting conditions generated by the absence of a carbon energy source in the medium play a determinative role in reprogramming the undifferentiated sporophytic callus in the gametophytic pathway.

Other studies (Kato, 1965*a*; Mehra, 1972; Sulklyan and Mehra, 1977; Kshirsagar and Mehta, 1978) have also established that under a variety of cultural conditions, a low level of sucrose or its absence thereof in the medium enables callus cells of sporophytic origin to regenerate stable gametophytic structures. The root callus of *Cyclosorus dentatus (= Christella dentata)* (Mehra and Palta, 1971) offers a striking example of modulation of single cells in the aposporous pathway by withholding sucrose from the medium. The callus obtained by culturing excised roots in a medium containing 2,4-D, yields by gentle shaking a fine suspension of single cells and cell clumps. Growth of single cells plated on a medium lacking sucrose is largely initiated by an asymmetric division as in a germinating spore, separating the rhizoid and protonema initials. While the rhizoid initial does not divide any further, the protonema initial faithfully recapitulates the ontogenetic sequence of a germinated spore to form a filamentous protonema, a planar gametophyte and finally the phenotypic elaboration of a mature prothallus bearing sex organs (Fig. 14.4). These observations confirm and give precision to the results on the

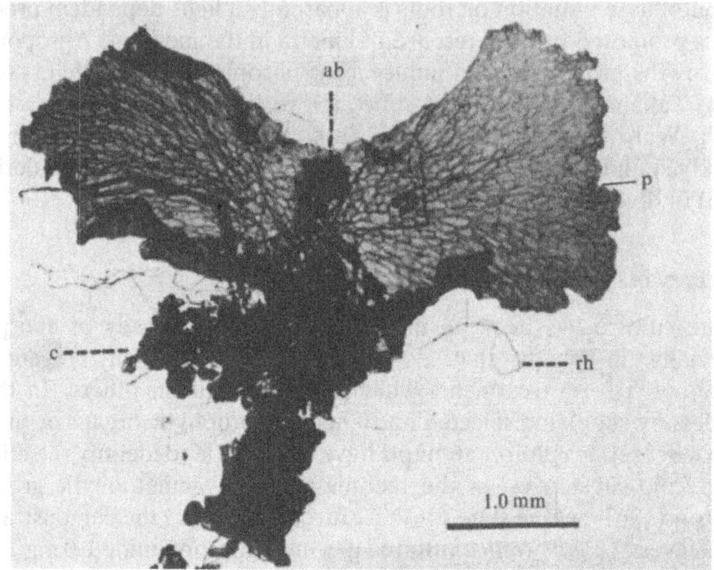

Figure 14.3. Dedifferentiation of sporophytic callus into gametophytes. A heart-shaped gametophyte formed from a leaf callus of *Pteris cretica*. ab, apogamous bud; c, callus; p, prothallus; rh, rhizoid (From Bristow, 1962).

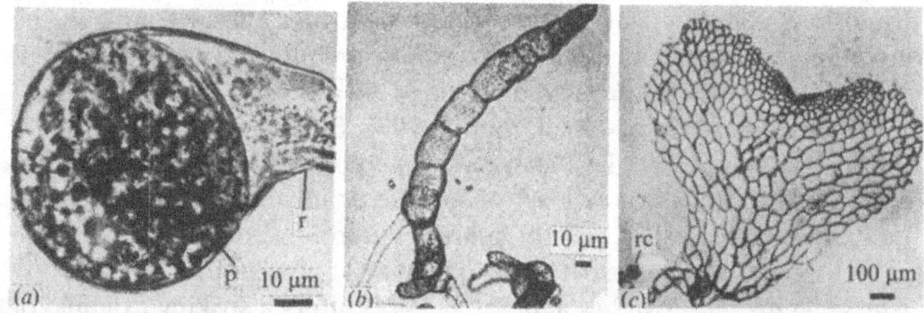

Figure 14.4. Dedifferentiation of single cells of the root callus of *Cyclosorus dentatus* (= *Christella dentata*) into gametophytes. (*a*) First division of the cell to form rhizoid (r) and protonema (p) initials. (*b*) Formation of the filamentous gametophyte. (*c*) A heart-shaped gametophyte; rc, the original root callus cells. (From Mehra and Palta, 1971.)

effect of nutrient-limiting conditions on the induction of apospory on sporophytic tissues and organs of other ferns.

A final contribution of the tissue culture approach is the induction of apospory in a non-filicalean fern. Whittier (1978) showed that when leaves of haploid *Botrychium alatum* are cultured in a low sucrose-containing medium, massive, white, fleshy, aposporous gametophytes, identical to the normal gametophytes of this species, arise from the petiole and leaf blades during a three to six month culture period. Results such as these suggest that the absence of reliable reports of apospory in other non-filicalean ferns is due to the failure to formulate suitable culture conditions to foster this morphogenetic event.

Origin and development of aposporous gametophytes

The formation of an aposporous regenerate is an extended process that begins in one or more cells of a donor sporophyte organ. Although an epidermal origin of apospory has been suggested in several plants (Beyerle, 1932; Lawton, 1932, 1936; Sardabhikari, 1936; Morlang, 1967; Takahashi, 1969; Mehra and Sulklyan, 1969), only in a few cases has the ontogenetic sequence of events been reconstructed. As shown in *Microgramma vacciniifolia* (Polypodiaceae; Hirsch, 1976), apospory is heralded by the initiation of autoradiographically detectable DNA synthesis in isolated cells of the leaf epidermis which have long ceased to divide. Subsequently, a filamentous protonema is formed to the exterior of the leaf by the division of the epidermal cell. The filament is transformed into a planar gametophyte when the plane of division in the terminal or subterminal cell changes; in other cases, gametophytic outgrowths are found to be planar from the beginning. This latter situation is also true of *Nephrodium pseudo-mas (=Dryopteris affinis)* (Digby, 1905) and *Osmunda javanica* (Sarbadhikari, 1936). *Pteridium aquilinum* is an example in which aposporous growths are elicited from the epidermal as well as from the subepidermal cells of expanded leaves by various routes (Takahashi, 1962). When aposporous regenerates arise from the margins of cultured juvenile leaves of *P. aquilinum*, their origin is traceable to single cells or to a group of as many as ten or more adjacent cells. Depending upon the plane of division of the single progenitor cells, either filamentous or three-dimensional outgrowths are formed initially (Sheffield, 1984; Sheffield and Bell, 1981*a*). In contrast to the single-celled origin of aposporous gametophytes in filicalean ferns, origin of tuberous gametophytes from cultured leaves of *B. alatum* involves many cells of the epidermis and subepidermis (Whittier, 1978).

During the formation of aposporous gametophytes on cultured roots of *P. aquilinum*, isolated cells in the outer layers of the root apex become intensely green in color, followed by bulging of the walls to form a protuberance. The protrusion which is delimited by a wall, sequentially forms a filament and a planar gametophyte by characteristic divisions (Munroe and Sussex, 1969). Comparative ultrastructural studies have shown that cells of normal and apospory-prone roots of *P. aquilinum* have essentially the same cytology. While most cells of the root degenerate, a few surviving ones that divide in the aposporous pathway display a striking complexity in structure, size and arrangement of plastids, mitochondria, ER, and ribosomes (Munroe and Bell, 1970). If the ultrastructural changes of cells committed to apospory are precisely identified from those of cells that follow a sporophytic program, we could gain some greatly needed insight into the cellular basis of apospory.

Nuclear cytology of apospory

Digby (1905) first studied the nuclear cytology of apospory in *Nephrodium pseudo-mas* (= *Dryopteris affinis*) and showed that meiosis is abrogated and consequently there is no reduction in the chromosome number during transition of the sporophyte to the gametophyte. The regeneration of diploid gametophytes by apospory has been subsequently confirmed as a secure generalization by careful cytological studies in *Osmunda regalis* (Manton, 1932; Lawton, 1936), *O. javanica* (Sarbadhikari, 1939), *Cystopteris fragilis* (Lawton, 1936), *Pteridium aquilinum* (Takahashi, 1962), and *Asplenium platyneuron* (Morlang, 1967) and by Feulgen microspectrophotometry of DNA contents of nuclei of normal and aposporously generated gametophytes of *Thelypteris palustris* (Bell and Richards, 1958). The results of these studies indicate that there is a certain orderliness in the broad cytological aspects of apospory in the different ferns.

One unexpected outcome of the cytological analysis of apospory is the possibility of obtaining polyploid series of gametophytes and sporophytes, if apospory is regularly followed by normal sexual recombination in successive generations. Thus, Manton (1932, 1950) showed that tetraploids found in a population of *O. regalis* sporophytes arose from aposporous gametophytes by fusion between diploid gametes, while triploids had their origin in chance fertilization of a diploid egg by a haploid sperm. The high incidence of apospory noted in certain species of *Asplenium* has led to the suggestion that fusion between diploid gametes might account for the production of fertile allotetraploids in the genus (Morlang, 1967). Tetraploid sporophytes have been regenerated from aposporous gametophytes of *Aspidium marginale* (=*Dryopteris marginalis*), *Woodwardia virginica* (Lawton, 1932), *Cystopteris fragilis* (Lawton, 1936), and *Pteridium aquilinum* (Takahashi, 1962). Although both triploid and tetraploid sporophytes of many aposporous ferns regenerate gametophytes with the same chromosome complement, the inability of the latter to produce sex organs has blocked attempts at repeated induction of apospory in successive generations. Production of octoploid sporophytes by fertilization between gametes from tetraploid gametophytes and subsequent aposporous regeneration of octoploid gametophytes has been reported in *Todea barbara, Onoclea sensibilis* (Wetmore *et al.*, 1963), and *Pteris vittata* (Mehra, 1972; Palta and Mehra, 1986).

Factors influencing apospory

Questions concerning the factors regulating apospory are of interest, since certain physical, environmental, and nutritional changes have been invoked from time to time as favoring this event. Under propitious conditions some cells in the sporophyte presumably act as inducers and others as responders to give rise to structures ranging from a tubular rhizoid to a planar gametophyte and everything in between. What are the conditions necessary for establishing this dialogue between cells? Although an answer to this question has been sought periodically since the discovery of apospory, no set of conditions which regularly and reproducibly induce apospory has been recognized. In the following paragraphs, the importance of four sets of factors, namely, age of the donor sporophyte, coordination between cells, senescence of neighboring cells, and unfavorable nutrient conditions, will be considered.

It is a general experience that apospory as a developmental episode peaks in the early formed fronds of a sporophyte and diminishes in later formed ones (Takahashi, 1962; Morel, 1963; Mehra and Sulklyan, 1969; Cheema, 1979; Sheffield and Bell, 1981*a*; von Aderkas, 1986). This occurs despite the fact that both early and later formed fronds are histologically similar and possess the same structural configuration. The pattern of decreasing developmental expression of progressively older fronds of a donor sporophyte suggests that the sporophytic form is irreversibly established in the frond as it attains physiological maturity.

A developmental approach assumes that integration of tissues within an organ is due to coordination between the individual cells, failure of which results in the expression of independent programs by single cells or groups of cells. In a fern frond, breakdown of coordination between cells may be due to any number of factors, but most commonly is due to the natural death of cells. This might free the remaining cells to regenerate gametophytic structures (Takahashi, 1962; Wetmore *et al.*, 1963). This assumption is however insufficient to explain the origin of aposporous regenerates from cells which are not physiologically isolated from the rest of the sporophytic organ.

It has been emphasized (for example, Munroe and Sussex, 1969; Munroe and Bell, 1970) that aposporous gametophytes have their origin in viable cells of an otherwise senescent organ. Senescence is an autolytic process resulting in cell destruction by enzyme action. Although it is conceivable that products of cell deterioration and senescence may activate the capacity for division and growth of the surviving cells, there is minimal evidence to support this view. Moreover, the validity of this assumption is also questionable in the light of other studies. According to Hirsch (1975), when leaves of *Microgramma vacciniifolia* are cultured in a medium enriched with high levels of sucrose, they undergo necrosis within a week after culture; however, those grown in a medium containing reduced levels of sucrose regenerate aposporous outgrowths without revealing any excessive symptoms of senescence. From this work, accelerated senescence appears to inhibit apospory rather than cause it. Along the same lines, Sheffield and Bell (1981*a*) have shown that leaves of *Pteridium aquilinum* produce aposporous gametophytes as early as three days after culture before any signs of senescence set in.

As described earlier, tissue culture experiments have raised the possibility that a trigger for apospory is starvation, that is, a depletion of carbon energy source that halts growth. Generally, a medium containing moderate amounts of sucrose as the carbon energy source is known to favor differentiation of sporophytic structures, while in the absence of sucrose, gametophytic growth predominates. There is also mounting evidence to indicate that some kind of nutritional gradients in the responding organ may regulate apospory *in vivo*. For example, aposporous gametophytes are formed on intact leaves of *Ceratopteris pteridoides* only after the plant is starved for a prolonged period resulting in leaf senescence (Cheema, 1979). Prolific growth of aposporous gametophytes occurs on intact leaves of *P. aquilinum* whose active vascular flow has ceased by occlusion of its tracheary elements (Sheffield and Bell, 1981*b*; Sheffield, Bell and Laird, 1982). A major aspect of the analysis of how the stoppage of traffic from the growing sporophyte activates the potential for apospory in an attached leaf must concern itself with a regulatory role for certain substances as suppressants of morphogenetic expression.

This brief summary indicates that approaches thus far made have not contributed substantially to our understanding of the factors that control apospory in ferns. Apparently, age of the donor sporophyte, senescence of cells, and nutritional deprivation affect the growth of regenerates, but whether these factors singly or in combination supply the trigger for apospory is another matter. It is clear that apart from the usual nutrients that nourish an organ of the sporophyte, there is an array of regulatory substances which account for the diversity of its growth patterns. At the molecular level, the transformation of the sporophytic cell into a gametophytic entity requires cell-specific programs of gene transcription and translation that account for morphogenetic expressions as one life style is replaced by another. With these thoughts in mind, Sheffield and Bell (1981*a*) have proposed a model for apospory in *P. aquilinum*. According to this model, regulatory substances from the sporophyte transmitted through the vascular system maintain genes for sporophytic growth in the frond in a state of readiness, at the same time as genetic information for gametophytic growth is silenced. Cessation of vascular flow provides the cells with reprogramming cues necessary to turn on gene systems for gametophytic growth. The failure of successively older fronds on a sporophyte to regenerate aposporously has been attributed to the progressive dilution of gametophytic proteins, so that derepression of genes to code for these proteins becomes increasingly difficult; more probably, it is due to decay of the genes with the age of the fronds. Attractive as this model seems to be, a test of its validity is not easy.

Apogamy, apospory and alternation of generations

As reproductive strategies, apogamy and apospory serve to circumvent the normal life cycle of ferns involving alternation of generations between an unreduced sporophyte and a reduced gametophyte. Consequently, they have an important bearing on the life cycle because they interfere with syngamy and meiosis and thereby provide opportunities for changes in chromosome numbers.

Since apogamous sporophytes are formed vegetatively from gametophytes without fertilization, they have a single set of chromosomes which has to be doubled to cause pairing during meiosis. The cytological change in the apogamous sporophyte that results in the doubling of chromosome number occurs during sporogenesis. This compensating event obviously accounts for a subsequent normal meiosis and the establishment of an alternation of generation between a stable sporophyte and a stable gametophyte. Thus, the basic cytological change in the apogamous sporophyte is geared to the production of spores with an unreduced chromosome number.

As already emphasized, following an aposporous episode, both gametophyte and sporophyte end up with the same chromosome number in their cells. No compensating nuclear event ever takes place in the basically diploid gametophyte which thus alternates with a diploid sporophyte.

In the typical fern life cycle, illustrative of a heteromorphic alternation of generations, the sporophyte and gametophyte are morphologically so dissimilar that they sometimes seem like two different organisms. This occurs irrespective of whether they carry the same or different chromosome number. From this it follows that morphological expressions of the gametophyte and sporophyte are not due to the number of gene copies present in the nucleus. The very occurrence of apogamy and apospory constitutes additional evidence favoring this line of argument.

If differences in chromosome number or in the quantity of DNA do not explain the basis for the two forms of growth in the fern life cycle, what other factors might contribute to the divergence in morphology of the gametophyte and sporophyte? One is the distinctive environments in which the gametophytic (spore) and sporophytic (zygote) generations are born and nurtured. As is quite evident by now, in nature and in the laboratory, the spore germinates to form the gametophyte without any restraint, while the zygote normally develops within the confines of the archegonium against a turgor generated by the surrounding jacket cells. To analyze the role of the archegonial environment on the beginnings of gametophyte development, spores of *P. aquilinum* were implanted in the archegonial cushion of the gametophyte, but the results were ambiguous in terms of the nature of the outgrowths that appeared and their origin (Bell, 1959). On the other hand, surgical removal of undivided zygotes from archegonia and their culture in an enriched nutrient medium led to the development of prothalloid structures which showed no tendency to organize as sporophytes (DeMaggio and Wetmore, 1961). The similarity between spore germination and zygote development when both occur in an unrestrained environment is remarkable in view of their difference in ploidy levels. This suggests a morphogenetic role for the physical and physiological environment of the archegonium in molding the zygote into an embryo.

Another view has attributed gene activation in the egg during oogenesis and in the spore during sporogenesis to account for the distinct morphology of the sporophyte and gametophyte. This view derives its strength from electron microscopic examination of oogenesis which has revealed profound changes in the cytoplasmic organization of the differentiating egg (see Chapter 12). Presumably, activation of different sets of genes accompanied by corresponding changes in the egg and spore accounts for apospory and apogamy, respectively.

In summary, current speculations on the basis for differences between the sporophyte and gametophyte in ferns center on environmental influences and gene activation. Evidence presently available falls short of a general proof of either proposition. Apart from causing transient perturbations in the chromosome number, a switch in the life cycle involving apogamy or apospory does not affect the basic heteromorphic alternation of generations.

General comments

Despite the significance of apospory to the concept of alternation of generations in the fern life cycle, the information presently available on apospory is limited and scattered. As a consequence, it is only realistic to recognize that our understanding of the physiological and developmental aspects of aposporous regeneration remains hazy. So, to some extent, this chapter is a thinly-veiled plea for more investigations on apospory.

To gain further insight into apospory from physiological and developmental perspectives, we have to learn more about the changes that occur in the target cells of the sporophyte during their dedifferentiation. An in-depth ultrastructural description of the events of dedifferentiation in the cells of a great number of aposporus systems will be valuable in this context. Also needed are data on the hormonal and metabolic changes in the donor sporophyte organ during the early stages of aposporous growth. Thus equipped, we will be in a position to answer the central question concerned with the mechanism of the origin of aposporous structures from sporophytic organs.

References

Agnew, N., McCabe, A., and Smith, D. L. (1984). Photocontrol of spore germination in *Polypodium vulgare* L. *New Phytol.* **96,** 167–78.

Albaum, H. G. (1938*a*). Normal growth, regeneration, and adventitious outgrowth formation in fern prothallia. *Am. J. Bot.* **25,** 37–44.

Albaum, H. G. (1938*b*). Inhibitions due to growth hormones in fern prothallia and sporophytes. *Am. J. Bot.* **25,** 124–33.

Allen, R. F. (1914). Studies in spermatogenesis and apogamy in ferns. *Trans. Wisconsin Acad. Arts, Sci. Lett.* **17,** 1–56.

Andersson-Kottö, I. (1929). A genetical investigation in *Scolopendrium vulgare. Hereditas* **12,** 109–78.

Andersson-Kottö, I. (1932). Observations on the inheritance of apospory and alternation of generations. *Svensk. Bot. Tidsk.* **26,** 99–106.

Andersson-Kottö, I., and Gairdner, A. E. (1936). The inheritance of apospory in *Scolopendrium vulgare. J. Genet.* **32,** 189–228.

Atkinson, L. R. (1960). A new germination pattern for the Hymenophyllaceae. *Phytomorphology* **10,** 26–36.

Atkinson, L. R. (1962). The Schizaeaceae: The gametophyte of *Anemia. Phytomorphology* **12,** 264–88.

Atkinson, L. R. (1975). The gametophyte of five old world Thelypteroid ferns. *Phytomorphology* **25,** 38–54.

Atkinson, L. R., and Stokey, A. G. (1964). Comparative morphology of the gametophyte of homosporous ferns. *Phytomorphology* **14,** 51–70.

Bähre, R. (1975*a*). Zur Regulation des Protonemawachstums von *Athyrium filix-femina* (L.) Roth. I. CO_2-Äthylen-Antagonismus. *Z. Pflanzenphysiol.* **76,** 243–7.

Bähre, R. (1975*b*). Zur Regulation des Protonemawachstums von *Athyrium filix-femina* (L.) Roth. II. Acetylcholin als Äthylenantagonist. *Z. Pflanzenphysiol.* **76,** 248–51.

Bähre, R. (1976). Zur Regulation des Protonemawachstums von *Athyrium filix-femina* (L.) Roth. III. Wirkung von Auxin (IAA) und Antiauxin (PCIB) unter verschiedenen Lichtbedingungen. *Z. Pflanzenphysiol.* **77,** 323–35.

Bähre, R. (1977). Zur Regulation des Protonemawachstums von *Athyrium filix-femina* (L.) Roth. IV. Wirkung cholinerger Substanzen in Gegenwart eines Antiauxins (PCIB). *Z. Pflanzenphysiol.* **81,** 278–82.

Baron-Ferrand, M., Kuligowski, J., Chenou, E., and Tourte, Y. (1984). Effets de la colchicine sur l'oogenèse et sur les processus de fécondation chez une plante. *Ann. Sci. Nat. Bot.* (Ser. 13) **6,** 81–91.

Bassel, A. R., Kuehnert, C. C., and Miller, J. H. (1981). Nuclear migration and asymmetric cell division in *Onoclea sensibilis* spores: An ultrastructural and cytochemical study. *Am. J. Bot.* **68,** 350–60.

Bassel, A. R., and Miller, J. H. (1982). The effects of centrifugation on asymmetric cell division and differentiation of fern spores. *Ann. Bot.* **50,** 185–98.

Beisvåg, T. (1970). An electron microscopic investigation of the young gametophyte of the fern *Blechnum spicant* (L.) Roth. *Grana* **10,** 121–35.

Bell, P. R. (1958). Variations in the germination-rate and development of fern spores in culture. *Ann. Bot.* **22,** 503–11.

Bell, P. R. (1959). The experimental investigation of the pteridophyte life cycle. *J. Linn. Soc. (Bot).* **56,** 188–203.

Bell, P. R. (1960*a*). The morphology and cytology of sporogenesis of *Trichomanes proliferum* Bl. *New Phytol.* **59,** 53–9.

Bell, P. R. (1960*b*). Interaction of nucleus and cytoplasm during oogenesis in *Pteridium aquilinum* (L.) Kuhn. *Proc. Roy. Soc. Lond.* **153B,** 421–32.

Bell, P. R. (1969). The cytoplasmic vesicles of the female reproductive cells of *Pteridium aquilinum. Z. Zellforsch.* **96,** 49–62.

Bell, P. R. (1974*a*). Nuclear sheets in the egg of a fern, *Dryopteris filix-mas. J. Cell Sci.* **14,** 69–83.

Bell, P. R. (1974*b*). Microtubules in relation to flagellogenesis in *Pteridium* spermatozoids. *J. Cell Sci.* **15,** 99–111.

Bell, P. R. (1974*c*). The origin of the multilayered structure in the spermatozoid of *Pteridium aquilinum. Cytobiologie* **8,** 203–12.

Bell, P. R. (1975). Observations on the male nucleus during fertilization in the fern *Pteridium aquilinum. J. Cell Sci.* **17,** 141–53.

Bell, P. R. (1978). A microtubule-nuclear envelope complex in the spermatozoid of *Pteridium. J. Cell Sci.* **29,** 189–95.

Bell, P. R. (1979*a*). Demonstration of succinic dehydrogenase in mitochondria of fern egg cells at electron microscope level. *Histochemistry* **62:** 85–91.

Bell, P. R. (1979*b*). Gametogenesis and fertilization in ferns. In: *The Experimental Biology of Ferns,* ed. A. F. Dyer, pp. 471–503. London: Academic Press.

Bell, P. R. (1980). Nucleocytoplasmic interaction during maturation of the egg of the fern *Histiopteris incisa* (Thunb.) J. Smith. *Ann. Bot.* **45,** 475–81.

Bell, P. R. (1981). Megasporogenesis in a heterosporous fern: Features of the organelles in meiotic cells and young megaspores. *J. Cell Sci.* **51,** 109–19.

Bell, P. R. (1982). Tubular elements in plastids in the female gamete of a fern, *Pteris ensiformis. Eur. J. Cell Biol.* **26,** 303–5.

Bell, P. R. (1983). Nuclear bodies in the maturing egg cell of a fern, *Pteridium aquilinum. J. Cell Sci.* **60,** 109–16.

Bell, P. R. (1985). Maturation of the megaspore in *Marsilea vestita. Proc. Roy. Soc. Lond.* **233B,** 485–94.

Bell, P. R. (1986). Features of egg cells of living representatives of ancient families of ferns. *Ann. Bot.* **57,** 613–21.

Bell, P. R., and Duckett, J. G. (1976). Gametogenesis and fertilization in *Pteridium. Bot. J. Linn. Soc.* **73,** 47–78.

Bell, P. R., Duckett, J. G., and Myles, D. G. (1971). The occurrence of a multilayered structure in the motile spermatozoids of *Pteridium aquilinum. J. Ultrastr. Res.* **34,** 181–9.

Bell, P. R., and Mühlethaler, K. (1962*a*). A membrane peculiar to the egg in the gametophyte of *Pteridium aquilinum. Nature (London)* **195,** 198.

Bell, P. R., and Mühlethaler, K. (1962*b*). The fine structure of the cells taking part in oogenesis in *Pteridium aquilinum* (L.) Kuhn. *J. Ultrastr. Res.* **7,** 452–66.

Bell, P. R., and Mühlethaler, K. (1964*a*). The degeneration and reappearance of mitochondria in the egg cells of a plant. *J. Cell Biol.* **20,** 235–48.

Bell, P. R., and Mühlethaler, K. (1964*b*). Evidence for the presence of deoxyribonucleic acid in the organelles of the egg cells of *Pteridium aquilinum*. *J. Mol. Biol.* **8,** 853–62.

Bell, P. R., and Pennell, R. I. (1987). The origin and composition of nucleolus-like inclusions in the cytoplasm of fern egg cells. *J. Cell Sci.* **87,** 283–90.

Bell, P. R., and Richards, B. M. (1958). Induced apospory in polypodiaceous ferns. *Nature (London)* **182,** 1748–9.

Bell, P. R., and Zafar, A. H. (1961). Changes in the level of the protein nitrogen during growth of the gametophyte and the initiation of the sporophyte of *Dryopteris borreri* Newm. *Ann. Bot.* **25,** 531–46.

Bergfeld, R. (1963*a*). Die Wirkung von hellroter und blauer Strahlung auf die Chloroplastenbildung. *Z. Naturforsch.* **18b,** 328–31.

Bergfeld, R. (1963*b*). Die Beeinflussung der Zellkerne in den Vorkeimen von *Dryopteris filix-mas* durch rote und blaue Strahlung. *Z. Naturforsch.* **18b,** 557–62.

Bergfeld, R. (1964*a*). Der Einfluss roter und blauer Strahlung auf die Ausbildung der Chloroplasten bei gehemmter Proteinsynthese. *Z. Naturforsch.* **19b,** 1076–8.

Bergfeld, R. (1964*b*). Die lichtabhängige Ausbildung der Zellkerne in den Vorkeimen von *Dryopteris filix-mas* bei Hemmung der Proteinsynthese. *Z. Naturforsch.* **19b,** 1142–6.

Bergfeld, R. (1965). Zellteilung und Morphogenese der Vorkeime von *Dryopteris filix-mas* (L.) Schott in hellroter und blauer Strahlung bei Hemmung der Proteinsynthese. *Z. Naturforsch.* **20b,** 591–4.

Bergfeld, R. (1967). Kern- und Nucleolusausbildung in den Gametophytenzellen von *Dryopteris filix-mas* (L.) Schott bei Umsteuerung der Morphogenese. *Z. Naturforsch.* **22b,** 972–6.

Bergfeld, R. (1968). Chloroplastenausbildung und Morphogenese der Gametophyten von *Dryopteris filix-mas* (L.) Schott nach Applikation von Chloramphenicol und Actidion (Cycloheximid). *Planta* **81,** 274–9.

Bergfeld, R. (1970). Feinstruktur der Chloroplasten in den Gametophytenzellen von *Dryopteris filix-mas* (L.) Schott nach Einwirkung hellroter und blauer Strahlung. *Z. Pflanzenphysiol.* **63,** 55–64.

Beyerle, R. (1932). Untersuchungen über die Regeneration von Farnprimärblättern. *Planta* **16,** 622–65.

Bhardwaja, T. N., and Sen, S. (1966). Effect of temperature on the viability of spores of the water fern *Marsilea*. *Sci. & Cult.* **32,** 47–8.

Bierhorst, D. W. (1958). Observations on the gametophytes of *Botrychium virginianum* and *B. dissectum*. *Am. J. Bot.* **45,** 1–9.

Bierhorst, D. W. (1966). The fleshy, cylindrical, subterranean gametophyte of *Schizaea melanesica*. *Am. J. Bot.* **53,** 123–33.

Bierhorst, D. W. (1967). The gametophyte of *Schizaea dichotoma*. *Am. J. Bot.* **54,** 538–49.

Bierhorst, D. W. (1968*a*). Observations on *Schizaea* and *Actinostachys* spp., including *A. oligostachys,* sp. nov. *Am. J. Bot.* **55,** 87–108.

Bierhorst, D. W. (1968*b*). On the Stromatopteridaceae (Fam. nov.) and on the Psilotaceae. *Phytomorphology* **18,** 232–68.

Bierhorst, D. W. (1971). *Morphology of Vascular Plants*. New York: The MacMillan Co.

Bierhorst, D. W. (1975*a*). Gametophytes and embryos of *Actinostachys pennula, A. wagneri*, and *Schizaea elegans,* with notes on other species. *Am. J. Bot.* **62,** 319–35.

Bierhorst, D. W. (1975*b*). The apogamous life cycle of *Trichomanes pinnatum* –

A confirmation of Klekowski's predictions on homoeologous pairing. *Am. J. Bot.* **62,** 448–56.

Bilderback, D. E. (1978*a*). The development of the sporocarp of *Marsilea vestita. Am. J. Bot.* **65,** 629–37.

Bilderback, D. E. (1978*b*). The ultrastructure of the developing sorophore of *Marsilea vestita. Am. J. Bot.* **65,** 638–45.

Bilderback, D. E., Bilderback, D. E., Jahn, T. L., and Fonseca, J. R. (1974). The locomotor behavior of *Lygodium* and *Marsilea* sperm. *Am. J. Bot.* **61,** 888–90.

Bloom, W. W. (1955). Comparative viability of sporocarps of *Marsilea quadrifolia* L. in relation to age. *Trans. Illinois State Acad. Sci.* **47,** 72–6.

Bloom, W. W. (1962). Some factors influencing rhizoid formation in female gametophytes of the Marsileaceae. *Proc. Indiana Acad. Sci.* **72,** 118–9.,

Bloom, W. W., and Nichols, K. E. (1972*a*). Rhizoid formation in megagametophytes of *Marsilea* in response to growth substances. *Am. Fern J.* **62,** 24–46.

Bloom, W. W., and Nichols, K. E. (1972*b*). Rhizoid initiation in relation to gravitation presentation time in *Marsilea* megagametophytes. *Proc. Indiana Acad. Sci.* **82,** 109–12.

Bonnet, A. L. M. (1957). Contribution à l'étude des Hydroptéridées. III. Recherches sur *Azolla filiculoides* Lamk. *Rev. Cytol, Biol. Veg.* **28,** 1–88.

Bordonneau, M., and Tourte, Y. (1987). Potentialités ontégénetique des tissus gamétophytiques chez le *Pilularia globulifera* L. *Compt. Rend. Acad. Sci. Paris* Ser. III **304,** 437–40.

Boterberg, A. (1956). Genèse et différenciation des parois sporales chez *Marsilea diffusa* Lepr. *La Cellule* **58,** 79–106.

Bower, F. O. (1923–28). *The Ferns (Filicales).* Vols. I–III. Cambridge: University Press.

Braithwaite, A. F. (1964). A new type of apogamy in ferns. *New Phytol.* **63,** 293–305.

Braithwaite, A. F. (1969). The cytology of some Hymenophyllaceae from the Solomon Islands. *Br. Fern Gaz.* **10,** 81–91.

Braithwaite, A. F. (1975). Cytotaxonomic observations on some Hymenophyllaceae from the New Hebrides, Fiji and New Caledonia. *Bot. J. Linn. Soc.* **71,** 167–89.

Breslavets, L. P. (1951). Changes in the shape and size of plastids in the prothalli of ferns and horsetails induced by X-rays. *Doklady Akad. Nauk. SSSR.* **78,** 1235–8.

Breslavets, L. P. (1952). Effect of a high dose of X-rays on the prothalli of a fern *(Dryopteris spinulosum). Doklady Akad. Nauk. SSSR.* **87,** 143–6.

Bristow, J. M. (1962). The controlled *in vitro* differentiation of callus derived from a fern, *Pteris cretica* L., into gametophytic or sporophytic tissues. *Develop. Biol.* **4,** 361–75.

Brokaw, C. J. (1958). Chemotaxis of bracken spermatozoids. Implications of electrochemical orientation. *J. Expt. Biol.* **35,** 197–212.

Brokaw, C. J. (1959). Random and oriented movements of bracken spermatozoids. *J. Cell. Comp. Physiol.* **54,** 95–101.

Bünning, E. (1958). Polarität und inäquale Teilung des pflanzlichen Protoplasten. *Protoplasmatologia* VIII **9a,** 1–86.

Bünning, E. and Etzold, H. (1958). Uber die Wirkung von polarisiertem Licht auf keimende Sporen von Pilzen, Moosen und Farnen. *Ber. Deut. Bot. Ges.* **71,** 304–6.

Bünning, E., and Mohr, H. (1955). Das Aktionsspektrum des Lichteinflusses auf die Keimung von Farnsporen. *Naturwissenschaften* **42,** 212.

Bürcky, K. (1977*a*). Antheridiogene in *Anemia phyllitidis* L. Sw. (Schizaeaceae). 1. Zeitverlauf der Antheridiogensynthese. *Z. Pflanzenphysiol.* **84,** 167–71.

Bürcky, K. (1977*b*). Antheridiogene in *Anemia phyllitidis* L. Sw. (Schizaeaceae). 2. Gaschromatographischer Nachweis des Antheridiogens A_{An1}. *Z. Pflanzenphysiol.* **84,** 173–8.

Bürcky, K. (1977*c*). Das Vorkommen von Abscisinsäure in *Anemia phyllitidis* L. Sw. (Schizaeaceae) während der Sporenreifung. *Z. Pflanzenphysiol.* **85,** 181–3.

Bürcky, K. (1977*d*). Gibberellinaktivität verschiedener Schizaeaceen-Antheridiogene im Zwergerbsentest. *Z. Naturforsch.* **32c,** 652–3.

Bürcky, K., Gemmrich, A., and Schraudolf, H. (1978). Gas chromatographic evidence of the antheridiogen of *Lygodium japonicum* (A_{Ly}) (Schizaeaceae). *Experientia* **34,** 718.

Burns, R. G., and Ingle, J. (1968). The induction of biplanar growth in fern gametophytes in the presence of RNA base analogues. *Plant Physiol.* **43,** 1987–90.

Burns, R. G., and Ingle, J. (1970). The relationship between the kinetics of ribonucleic acid accumulation and the morphological development of the fern gametophyte, *Dryopteris borreri. Plant Physiol.* **46,** 423–8.

Calvert, H. E., Perkins, S. A., and Peters, G. A. (1983). Sporocarp structure in the heterosporous water fern *Azolla mexicana* Presl. In: *Scanning Electron Microscopy,* 1983, III, pp. 1499–510. O'Hare: SEM Inc.

Carlson, P. S. 1969. Production of auxotrophic mutants in ferns. *Genet. Res.* **14,** 337–9.

Caruso, J. L. (1973). Influence of phenylboric acid on germinating fern spores. *Can. J. Bot.* **51,** 1998–2000.

Cave, C. F., and Bell, P. R. (1973). The cytochemistry of the walls of the spermatocytes of *Ceratopteris thalictroides. Planta* **109,** 99–104.

Cave, C. F., and Bell, P. R. (1974*a*). The synthesis of ribonucleic acid and protein during oogenesis in *Pteridium aquilinum. Cytobiologie* **9,** 331–43.

Cave, C. F., and Bell, P. R. (1974*b*). The nature of the membrane around the egg of *Pteridium aquilinum* (L.) Kuhn. *Ann. Bot.* **38,** 17–21.

Cave, C. F., and Bell, P. R. (1975). Evidence for the association of acyl transferases with the production of nuclear evaginations in maturing eggs of the fern *Dryopteris filix-mas. Histochemistry* **44,** 57–65.

Cave, C. F., and Bell, P. R. (1979). The effect of colchicine on spermatogenesis in a fern, *Pteridium aquilinum* (L.) Kuhn. *Ann. Bot.* **44,** 407–15.

Chapman, R. H., Klekowski, E. J., Jr., and Selander, R. K. (1979). Homoeologous heterozygosity and recombination in the fern *Pteridium aquilinum. Science* **204,** 1207–9.

Charlton, F. B. (1938). Formative effects of radiation upon fern prothallia. *Am. J. Bot.* **25,** 431–42.

Cheema, H. K. (1979). Morphogenetic studies on the induction of apospory in *Ceratopteris pteridoides* (Hook.) Hieron, in aseptic culture. *Indian J. Expt. Biol.* **17,** 1403–5.

Cheema, H. K. (1980). *In vitro* morphogenetic studies on the gametophyte of *Anogramma leptophylla* (Sw.) Link. *Curr. Sci.* **49,** 593–4.

Chen, C. Y., and Ikuma, H. (1979). Photocontrol of the germination of *Onoclea* spores. V. Analysis of germination processes by means of temperature. *Plant Physiol.* **63,** 704–8.

Cheng, C. Y., and Schraudolf, H. (1974). Nachweis von Abscisinsäure in Sporen und jungen Prothallien von *Anemia phyllitidis* L. Sw. *Z. Pflanzenphysiol.* **71,** 366–9.

Chia, S. E., and Raghavan, V. (1982). Abscisic acid effects on spore germination and protonemal growth in the fern, *Mohria caffrorum. New Phytol.* **92,** 31–7.

Chiang, Y. L. (1976). On the sex of the prothallia of *Ceratopteris pteridoides* (Hook).) Hieron. *Taiwania* **21,** 134–7.

Clutter, M. E., and Sussex, I. M. (1965). Meiosis and sporogenesis in excised fern leaves grown in sterile culture. *Bot. Gaz.* **126,** 72–8.

Cohen, D., and Crotty, W. J. (1979). ^{3}H-uridine incorporation in the premitotic stage of rhizoid cell differentiation in *Pteris vittata* L. *Am. J. Bot.* **66,** 179–82.

Cohen, H. P., and DeMaggio, A. E. (1986). Biochemistry of fern spore germination: Protease activity in ostrich fern spores. *Plant Physiol.* **80,** 992–6.

Conway, E. (1949). The autecology of bracken [*Pteridium aquilinum* (L.) Kuhn.]: The germination of the spore, and the development of the prothallus and the young sporophyte. *Proc. Roy. Soc. Edin.* **63B,** 325–43.

Cooke, T. J., and Paolillo, D. J., Jr. (1979*a*). The photobiology of fern gametophytes. I. The phenomenon of red/far-red and yellow/far-red photoreversibility. *J. Expt. Bot.* **30,** 71–80.

Cooke, T. J., and Paolillo, D. J., Jr. (1979*b*). The photobiology of fern gametophytes. II. The photocontrol of filamentous growth and its implications for the photocontrol of the transition to two-dimensional growth. *Am. J. Bot.* **66,** 376–85.

Cooke, T. J., and Paolillo, D. J., Jr. (1980*a*). Dark growth and the associated phenomenon of age-dependent photoresponses in fern gametophytes. *Ann. Bot.* **45,** 693–702.

Cooke, T. J., and Paolillo, D. J., Jr. (1980*b*). The control of the orientation of cell divisions in fern gametophytes. *Am. J. Bot.* **67,** 1320–33.

Cooke, T. J., and Racusen, R. H. (1982). Cell expansion in the filamentous gametophyte of the fern *Onoclea sensibilis* L. *Planta* **155,** 449–58.

Cooke, T. J., Racusen, R. H., and Briggs, W. R. (1983). Initial events in the tip-swelling response of the filamentous gametophyte of *Onoclea sensibilis* L. to blue light. *Planta* **159,** 300–7.

Cooke, T. J., Racusen, R. H., Hickok, L. G., and Warne, T. R. (1987). The photocontrol of spore germination in the fern *Ceratopteris richardii. Plant Cell Physiol.* **28,** 753–9.

Copeland, E. B. (1947). *Genera Filicum. The Genera of Ferns.* Waltham: Chronica Botanica.

Corey, E. J., and Myers, A. G. (1985). Total synthesis of (+)-antheridium-inducing factor (A_{An},2) of the fern *Anemia phyllitidis.* Clarification of stereochemistry. *J. Am. Chem. Soc.* **107,** 5574–6.

Corey, E. J., Myers, A. G., Takahashi, N., Yamane, H., and Schraudolf, H. (1986). Constitution of antheridium-inducing factor of *Anemia phyllitidis. Tetrahedron Lett.* **27,** 5083–4.

Courbet, H. (1955). Influence de la concentration ionique du milieu sur la germination des spores et la croissance des prothalles de Filicales en culture aseptique. *Compt. Rend. Acad. Sci. Paris* **241,** 441–3.

Courbet, H. (1957). Action de quelques sucres sur la germination des spores et la croissance des prothalles d*Áthyrium filix femina* Roth en culture aseptique. *Compt. Rend. Acad. Sci. Paris.* **244,** 107–10.

Courbet, H. (1963). Les spores des fougeres. *Bull. Acad. Soc. Lorraines Sci.* **3,** 53–65.

Courbet, H., and Metche, M. (1971). Dosage de l'azote minéralisable et des aminés libres des spores de quelques fougères. *Compt. Rend. Acad. Sci. Paris* **272D,** 1904–7.

Cousens, M. I. (1975). Gametophyte sex expression in some species of *Dryopteris. Am. Fern J.* **65,** 39–42.

Cousens, M. I. (1979). Gametophyte ontogeny, sex expression, and genetic load as measures of population divergence in *Blechnum spicant. Am. J. Bot.* **66,** 116–32.

Cousens, M. I., and Horner, H. T., Jr. (1970). Gametophyte ontogeny and sex expression in *Dryopteris ludoviciana. Am. Fern J.* **60,** 13–27.

Crabbe, J. A., Jermy, A. C., and Mickel, J. T. (1975). A new generic sequence for the pteridophyte herbarium. *Fern Gaz.* **11,** 141–62.

Cran, D., and Dyer, A. F. (1975). The effect of a change in light quality on plastids of protonemata of *Dryopteris borreri. Plant Sci. Lett.* **5,** 57–65.

Cran, D. G. (1979). The ultrastructure of fern gametophyte cells. In: *The Experimental Biology of Ferns,* ed. A. F. Dyer, pp. 171–212. London: Academic Press.

Cran, D. G., and Dyer, A. F. (1973). Membrane continuity and associations in the fern *Dryopteris borreri. Protoplasma* **76,** 103–8.

Crist, K. C., and Farrar, D. R. (1983). Genetic load and long-distance dispersal in *Asplenium platyneuron. Can. J. Bot.* **61,** 1809–14.

Crotty, W. J. (1967). Rhizoid cell differentiation in the fern gametophyte of *Pteris vittata. Am. J. Bot.* **54,** 105–17.

Crotty, W. J., and Ledbetter, M. C. (1973). Membrane continuities involving chloroplasts and other organelles in plant cells. *Science* **182,** 839–41.

Czaja, A. T. (1921). Uber Befruchtung, Bastardierung und Geschlechtertrennung bei Prothallien homosporer Farne. *Z. Bot.* **13,** 545–89.

Davidonis, G. H., and Ruddat, M. (1973). Allelopathic compounds, thelypterin A and B in the fern *Thelypteris normalis. Planta* **111,** 23–32.

Davidonis, G. H., and Ruddat, M. (1974). Growth inhibition in gametophytes and oat coleoptiles by thelypterin A and B released from roots of the fern *Thelypteris normalis. Am. J. Bot.* **61,** 925–30.

Davie, J. H. (1951). The development of the antheridium in the Polypodiaceae. *Am. J. Bot.* **38,** 621–8.

Davis, B. D. (1968*a*). Effect of light quality on the transition to two-dimensional growth by gametophytes of *Pteridium aquilinum. Bull. Torrey Bot. Cl.* **95,** 31–6.

Davis, B. D. (1968*b*). The transition from filamentous to two-dimensional growth in fern gametophytes. I. The requirement for protein synthesis in gametophytes of *Pteridium aquilinum. Am. J. Bot.* **55,** 532–40.

Davis, B. D. (1968*c*). Is riboflavin the photoreceptor in the induction of two-dimensional growth in fern gametophytes? *Plant Physiol.* **43,** 1165–7.

Davis, B. D. (1969). The transition from filamentous to two-dimensional growth in fern gametophytes. II. Kinetic studies on *Pteridium aquilinum. Am. J. Bot.* **56,** 1048–53.

Davis, B. D. (1970). Cycloheximide: Nonspecific inhibition of two-dimensional growth in gametophytes of *Pteridium aquilinum. Bull. Torrey Bot. Cl.* **97,** 53–8.

Davis, B. D. (1971). The transition from filamentous to two-dimensional growth in fern gametophytes. III. Interaction of cell elongation and cell division. *Am. J. Bot.* **58,** 212–7.

Davis, B. D. (1975). Bending growth in fern gametophyte protonema. *Plant Cell Physiol.* **16,** 537–41.

Davis, B. D., Chen, J. C. W., and Philpott, M. (1974). The transition from

filamentous to two-dimensional growth in fern gametophytes. IV. Initial events. *Am. J. Bot.* **61,** 722–9.
DeBary, A. (1878). Ueber apogame Farne und die Erscheinung der Apogamie im Allgemeinen. *Bot. Zeit.* **36,** 449–87.
Deimling, A. v., and Mohr, H. (1967). Eine Analyse der durch Blaulicht bewirkten Steigerung der Proteinsynthese bei Farnvorkeimen auf der Ebene der Aminosäuren. *Planta* **76,** 269–84.
DeMaggio, A., Wetmore, R., and Morel, G. (1963). Induction de tissu vasculaire dans le prothalle de fougère. *Compt. Rend. Acad. Sci. Paris* **256,** 5196–9.
DeMaggio, A. E. (1968). Meiosis *in vitro*: Sporogenesis in cultured fern plants. *Am. J. Bot.* **55,** 915–22.
DeMaggio, A. E. (1972). Induced vascular tissue differentiation in fern gametophytes. *Bot. Gaz.* **133,** 311–7.
DeMaggio, A. E. (1977). Cytological aspects of reproduction in ferns. *Bot. Rev.* **43,** 427–48.
DeMaggio, A. E., Greene, C., and Stetler, D. A. (1980). Biochemistry of fern spore germination. Glyoxylate and glycolate cycle activity in *Onoclea sensibilis* L. *Plant Physiol.* **66,** 922–4.
DeMaggio, A. E., Greene, C., Unal, S., and Stetler, D. A. (1979). Microbodies in germinating fern spores: Evidence for glyoxysomal activity. *Science* **206,** 580–2.
DeMaggio, A. E., and Krasnoff, M. (1980). Peroxisomal enzyme activity in *Todea barbara* gametophytes and sporophytes. *Phytochemistry* **19,** 1339–41.
DeMaggio, A. E., and Lambrukos, J. (1974). Polyploidy and gene dosage effects on peroxidase activity in ferns. *Biochem. Genet.* **12,** 429–40.
DeMaggio, A. E., and Raghavan, V. (1972). Germination of bracken fern spore. Electrophoretic analysis of proteins synthesized during initiation and elongation of the rhizoid. *Expt. Cell Res.* **73,** 182–6.
DeMaggio, A. E., and Raghavan, V. (1973). Photomorphogenesis and nucleic acid metabolism in fern gametophytes. *Adv. Morphogenesis* **10,** 227–63.
DeMaggio, A. E., and Stetler, D. A. (1971). Polyploidy and gene dosage effects on chloroplasts of fern gametophytes. *Expt. Cell Res.* **67,** 287–94.
DeMaggio, A. E., and Stetler, D. A. (1980). Storage products in spores of *Onoclea sensibilis* L. *Am. J. Bot.* **67,** 452–5.
DeMaggio, A. E., and Wetmore, R. H. (1961). Morphogenetic studies on the fern *Todea barbara*. III. Experimental embryology. *Am. J. Bot.* **48,** 551–65.
Demalsy, P. (1953). Etudes sur les Hydroptéridales. III. Le sporophyte d'*Azolla nilotica. La Cellule* **56,** 5–60.
Digby, L. (1905). On the cytology of apogamy and apospory. II. Preliminary note on apospory. *Proc. Roy. Soc. Lond.* **76B,** 463–7.
Donaher, D. J., and Partanen, C. R. (1971). The role of light in the interrelated processes of morphogenesis and photosynthesis in the fern gametophyte. *Physiol. Plantarum* **25,** 461–8.
Doonan, J. H., Lloyd, C. W., and Duckett, J. G. (1986). Anti-tubulin antibodies locate the blepharoplast during spermatogenesis in the fern *Platyzoma microphyllum* R.Br.: A correlated immunofluorescence and electron-microscopic study. *J. Cell Sci.* **81,** 243–65.
Döpp, W. (1932). Die Apogamie bei *Aspidium remotum* Al. Br. *Planta* **17,** 86–152.
Döpp, W. (1933). Weitere Untersuchungen an apogamen Farnen. I. *Aspidium filix-mas* SW. var. *crist. hort. Ber. Deut. Bot Ges.* **51,** 341–7.
Döpp, W. (1937). Gestaltung und Organbidung innerhalb der Gametophyt-

generation der Polypodiaceen unter besonderer Berücksichtigung genetischer Gesichtpunkte. *Beitr. Biol. Pflanzen* **24,** 201–38.

Döpp, W. (1939). Cytologische und genetische Untersuchungen innerhalb der Gattung *Dryopteris. Planta* **29,** 481–533.

Döpp, W. (1950). Eine die Antheridienbildung bei Farnen fördernde Substanz in den Prothallien von *Pteridium aquilinum* (L.) Kuhn. *Ber. Deut. Bot. Ges.* **63,** 139–47.

Döpp, W. (1955*a*). Hemmung der Antheridienbildung an den Prothallien von *Pteridium aquilinum* durch 2,3,5-Trijodobenzoesäure. *Naturwissenschaften* **42,** 99.

Döpp, W. (1955*b*). Polyploidie und andere Erscheinungen bei Farnprothallien durch Wirkung von Gesarol und seinen Bestandteilen. *Naturwissenschaften* **42,** 99–100.

Döpp, W. (1959). Uber eine hemmende und eine fördernde Substanz bei der Antheridienbildung in den Prothallien von *Pteridium aquilinum. Ber. Deut. Bot. Ges.* **72,** 11–24.

Döpp, W. (1962). Weitere Untersuchungen über die Physiologie der Antheridienbildung bei *Pteridium aquilinum. Planta* **58,** 483–508.

Drumm, H., and Mohr, H. (1967*a*). Die Regulation der RNS-Synthese in Farngametophyten durch Licht. *Planta* **72,** 232–46.

Drumm, H., and Mohr, H. (1967*b*). Die Regulation der DNS-Synthese in Farngametophyten durch Licht. *Planta* **75,** 343–51.

Dubey, J. P., and Roy, S. K. (1985). A new antheridiogen from the fern *Pityrogramma calomelanos* (L.) Link. *Proc. Indian Acad. Sci. (Plant Sci).* **95,** 173–9.

Duckett, J. G. (1970). Sexual behaviour of the genus *Equisetum,* subgenus *Equisetum. Bot. J. Linn. Soc.* **63,** 327–52.

Duckett, J. G. (1972). Sexual behaviour of the genus *Equisetum,* subgenus *Hippochaete. Bot. J. Linn. Soc.* **65,** 87–108.

Duckett, J. G. (1975). Spermatogenesis in pteridophytes. In: *The Biology of the Male Gamete,* ed. J. G. Duckett and P. A. Racey, pp. 97–127. London: Academic Press.

Duckett, J. G. (1977). Towards an understanding of sex determination in *Equisetum:* An analysis of regeneration in gametophytes of the subgenus *Equisetum. Bot. J. Linn. Soc.* **74,** 215–42.

Duckett J. G. (1979*a*). Comparative morphology of the gametophytes of *Equisetum* subgenus *Hippochaete* and the sexual behaviour of *E. ramosissimum* subsp. *debile,* (Roxb.) Hauke, *E. hyemale* var. *affine* (Engelm.) A.A., and *E. laevigatum* A. Br. *Bot. J. Linn. Soc.* **79,** 179–203.

Duckett, J. G. (1979*b*). An experimental study of the reproductive biology and hybridization in the European and North American species *Equisetum. Bot. J. Linn. Soc.* **79,** 205–29.

Duckett, J. G., and Bell, P. R. (1971). Studies on fertilization in archegoniate plants. I. Changes in the structure of the spermatozoids of *Pteridium aquilinum* (L.) Kuhn during entry into the archegonium. *Cytobiologie* **4,** 421–36.

Duckett, J. G., and Bell, P. R. (1972). Studies on fertilization in archegoniate plants. II. Egg penetration in *Pteridium aquilinum* (L.) Kuhn. *Cytobiologie* **4,** 35–50.

Duckett, J. G., and Duckett, A. R. (1980). Reproductive biology and population dynamics of wild gametophytes of *Equisetum. Bot. J. Linn. Soc.* **80,** 1–40.

Duckett, J. G., Klekowski, E. J., Jr., and Hickok, L. G. (1979). Ultrastructural studies of mutant spermatozoids in ferns. I. The mature nonmotile spermatozoid of mutation 230X in *Ceratopteris thalictroides* (L.) Brongn. *Gamete Res.* **2,** 317–43.

Duckett, J. G., and Pang, W. C. (1984). The origins of heterospory: A comparative study of sexual behaviour in the fern *Platyzoma microphyllum* R.Br. and the horsetail *Equisetum giganteum*. L. *Bot. J. Linn. Soc.* **88,** 11–34.

Duncan, R. E. (1941). Apogamy in *Doodia caudata*. *Am. J. Bot.* **28,** 921–31.

Duncan, R. E. (1943). Origin and development of embryos in certain apogamous forms of *Dryopteris*. *Bot. Gaz.* **105,** 202–11.

Durand-Rivierès, R. (1960). Sur la culture *in vitro* de racine isolées de jeunes sporophytes de *Pteris longifolia* L. (Polypodiacées). *Compt. Rend. Acad. Sci. Paris* **250,** 2442–4.

Durand-Rivierès, R., and Fillon, F. (1968). Substances de croissance et diffèrenciation du prothalle de *Pteris longifolia* L. *Compt. Rend. Acad. Sci. Paris* **266D,** 471–3.

Dyar, J. J., and Shade, J. (1974). The influence of benzimidazole on the gametophyte of *Thelypteris felix-mas*. *Plant Physiol.* **53,** 668–8.

Dyer, A. F. (1979). The culture of fern gametophytes for experimental investigation. In: *The Experimental Biology of Ferns,* ed. A. F. Dyer, pp. 253–305. London: Academic Press.

Dyer, A. F., and Cran, D. G. (1976). The formation and ultrastructure of rhizoids on protonemata of *Dryopteris borreri* Newm. *Ann. Bot.* **40,** 757–65.

Dyer, A. F., and King, M. A. L. (1979). Cell division in fern protonemata. In: *The Experimental Biology of Ferns,* ed. A. F. Dyer, pp. 307–54. London: Academic Press.

Eakle, T. W. (1975). Photoperiodic control of spore germination of *Acrostichum aureum*. *Am. Fern J.* **65,** 94–5.

Edwards, M. E. (1977). Carbon dioxide and ethylene control of spore germination in *Onoclea sensibilis* L. *Plant Physiol.* **59,** 756–8.

Edwards, M. E., and Miller, J. H. (1972*a*). Growth regulation by ethylene in fern gametophytes. II. Inhibition of cell division. *Am. J. Bot.* **59,** 450–7.

Edwards, M. E., and Miller, J. H. (1972*b*). Growth regulation by ethylene in fern gametophytes. III. Inhibition of spore germination. *Am. J. Bot.* **59,** 458–65.

Elmore, H. W., and Adams, R. J. (1976). Scanning electron microscopic observations on the gametophyte and sperm of the bracken fern, *Pteridium aquilinum* (L.) Kuhn. *New Phytol.* **76,** 519–22.

Elmore, H. W., and Whittier, D. P. (1973). The role of ethylene in the induction of apogamous buds in *Pteridium* gametophytes. *Planta* **111,** 85–90.

Elmore, H. W., and Whittier, D. P. (1974). Ethylene and carbohydrate requirements for apogamous bud induction in *Pteridium* gametophytes. *Can. J. Bot.* **52,** 2089–96.

Elmore, H. W., and Whittier, D. P. (1975*a*). The involvement of ethylene and sucrose in the inductive and developmental phases of apogamous bud formation in *Pteridium* gametophytes. *Can. J. Bot.* **53,** 375–81.

Elmore, H. W., and Whittier, D. P. (1975*b*). Ethylene production and ethylene-induced apogamous bud formation in nine gametophytic strains of *Pteridium aquilinum* (l.) Kuhn. *Ann. Bot.* **39,** 965–71.

Emigh, V. D., and Farrar, D. R. (1977). Gemmae: A role in sexual reproduction in the fern genus *Vittaria*. *Science* **198,** 297–8.

Endo, M., Nakanishi, K., Näf, U., Mc Keon, W., and Walker, R. (1972).

Isolation of the antheridiogen of *Anemia phyllitidis*. *Physiol. Plantarum* **26,** 183–5.

Endress, A. G. (1974). Spore germination of *Ceratopteris thalictroides* (L.) Brongn. *Ann. Bot.* **38,** 877–81.

Estes, L. W. (1963). Morphological effects of ultraviolet radiation on the prothalli of *Onoclea sensibilis* L. *Phytomorphology* **13,** 284–9.

Esteves, L. M., Felippe, G. M., and Melhem, T. S. (1985). Germination and morphology of spores of *Trichipteris corcovadensis*. *Am. Fern J.* **75,** 92–102.

Etzold, H. (1965). Der Polarotropismus und Phototropismus der Chloronemen von *Dryopteris filix-mas* (L.) Schott. *Planta* **64,** 254-80.

Etzold, H., and Jaffe, L. (1963). Die Polaritätsinduktion bei der *Equisetum* spore durch polarisiertes Licht und partielle Belichtung. *Expt. Cell Res.* **29,** 188–93.

Evans, A. M. (1964). Ameiotic alternation of generations: A new life cycle in the ferns. *Science* **143,** 261–3.

Evans, A. M. (1968). Interspecific relationships in *Polypodium pectinatum-plumula* complex. *Ann. Mo. Bot. Gard.* **55,** 193–293.

Evans, L. S., and Bozzone, D. M. (1977). Effect of buffered solutions and sulfate on vegetative and sexual development in gametophytes of *Pteridium aquilinum*. *Am. J. Bot.* **64,** 897–902.

Evans, L. S., and Bozzone, D. M. (1978). Effect of buffered solutions and various anions on vegetative and sexual development in gametophytes of *Pteridium aquilinum*. *Can. J. Bot.* **56,** 779–85.

Evans, L. S., and Conway, C. A. (1980). Effects of acidic solutions on sexual reproduction of *Pteridium aquilinum*. *Am. J. Bot.* **67,** 866–75.

Faivre, M. (1969*a*). Action de la kinétine sur le gamétophyte de *Gymnogramme calomelanos* (Filicinées, Leptosporangiées). *Compt. Rend. Acad. Sci. Paris* **269D,** 32–5.

Faivre, M. (1969*b*). Morphologie des gamétophytes du *Gymnogramme calomelanos* (Filicinées, Leptosporangiées) cultivées en présence de kinétine, glucose et saccharose. *Compt. Rend. Acad. Sci. Paris* **269D,** 693–6.

Faivre, M. (1970). Etude expérimentale sur la régénération de gamétophytes à partir des cellules de filaments germinatifs issus de spores de *Gymnogramme calomelanos* (Filicinées, Leptosporangiées). *Compt. Rend. Acad. Sci. Paris* **270D,** 1254–7.

Faivre, M. (1975). Etude du comportement et de la régénération du jeune gamétophyte du *Gymnogramme calomelanos* (Leptosporangiées-Filicinées) en fonction de divers facteurs. *Rev. Gen. Bot.* **82,** 5–91.

Faivre, M., Kuligowski, J., and Tourte, Y. (1982). Etude experimentale de la fecondation chez une pteridophyte: Evenements nucleaires et cytoplasmiques *Cytobios* **35,** 195–208.

Faivre-Baron, M. (1977*a*). Étude cytophysiologique de la régénération du jeune gamétophyte d'une fougère: Le *Gymnogramme calomelanos* L. I. Caractères infrastructuraux particuliers aux cellules du filament germinatif, en conditions de cultures contrôles. *Beitr. Biol. Pflanzen* **53,** 103–20.

Faivre-Baron, M. (1977*b*). Etude cytophysiologique de la régénération du jeune gamétophyte d'une fougère: Le *Gymnogramme calomelanos* L. II. Caractères infrastructuraux des cellules du filament germinatif au cours des phenomènès de régénération. *Beitr. Biol. Pflanzen* **53,** 127–42.

Faivre-Baron, M. (1977*c*). Incorporation de thymidine et leucine tritées dans les gamétophytes d'une fougère: Le *Gymnogramme calomelanos* L. à différents stades de leurs développements. *Beitr. Biol. Pflanzen* **53,** 411–20.

Faivre-Baron, M. (1978*a*). Etude des mécanismes de corrélations d'inhibition chez le jeune gamétophyte du *Gymnogramme calomelanos* L. *Biochem. Physiol. Pflanzen* **172,** 79–91.

Faivre-Baron, M. (1978*b*). Etude du fonctionnement des zones organogènes du gamétophyte du *Gymnogramme calomelanos* L. (Pteridophytes, Filicinées). *Flora* **167,** 315–28.

Faivre-Baron, M. (1979*a*). Peroxysomes in the gametophytic cells of a fern: *Gymnogramme calomelanos* L. *Beitr. Biol. Pflanzen* **55,** 129–42.

Faivre-Baron, M. (1979*b*). Intervention de substances à groupements sulfhydryles dans la morphogenese prothallienne d'une fougère. *Biochem. Physiol. Pflanzen* **174,** 696–706.

Faivre-Baron, M. (1980). Action de deux inhibiteurs de la synthèse protéique sur la morphologie prothallienne d'une fougère. *Flora* **169,** 467–75.

Faivre-Baron, M. (1981). Effets de l'acide indolyl-acétique exogène et son transport dans le gamétophyte d'une fougère. *Z. Pflanzenphysiol.* **101,** 263–75.

Falk, H., and Steiner, A. M. (1968). Phytochrome-mediated polarotropism: An electron microscopical study. *Naturwissenschaften* **55,** 500.

Farlow, W. G. (1874). An asexual growth from the prothallus of *Pteris cretica. Quart. J. Micros. Sci.* **14,** 266–72.

Farmer, J. B., and Digby, L. (1907). Studies in apospory and apogamy in ferns. *Ann. Bot.* **21,** 161–99.

Farrar, D. R. (1967). Gametophytes of four tropical fern genera reproducing independently of their sporophytes in the southern Appalachians. *Science* **155,** 1266–7.

Farrar, D. R. (1974). Gemmiferous fern gametophytes - Vittariaceae. *Am. J. Bot.* **61,** 146-55.

Farrar, D. R. (1978). Problems in the identity and origin of the Appalachian *Vittaria* gametophyte, a sporophyteless fern of the eastern United States. *Am. J. Bot.* **65,** 1–12.

Farrar, D. R., and Wagner, W. H., Jr. (1968). The gametophyte of *Trichomanes holopterum* Kunze. *Bot. Gaz.* **129,** 210–9.

Fechner, A., and Schraudolf, H. (1982). Isolation and characterization of poly (adenylic acid)-containing ribonucleic acid from dry spores of *Anemia phyllitidis* L. Sw. *Z. Pflanzenphysiol.* **108,** 419–28.

Fechner, A., and Schraudolf, H. (1984). Translation and transcription in imbibed and germinating spores of *Anemia phyllitidis* L. Sw. *Planta* **161,** 451–8.

Fechner, A., and Schraudolf, H. (1986). Aphidicolin inhibition of DNA synthesis and germination in spores of *Anemia phyllitidis* L. Sw. *Plant Physiol.* **81,** 714–6.

Fellenberg-Kressel, M. (1969). Untersuchungen über die Archegonien- und Antheridienbildung bei *Microlepia speluncae* (L.) Moore in Abhängigkeit von inneren und äusseren Faktoren. *Flora* **160,** 14–39.

Feller, M.-J. (1953). Sporocarpe et sporogenése chez *Marsilea hirsuta* R. Br. *La Cellule* **55,** 305–77.

Fisher, R. W., and Miller, J. H. (1975). Growth regulation by ethylene in fern gametophytes. IV. Involvement of photosynthesis in overcoming ethylene inhibition of spore germination. *Am. J. Bot.* **62,** 1104–11.

Fisher, R. W., and Miller J. H. (1978). Growth regulation by ethylene in fern gametophytes. V. Ethylene and the early events of spore germination. *Am. J. Bot.* **65,** 334–9.

Fisher, R. W., and Shropshire, W., Jr. (1979). Reversal by light of ethylene-induced inhibition of spore germination in the sensitive fern *Onoclea sensibilis*. An action spectrum. *Plant Physiol.* **63,** 984–8.

Föhr, K. J., Enssle, M., and Schraudolf, H. (1987). Calmodulin-like protein from fern *Anemia phyllitidis* L. Sw. *Planta* **171,** 127–9.

Fraser, T. W., and Smith, D. L. (1974). Young gametophytes of the fern *Polypodium vulgare* L. An ultrastructural study. *Protoplasma* **82,** 19–32.

Froeschel, P. (1953). Remstoffen bij de lagere plantaardige organismen. *Natuurw. Tijdschr.* **35,** 70–5.

Furuya, M. (1978). Photocontrol of developmental processes in fern gametophytes. *Bot. Mag. (Tokyo), Special Issue* **1,** 219–42.

Furuya, M. (1983). Photomorphogenesis in ferns. *Encyl. Plant Physiol. (New Series)* **16B,** 569–600.

Furuya, M., Ito, M., and Sugai, M. (1967). Photomorphogenesis in *Pteris vittata. Jap. J. Expt. Morph.* **21,** 398–408.

Furuya, M., Kadota, A., and Uematsu-Kaneda, H. (1982). Percent P_{FR}-dependent germination of spores in *Pteris vittata. Plant Cell Physiol.* **23,** 1213–17.

Ganders, F. R., (1972). Heterozygosity for recessive lethals in homosporous fern populations: *Thelypteris plaustris* and *Onoclea sensibilis*. *Bot. J. Linn. Soc.* **65,** 211–21.

Gantt, E., and Arnott, H. J. (1963). Chloroplast division in the gametophyte of the fern *Matteuccia struthiopteris* (L.) Todaro. *J. Cell Biol.* **19,** 446–8.

Gantt, E., and Arnott, H. J. (1965). Spore germination and development of the young gametophyte of the ostrich fern *(Matteuccia struthiopteris). Am. J. Bot.* **52,** 82–94.

Gastony, G. J. (1979). Spore morphology in the Cyatheaceae. III. The genus *Trichipteris. Am. J. Bot.* **66,** 1238–60.

Gastony, G. J., and Darrow, D. C. (1983). Chloroplastic and cytosolic isozymes of the homosporous fern *Athyrium filix-femina* L. *Am. J. Bot.* **70,** 1409–15.

Gastony, G. J., and Gottlieb, L. D. (1982). Evidence for genetic heterozygosity in a homosporous fern. *Am. J. Bot.* **69,** 634-7.

Gastony, G. J., and Gottlieb, L. D. (1985). Genetic variation in the homosporous fern *Pellaea andromedifolia. Am. J. Bot.* **72,** 257–67.

Gastony, G. J., and Haufler, C. H. (1976). Chromosome numbers and apomixis in the fern genus *Bommeria* (Gymnogrammaceae). *Biotropica* **8,** 1-11.

Gastony, G. J., and Tryon, R. M. (1976). Spore morphology in the Cyatheaceae. II. The genera *Lophosoria, Metaxya, Sphaeropteris, Alsophila,* and *Nephelea. Am. J. Bot.* **63,** 738–58.

Gemmrich, A. R. (1975). Prolindehydrogenase aus keimenden Farnsporen. *Phytochemistry* **14,** 353–7.

Gemmrich, A. R. (1977*a*). Mobilization of reserve lipids in germinating spores of the fern *Anemia phyllitidis* L. *Plant Sci. Lett.* **9,** 301–7.

Gemmrich, A. R. (1977*b*). Fatty acid composition of fern spore lipids. *Phytochemistry* **16,** 1044-6.

Gemmrich, A. R. (1979*a*). Lipid synthesis in imbibed spores of the fern *Anemia phyllitidis* L. Sw. *Z. Pflanzenphysiol.* **91,** 317–24.

Gemmrich, A. R. (1979*b*). Isocitrate lyase in germinating spores of the fern *Anemia phyllitidis. Phytochemistry* **18,** 1143–6.

Gemmrich, A. R. (1980). Developmental changes in microbody enzyme activities in germinating spores of the fern *Pteris vittata. Z. Pflanzenphysiol.* **97,** 153–60.

Gemmrich A. R. (1981). Ultrastructural and enzymatic studies on the development of microbodies in germinating spores of the fern *Anemia phyllitidis*. *Z. Pflanzenphysiol.* **102,** 69–80.

Gemmrich A. R. (1982). Effect of red light and gibberellic acid on lipid metabolism in germinating spores of the fern *Anemia phyllitidis*. *Physiol. Plantarum* **54,** 58–62.

Gemmrich A. R. (1986*a*). Antheridiogenesis in the fern *Pteris vittata*. I. Photocontrol of antheridium formation. *Plant. Sci.* **43,** 135–40.

Gemmrich, A. R. (1986*b*). Antheridiogenesis in the fern *Pteris vittata*. II. Hormonal control of antheridium formation. *J. Plant Physiol.* **125,** 157–66.

Gifford, E. M., Jr., and Brandon, D. D. (1978). Gametophytes of *Botrychium multifidum* as grown in axenic culture. *Am. Fern J.* **68,** 71–5.

Goebel, K. (1905). Kleinere Mitteilungen. *Flora* **95,** 232–50.

Gratzy-Wardengg, S. A. E. (1929). Osmotische Untersuchungen an Farnprothallien. *Planta* **7,** 307–39.

Greany, R. H., and Miller, J. H. (1976). An interpretation of dose-response curves for light-induced cell elongation in fern protonemata. *Am. J. Bot.* **63,** 1031–7.

Grill, R. (1987). Induction of two-dimensional growth by red and green light in the fern *Anemia phyllitidis* L. Sw. *J. Plant Physiol.* **131,** 363–71,

Grill, R., and Schraudolf, H. (1981). *In vivo* phytochrome difference spectrum from dark grown gametophytes of *Anemia phyllitidis* L. Sw. treated with norflurazon. *Plant Physiol.* **68,** 1–4.

Guervin, C. (1968). Les 3 étapes de l'édifcation du prothalle du *Gymnogramme sulphurea* Desv. *Compt. Rend. Acad. Sci. Paris* **266D,** 343–5.

Guervin, C. (1971). Contribution à l'étude d'une Polypodiacée: Le *Gymnogramme sulphurea* Desv. I. Cyto-morphogénèse du gamètophyte. *Rev. Gen. Bot.* **78,** 5–51.

Guervin, C. (1972). Contribution à l'étude d'une Polypodiacée: Le *Gymnogramme sulphurea* Desv. III. Etude des lipides. *Rev. Gen. Bot.* **79,** 109–37.

Guervin, C., and Laroche, J. (1967). Cultures d'*Equisetum arvense* L. et de *Gymnogramme sulphurea* Desv.: Importance du facteur pH. *Compt. Rend. Acad. Sci. Paris* **264D,** 330–3.

Guervin, C., Laroche, J., and Le Coq, C. (1979). Le génome et l'expression sexuelle chez l'*Equisetum arvense* L. II. Comportement du gamétophyte cultivé sur milieu témoin additionné de phytohormones. *Rev. Gen. Bot.* **86,** 3–20.

Guiragossian, H. A., and Koning, R. E. (1986). Induction of spore germination in *Schizaea pusilla* (Schizaeaceae). *Am. J. Bot.* **73,** 1588–94.

Gullvåg, B. M. (1968*a*). On the fine structure of the spores of *Equisetum fluviatile* var. *verticillatum* studied in the quiescent, germinated and nonviable state. *Grana Palynol.* **8,** 23–69.

Gullvåg, B. M. (1968*b*). Fine structure of the plastids and possible ways of distribution of the chloroplast products in some spores of archegoniatae. *Phytomorphology* **18,** 520–35.

Gullvåg, B. M. (1969). Primary storage products of some pteridophyte spores – A fine structural study. *Phytomorphology* **19,** 82–92.

Gullvåg, B. M. (1971*a*). Microbodies in the spore of *Equisetum:* A fixation study. *Grana* **11,** 36–40.

Gullvåg, B. M. (1971*b*). The fine structure of pteridophyte spores before and during germination. In: *Pollen and Spore Morphology* IV, ed. G. Erdtman and P. Sorsa, pp. 252–295. Stockholm: Almquist & Wiksell.

Haigh, M. V., and Howard, A. (1970). Mutations affecting cell morphology in *Osmunda regalis. J. Hered.* **61,** 285–7.
Haigh, M. V., and Howard, A. (1973*a*). Effect of X rays on the growth of the first rhizoid in spores of *Osmunda regalis. Radiation Bot.* **13,** 37–45.
Haigh, M. V., and Howard, A. (1973*b*). Radiation-induced tumorous outgrowths in young gametophytes of *Osmunda regalis. Radiation Bot.* **13,** 111–19.
Hannaford, J. E., and DeMaggio, A. E. (1975). Cytological studies in a polyploid series of the fern, *Todea barbara.* In: *Form, Structure and Function in Plants,* ed. H. Y. Mohan Ram, J. J. Shah and C. K. Shah, pp. 59–68. Meerut: Sarita Prakashan.
Hartmann, K. M., Menzel, H., and Mohr, H. (1965). Ein Beitrag zur Theorie der polarotropischen und phototropischen Krümmung. *Planta* **64,** 363–75.
Hartt, C. E. (1925). Conditions for germination of spores of *Onoclea sensibilis. Bot. Gaz.* **79,** 427–40.
Harvey, W. H., and Caponetti, J. D. (1972). *In vitro* studies on the induction of sporogenous tissue on leaves of cinnamon fern. I. Environmental factors. *Can. J. Bot.* **50,** 2673–82,
Harvey, W. H., and Caponetti, J. D. (1973). *In vitro* studies on the induction of sporogenous tissue on leaves of cinnamon fern. II. Some aspects of carbohydrate metabolism. *Can. J. Bot.* **51,** 341–9.
Haufler, C. H., and Adams, W. W., III. (1982). Early gametophyte ontogeny of *Gleichenia bifida* (Willd.) Spreng.: Phylogenetic and ecological implications. *Am. J. Bot.* **69,** 1560–5.
Haufler, C. H., and Gastony, G. J. (1978). Antheridiogen and the breeding system in the fern genus *Bommeria. Can. J. Bot.* **56,** 1594–601.
Haufler, C. H., and Ranker, T. A. (1985). Differential antheridiogen response and evolutionary mechanisms in *Cystopteris. Am. J. Bot.* **72,** 659–65.
Haufler, C. H., and Soltis, D. E. (1984). Obligate outcrossing in a homosporous fern: Field confirmation of a laboratory prediction. *Am. J. Bot.* **71,** 878–81.
Haufler, C. H., and Soltis, D. E. (1986). Genetic evidence suggests that homosporous ferns with high chromosome numbers are diploid. *Proc. Natl. Acad. Sci. USA* **83,** 4389–93.
Hauke, R. L. (1971). The effect of light quality and intensity on sexual expression in *Equisetum* gametophytes. *Am. J. Bot.* **58,** 373–7.
Hauke, R. L. (1977). Experimental studies on growth and sexual determination in *Equisetum* gametophytes. *Am. Fern J.* **67,** 18–31.
Haupt, W. (1957). Die Induktion der Polarität bei der Spore von *Equisetum. Planta* **49,** 61–90.
Haupt, W. (1958). Über den Primärvorgang bei der polarisierenden Wirkung des Lichtes auf keimende *Equisetum*-Sporen. *Planta* **51,** 74–83.
Haupt, W. (1985). Effects of nutrients and light pretreatment on phytochrome-mediated fern-spore germination. *Planta* **164,** 63–8.
Haupt, W., and Björn, L. O. (1987). No action dichroism for light-controlled fern spore germination. *J. Plant Physiol.* **129,** 119–28.
Haupt, W., and Filler, E. (1986). Sequential photoregulation of fern-spore germination. *J. Plant Physiol.* **125,** 409–16.
Haupt, W., and Meyer zu Bentrup, F. W. (1961). Versuche zur Polaritätsinduktion durch Licht bei *Equisetum*sporen und *Fucus*zygoten. *Naturwissenschaften* **48,** 723.
Hayami, J., Kadota, A., and Wada, M. (1986). Blue light-induced phototropic response and the intracellular photoreceptive site in *Adiantum* protonemata. *Plant Cell Physiol.* **27,** 1571–7.

Heilbronn, A. (1932). Polyploidie und Generationswechsel. *Ber. Deut. Ges. Bot.* **50,** 289–99.

Hepler, P. K. (1976). The blepharoplast of *Marsilea:* Its *de novo* formation and spindle association. *J. Cell Sci.* **21,** 361–90.

Herd, Y. R., Cutter, E. G., and Watanabe, I. (1985). A light and electron microscopic study of microsporogenesis in *Azolla microphylla. Proc. Roy. Soc. Edin.* **86B,** 53–8.

Herd, Y. R., Cutter, E. G., and Watanabe, I. (1986). An ultrastructural study of postmeiotic development in the megasporocarp of *Azolla microphylla. Can. J. Bot.* **64,** 822–33.

Heun, A. L. (1939). The cytology of apogamy in *Pteris cretica* Linn. var. *albolineata* Hort. *Bull. Torrey Bot. Cl.* **66,** 549–62.

Hevly, R. H. (1963). Adaptations of Cheilanthoid ferns to desert environments. *J. Arizona Acad. Sci.* **2,** 164–75.

Hickok, L. G. (1977*a*). The cytology and derivation of a temperature-sensitive meiotic mutant in the fern *Ceratopteris. Am. J. Bot.* **64,** 552–63.

Hickok, L. G. (1977*b*). An apomictic mutant for sticky chromosomes in the fern *Ceratopteris. Can. J. Bot.* **55,** 2186–95.

Hickok. L. G. (1978*a*). Homoeologous chromosome pairing and restricted segregation in the fern *Ceratopteris. Am. J. Bot.* **65,** 516–21.

Hickok, L. G. (1978*b*). Homoeologous chromosome pairing: Frequency differences in inbred and intraspecific hybrid polyploid ferns. *Science* **202,** 982–4.

Hickok, L. G. (1979). Apogamy and somatic restitution in the fern *Ceratopteris. Am. J. Bot.* **66,** 1074–8.

Hickok, L. G. (1983). Abscisic acid blocks antheridiogen-induced antheridium formation in gametophytes of the fern *Ceratopteris. Can. J. Bot.* **61,** 888–92.

Hickok, L. G. (1985). Abscisic acid resistant mutants in the fern *Ceratopteris:* Characterization and genetic analysis. *Can. J. Bot.* **63,** 1582–5.

Hickok, L. G., and Kiriluk, R. M. (1984). Effects of auxins on gametophyte development and sexual differentiation in the fern *Ceratopteris thalictroides* (L.) Brongn. *Bot. Gaz.* **145,** 37–42.

Hickok, L. G., and Klekowski, E. J., Jr. (1973). Abnormal reductional and nonreductional meiosis in *Ceratopteris:* Alternatives to homozygosity and hybrid sterility in homosporous ferns. *Am. J. Bot.* **60,** 1010–22.

Hickok, L. G., and Schwarz, O. J. (1986*a*). An *in vitro* whole plant selection system: Paraquat tolerant mutants in the fern *Ceratopteris. Theor. Appl. Genet.* **72,** 302–6.

Hickok, L. G., and Schwarz, O. J. (1986*b*). Paraquat tolerant mutants in *Ceratopteris:* Genetic characterization and reselection for enhanced tolerance. *Plant Sci.* **47,** 153–8.

Hickok, L. G., Warne, T. R., and Slocum, M. K. (1987). *Ceratopteris richardii:* Applications for experimental plant biology. *Am. J. Bot.* **74,** 1304–16.

Higinbotham, N. (1941). Development of the gametophytes and embryo of *Regnellidium diphyllum. Am. J. Bot.* **28,** 282–300.

Hirsch, A. M. (1975). The effect of sucrose on the differentiation of excised fern leaf tissue into either gametophytes or sporophytes. *Plant Physiol.* **56,** 390–93.

Hirsch, A. M. (1976). The development of aposporous gametophytes and regenerated sporophytes from epidermal cells of excised fern leaves: An anatomical study. *Am. J. Bot.* **63,** 263–71.

Holbrook-Walker, S. G., and Lloyd, R. M. (1973). Reproductive biology and gametophyte morphology of the Hawaiian fern genus *Sadleria* (Blechnaceae) relative to habitat diversity and propensity for colonization. *Bot. J. Linn. Soc.* **67,** 157–74.

Hotta, Y. (1960*a*). Role of nitrogenous compounds in the development of gametophyte of *Dryopteris erythrosora. Bot. Mag. (Tokyo)* **73,** 69–74.

Hotta, Y. (1960*b*). Morphological differentiation and growth substance in the gametophyte of *Dryopteris erythrosora. Bot. Mag. (Tokyo)* **73,** 191–4.

Hotta, Y., and Osawa, S. (1958). Control of differentiation in the fern gametophyte by amino acid analogs and 8-azaguanine. *Expt. Cell Res.* **15,** 85–94.

Hotta, Y., Osawa, S., and Sakaki, T. (1959). Ribonucleic acid and differentiation of the gametophyte of a polypodiaceous fern. *Develop. Biol.* **1,** 65–78.

Howard, A., and Haigh, M. V. (1968). Chloroplast aberrations in irradiated fern spores. *Mutation Res.* **6,** 263–80.

Howard, A., and Haigh, M. V. (1970). Radiation responses of fern spores during their first cell-cycle. *Int. J. Radiation Biol.* **18,** 147–58.

Howard, A., and Haigh, M. V. (1974). Apparent non-involvement of growth hormones in the induction by radiation of tumorous outgrowths in *Osmunda* prothalli. *Radiation Bot.* **14,** 59–61.

Howland, G. P. (1972). Changes in amounts of chloroplast and cytoplasmic ribosomal-RNAs and photomorphogenesis in *Dryopteris* gametophytes. *Physiol. Plantarum* **26,** 264–70.

Howland, G. P., and Boyd, E. L. (1974). Genetic control of photomorphogenesis. Isolation of nonfilamentous mutants after gamma irradiation of *Pteridium aquilinum* spores. *Radiation Bot.* **14,** 281–5.

Huckaby, C. S., Bassel, A. R. and Miller, J. H. (1982). Isolation of rhizoid and prothallial protoplasts from gametophytes of the fern, *Onoclea sensibilis. Plant Sci. Lett.* **25,** 203–8.

Huckaby, C. S., Kalantari, K., and Miller, J. H. (1982). Inhibition of *Onoclea sensibilis* spore germination by far-red light and cis-4-cyclohexene-1,2-dicarboximide. *Z. Pflanzenphysiol.* **105,** 375–8.

Huckaby, C. S., and Miller, J. H. (1984). Spore germination and rhizoid differentiation in *Onoclea sensibilis.* A two-dimensional electrophoretic analysis of the extant soluble proteins. *Plant. Physiol.* **74,** 656–62.

Huckaby, C. S., Nagmani, R., and Raghavan, V. (1981). Spore germination patterns in *Anogramma, Bommeria, Gymnopteris, Hemionitis* and *Pityrogramma. Am. Fern J.* **71,** 109–19.

Huckaby, C. S., and Raghavan, V. (1981*a*). Photocontrol of spore germination in the fern *Thelypteris kunthii. Physiol. Plantarum* **51,** 19–22.

Huckaby, C. S., and Raghavan, V. (1981*b*). Spore germination patterns in the ferns, *Cyathea* and *Dicksonia. Ann. Bot.* **47,** 397–403.

Huckaby, C. S., and Raghavan, V. (1981*c*). The spore-germination pattern of Thelypteroid ferns. *Am. J. Bot.* **68,** 517–23.

Hurel-Py, G. (1943). De l'influence de l'acide indole-β-acétique sur la culture des prothalles d'*Asplenium* en milieu aseptique. *Compt. Rend. Soc. Biol.* **137,** 57–8.

Hurel-Py, G. (1950*a*). Recherches préliminaires sur la culture aseptique des prothalles de Filicinées. *Rev. Gen. Bot.* **57,** 637–81.

Hurel-Py, G. (1950*b*). Recherches preliminaires sur la culture aseptiques des prothalles de Filicinées. *Rev. Gen. Bot.* **57,** 690–736.

Hurel-Py, G. (1955*a*). Action de quelques sucres sur la croissance des prothalles de fougères. *Compt. Rend. Acad. Sci. Paris* **240,** 1119–21.

Hurel-Py, G. (1955*b*). Action de quelques sucres sur la germination des spores d'*Alsophila australis* (Polypodiacées). *Compt. Rend. Acad. Sci. Paris* **241,** 1813–5.

Hurel-Py, G. (1969). Action du glucose et du fructose sur la croissance, le développement et le contenu en sucres solubles des prothalles de *Lygodium japonicum* (Sw.) en fonction de la durée d'éclairement. *Rev. Gen. Bot.* **76,** 531–71.

Hutchinson, S. A., and Fahim, M. (1958). The effects of fungi on the gametophytes of *Pteridium aquilinum* (L.) Kühn. *Ann. Bot.* **22,** 117–26.

Iqbal, J., and Schraudolf, H. (1977). Changes in protein biosynthesis during gibberellic acid induced induction and formation of antheridium in the fern *Anemia phyllitidis. Develop. Growth Diffn.* **19,** 85–92.

Iqbal, J., and Schraudolf, H. (1984*a*). Changes in isozymes shikimate dehydrogenase and peroxidase during gibberellic acid induced induction and formation of antheridium in the fern *Anemia phyllitidis. Pak. J. Bot.* **16,** 13–21.

Iqbal, J., and Schraudolf, H. (1984*b*). Tubulin synthesis during spermatogenesis in the fern *Anemia phyllitidis:* Demonstration with antitubulin antibodies. *Pak. J. Bot.* **16,** 101–7.

Isikawa, S. (1954). Light-sensitivity against the germination. I. 'Photoperiodism' of seeds. *Bot. Mag. (Tokyo)* **67,** 51–6.

Isikawa, S., and Oohusa, T. (1954). Effects of light upon the germination of spores of ferns. I. The relation of daily illuminating period to germination rate. *Bot. Mag. (Tokyo)* **67,** 193–7.

Isikawa, S., and Oohusa, T. (1956). Effects of light upon the germination of spores of ferns. II. Two light-periods of *Dryopteris crassirhizoma* Nakai. *Bot. Mag. (Tokyo)* **69,** 132–7.

Ito, M. (1960). Complete regeneration from single isolated cells of fern gametophyte. *Bot. Mag. (Tokyo).* **73,** 267.

Ito, M. (1962). Studies on the differentiation of fern gametophytes. I. Regeneration of single cells isolated from cordate gametophytes of *Pteris vittata. Bot. Mag. (Tokyo)* **75,** 19–27.

Ito, M. (1969). Intracellular position of nuclear division in protonema of *Pteris vittata. Embryologia* **10,** 273–83.

Ito, M. (1970). Light-induced synchrony of cell division in the protonema of the fern, *Pteris vittata. Planta* **90,** 22–31.

Ito, M. (1974). Effects of blue and red light on the cell division cycle in *Pteris* protonemata. *Plant Sci. Lett.* **3,** 351–5.

Jaffe, L., and Etzold, H. (1962). Orientation and locus of tropic photoreceptor molecules in spores of *Botrytis* and *Osmunda. J. Cell Biol.* **13,** 13–31.

Jarvis, S. J., and Wilkins, M. B. (1973). Photoresponses of *Matteuccia struthiopteris* (L.) Todaro. I. Germination. *J. Expt. Bot.* **24,** 1149–57.

Jayasekera, R. D. E., and Bell, P. R. (1971). The synthesis and distribution of ribonucleic acid in developing archegonia of *Pteridium aquilinum. Planta* **101,** 76–87.

Jayasekera, R. D. E., and Bell, P. R. (1972). The effect of thiouracil on the viability of eggs and embryogeny in *Pteridium aquilinum. Planta* **102,** 206–14.

Jayasekera, R. D. E., Cave, C. F., and Bell, P. R. (1972). The effect of thiouracil on the cytochemistry of the egg *Pteridium aquilinum. Cytobiologie* **6,** 253–60.

Johnson, D. M. (1985). New records for longevity of *Marsilea* sporocarps. *Am. Fern J.* **75,** 30–1.

Kadota, A., and Furuya, M. (1977). Apical growth of protonemata in *Adiantum capillus-veneris.* I. Red far-red reversible effect on growth cessation in the dark. *Develop. Growth Diffn.* **19,** 357–65.

Kadota, A., and Furuya, M. (1981). Apical growth of protonemata in *Adiantum capillus-veneris.* IV. Phytochrome-mediated induction in non-growing cells. *Plant Cell Physiol.* **22,** 629–38.

Kadota, A., Fushimi, Y., and Wada, M. (1986). Intracellular photoreceptive site for blue light-induced cell division in protonemata of the fern *Adiantum* – Further analysis by polarized light irradiation and cell centrifugation. *Plant Cell Physiol.* **27,** 989–95.

Kadota, A., Inoue, Y., and Furuya, M. (1986). Dichroic orientation of phytochrome intermediates in the pathway from P_R to P_{FR} as analyzed by double laser flash irradiations in polarotropism of *Adiantum* protonemata. *Plant Cell Physiol.* **27,** 867–73.

Kadota, A., Koyama, M., Wada, M., and Furuya, M. (1984). Action spectra for polarotropism in protonemata of the fern *Adiantum capillus-veneris. Physiol. Plantarum* **61,** 327–30.

Kadota, A., and Wada, M. (1986). Heart-shaped prothallia of the fern *Adiantum capillus-veneris* L. develop in the polarization plane of white light. *Plant Cell Physiol.* **27,** 903–10.

Kadota, A., Wada, M., and Furuya, M. (1979). Apical growth of protonemata in *Adiantum capillus-veneris.* III. Action spectra for the light effect on dark cessation of apical growth and the intracellular photoreceptive site. *Plant Sci. Lett.* **15,** 193–201.

Kadota, A., Wada, M., and Furuya, M. (1982). Phytochrome-mediated phototropism and different dichroic orientation of P_r and P_{fr} in protonemata of the fern *Adiantum capillus-veneris* L. *Photochem. Photobiol.* **35,** 533-6.

Kadota, A., Wada, M., and Furuya, M. (1985). Phytochrome-mediated polarotropism of *Adiantum capillus-veneris* L. protonemata as analyzed by microbeam irradiation with polarized light. *Planta* **165,** 30–6.

Kanamori, K. (1967). Origin and early development of apogamous embryos in the prothallia of *Dryopteris chinensis. J. Jap. Bot.* **42,** 111–8.

Kanamori, K. (1972). Apogamy in ferns with special reference to the apogamous embryogenesis. *Sci. Rep. Tokyo Kyoiku Daigaku Sec.* **B 15,** 111–31.

Kasemir, H., and Mohr, H. (1965). Die Regulation von Chlorophyll- und Proteingehalt in Farnvorkeimen durch sichtbare Strahlung. *Planta* **67,** 33–43.

Kato, J., and Purves, W. K., and Phinney, B. O. (1962). Gibberellin-like substances in plants. *Nature (London)* **196,** 687–8.

Kato, Y. (1955). Responses of plant cells to gibberellin. *Bot. Gaz.* **117,** 16–24.

Kato, Y. (1957*a*). Experimental studies on rhizoid-differentiation of certain ferns. *Phyton* **9,** 25–40.

Kato, Y. (1957*b*). Growth of the rhizoid and behavior of the nucleus in *Dryopteris erythrosora. Bot. Mag. (Tokyo)* **70,** 209–16.

Kato, Y. (1957*c*). The effects of colchicine and auxin on rhizoid formation of *Dryopteris erythrosora. Bot. Mag. (Tokyo)* **70,** 258–63.

Kato, Y. (1957*d*). Some experiments on the polarity of spores in *Dryopteris erythrosora* and *Equisetum arvense. Cytologia* **22,** 328–36.

Kato, Y. (1963). Physiological and morphogenetic studies of fern gametophytes in aseptic culture. I. Callus tissues from dark-cultured *Pteris vittata. Bot. Gaz.* **124,** 413–16.

Kato, Y. (1964*a*). Consequences of ultraviolet radiation on the differentiation and growth of fern gametophytes. *New Phytol.* **63,** 21–7.

Kato, Y. (1964*b*). Physiological and morphogenetic studies of fern gametophytes in aseptic culture. II. One- and two-dimensional growth in sugar media. *Bot. Gaz.* **125,** 33–7.

Kato, Y. (1964*c*). Physiological and morphogenetic studies of fern gametophytes by aseptic culture. III. Growth and differentiation of single cells isolated from callus tissues of *Pteris vittata. Cytologia* **29,** 79–85.

Kato, Y. (1965*a*). Physiological and morphogenetic studies of fern gametophytes and sporophytes in aseptic culture. IV. Controlled differentiation in leaf callus tissues. *Cytologia* **30,** 67–74.

Kato, Y. (1965*b*). Physiological and morphogenetic studies of fern gametophyte and sporophyte by aseptic culture. V. Further studies on one- and two-dimensional growth in gametophytes of *Pteris vittata. Bot. Mag. (Tokyo)* **78,** 149–55.

Kato, Y. (1965*c*). Physiological and morphogenetic studies of fern gametophyte and sporophyte by aseptic culture. VI. Notes on the alternation of generations. *Bot. Mag. (Tokyo)* **78,** 187–93.

Kato, Y. (1966). Responses of fern gametophytes to ultraviolet radiation. *Phyton* **23,** 23–7.

Kato, Y. (1967*a*). Physiological and morphogenetic studies of fern gametophytes and sporophytes in aseptic culture. VIII. Behaviour of *Pteris vittata* gametophytes on sugar and complex media in red light. *Planta* **77,** 127–34.

Kato, Y. (1967*b*). Effects of blue light on the survival of UV-irradiated fern gametophytic cells. *Cytologia* **32,** 519–23.

Kato, Y. (1968*a*). Physiological and morphogenetic studies of fern gametophytes and sporophytes by aseptic culture. IX. Spiral growth in callus cells of the fern, *Pteris vittata* L. *Cytologia* **33,** 331–5.

Kato, Y. (1968*b*). Physiological and morphogenetic studies of fern gametophytes and sporophytes in aseptic culture. X. Experiments on the induction of two-dimensional growth in gametophytes under red light. *Bot. Mag. (Tokyo)* **81,** 445–51.

Kato, Y. (1969*a*). Physiological and morphogenetic studies of fern gametophytes and sporophytes in aseptic culture. VII. Experimental modifications of dimensional growth in gametophytes of *Pteris vittata. Phytomorphology* **19,** 114–21.

Kato, Y. (1969*b*). Physiological and morphogenetic studies of fern gametophytes and sporophytes in aseptic culture. XI. Further studies on spiral growth in single cells from callus tissue and in rhizoid cells of *Pteris vittata. Phytomorphology* **19,** 279–85.

Kato, Y. (1970*a*). Physiological and morphogenetic studies of fern gametophytes and sporophytes in aseptic culture. XII. Sporophyte formation in dark cultured gametophyte of *Pteris vittata* L. *Bot. Gaz.* **131,** 205–10.

Kato, Y. (1970*b*). Studies on the formation, elongation and division of rhizoid cells in the fern, *Pteris vittata. Jap. J. Bot.* **20,** 213–36.

Kato, Y. (1973*a*). Control of rhizoid formation in the fern gametophyte. I. The effect of sugars. *Cytologia* **38,** 117–24.

Kato, Y. (1973*b*). Active charcoal and vermiculite: Effective agents on growth and morphogenesis of fern gametophytes. *Phytomorphology* **23,** 260–3.

Kawasaki, T. (1954*a*). Studies on the sexual generation of ferns. (1.) On the prothallia of *Cyathea boninsimensis* Copel. *J. Jap. Bot.* **29,** 201–4.

Kawasaki, T. (1954*b*). Studies on the sexual generation of ferns. (3). Effects of ultra short waves on spore-germination. *J. Jap. Bot.* **29,** 294–8.

Kelley, A. G., and Postlethwait, S. N. (1962). Effect of 2-chloroethyltrimethylammonium chloride on fern gametophytes. *Am. J. Bot.* **49,** 778–86.

Kermarrec, O., and Tourte, Y. (1984). Modification expérimentale de la différenciation du gamète femelle chez le *Marsilea vestita* (H. et G.). *Compt. Rend. Acad. Sci. Paris* Ser. III **299,** 709–14.

Khare, P. B. (1980). Studies on the reproductive biology of three homosporous ferns. *Brenesia* **18,** 191–200.

Khare, P. B. (1981). Effects of maleic hydrazide on gametophyte development in the fern *Cheilanthes farinosa. Kalikasan, Philipp. J. Biol.* **10,** 118–21.

Khare, P. B., and Roy, S. K. (1977). Maleic hydrazide effects on growth and differentiation of gametophyte in the fern *Dryopteris cochleata. Indian J. Expt. Biol.* **15,** 419–22.

Klekowski, E. J., Jr. (1969*a*). Reproductive biology of the Pteridophyta. II. Theoretical considerations. *Bot. J. Linn. Soc.* **62,** 347–59.

Klekowski, E. J., Jr. (1969*b*). Reproductive biology of the Pteridophyta. III. A study of the Blechnaceae. *Bot. J. Linn. Soc.* **62,** 361–77.

Klekowski, E. J., Jr. (1970*a*). Reproductive biology of the Pteridophyta. IV. An experimental study of mating systems in *Ceratopteris thalictroides* (L.) Brongn. *Bot. J. Linn. Soc.* **63,** 153–69.

Klekowski, E. J., Jr. (1970*b*). Evidence against self-incompatibility and for genetic lethals in the fern *Stenochlaena tenuifolia* (Desv.). *Bot. J. Linn. Soc.* **63,** 171–6.

Klekowski, E. J., Jr. (1970*c*). Populational and genetic studies of a homosporous fern – *Osmunda regalis. Am. J. Bot.* **57,** 1122–38.

Klekowski, E. J., Jr. (1971). Evidence for a duplicated locus in the fern *Osmunda regalis. J. Hered.* **62,** 367–70.

Klekowski, E. J., Jr. (1972). Evidence against self-incompatibility in the homosporous fern *Pteridium aquilinum. Evolution* **26,** 66–73.

Klekowski, E. J., Jr. (1973*a*). Genetic load in *Osmunda regalis* populations. *Am. J. Bot.* **60,** 146–54.

Klekowski, E. J., Jr. (1973*b*). Sexual and subsexual systems in homosporous pteridophytes: A new hypothesis. *Am. J. Bot.* **60,** 535–44.

Klekowski, E. J., Jr. (1976). Genetics of recessive lethality in the fern, *Osmunda regalis. J. Hered.* **67,** 146–8.

Klekowski, E. J., Jr. (1978). Screening aquatic ecosystems for mutagens with fern bioassays. *Environ. Health Persp.* **27,** 99–102.

Klekowski, E. J., Jr., and Baker, H. G. (1966). Evolutionary significance of polyploidy in the Pteridophyta. *Science* **153,** 305–7.

Klekowski, E. J., Jr., and Davis, E. L. (1977). Genetic damage to a fern population growing in a polluted environment: Segregation and description of gametophyte mutants. *Can. J. Bot.* **55,** 542–8.

Klekowski, E. J., Jr., and Hickok, L. G. (1974). Nonhomologous chromosome pairing in the fern *Ceratopteris. Am. J. Bot.* **61,** 422–32.

Klekowski, E. J., Jr., and Klekowski, E. (1982). Mutation in ferns growing in an environment contaminated with polychlorinated biphenyls. *Am. J. Bot.* **69,** 721–7.

Klekowski, E. J., Jr., and Lloyd, R. M. (1968). Reproductive biology of the Pteridophyta. I. General considerations and a study of *Onoclea sensibilis* L. *J. Linn. Soc. (Bot).* **60,** 315–24.

Knobloch, I. W. (1969). The spore pattern in some species of *Cheilanthes. Am. J. Bot.* **56,** 646–53.

Knobloch, I. W., Tai, W., and Ninan, T. A. (1973). The cytology of some species of the genus *Notholaena. Am. J. Bot.* **60,** 92–5.

Knudson, L. (1940). Permanent changes of chloroplasts induced by X rays in the gametophyte of *Polypodium aureum. Bot. Gaz.* **101,** 721–58.

Konar, R. N., and Kapoor, R. K. (1974). Embryology of *Azolla pinnata. Phytomorphology* **24,** 228–61.

Koop, H. U. (1973). Abbaustoffwechsel und Einbau von Bromdesoxyuridin in die Desoxyribonukleinsäure der Gametophyten von *Anemia phyllitidis. Protoplasma* **77,** 343–59.

Korn, R. W. (1974). Computer simulation of the early development of the gametophyte of *Dryopteris thelypteris* (L.) Gray. *Bot. J. Linn. Soc.* **68,** 163–71.

Koshiba, T., Minamikawa, T., and Wada, M. (1984). Hydrolytic enzyme activities in germinating spores of *Adiantum capillus-veneris* L. *Bot. Mag. (Tokyo)* **97,** 323–31.

Kotenko, J. L. (1985). Antheridial formation in *Onoclea sensibilis* L.: Genesis of the jacket cell walls. *Am. J. Bot.* **72,** 596–605.

Kotenko, J. L. (1986). Antheridium formation in *Onoclea sensibilis* L.: Cytoplasmic polarity and determination of wall positions. *Bot. Gaz.* **147,** 28–39.

Kotenko, J. L., Miller, J. H., and Robinson, A. I. (1987). The role of asymmetric cell division in pteridophyte cell differentiation. I. Localized metal accumulation and differentiation in *Vittaria* gemmae and *Onoclea* prothallia. *Protoplasma* **136,** 81–95.

Kott, L. S., and Britton, D. M. (1982). A comparative study of spore germination of some *Isoetes* species of northeastern North America. *Can. J. Bot.* **60,** 1679–87.

Kott, L. S., and Peterson, R. L. (1974). A comparative study of gametophyte development of the diploid and tetraploid races of *Polypodium virginianum. Can. J. Bot.* **52,** 91–6.

Kraiss, S., and Gemmrich, A. R. (1986). Photocontrol of chloroplast lipids in fern gametophytes. *Z. Naturforsch.* **41c,** 591–6.

Kressel, M. (1965). Antheridien- und Archegonienbildung bei *Microlepia speluncae* (L.) Moore. *Z. Pflanzenphysiol.* **53,** 366–8.

Kshirsagar, M. K., and Mehta, A. R. (1978). *In vitro* studies in ferns: Growth and differentiation in rhizome callus of *Pteris vittata. Phytomorphology* **28,** 50–8.

Kuligowski-Andres, J., Faivre, M., and Tourte, Y. (1982). Etude de la structure et du comportement du noyau spermatique lors de la fécondation chez une fougère, le *Marsilea vestita;* effet de la colchicine. *Compt. Rend. Acad. Sci. Paris* Ser. III **295,** 335–40.

Kuligowski-Andres, J., and Tourte, Y. (1978). Première étude, par autoradiographie, de la fécondation chez une Ptéridophyte. *Biol. Cell.* **31,** 101–8.

Kumar, G., and Roy, S. K. (1985). Effect of lycorine on germination and growth of prothallus of *Cheilanthes farinosa* Kaulf. *Indian J. Expt. Biol.* **23,** 356–8.

Kurita, S. (1961). Chromosome numbers of some Japanese ferns (II). *Bot. Mag. (Tokyo)* **74,** 395–401.

Labouriau, L. G. (1952). Contribution to the study of sporophyll morphogenesis in *Anemia* Sw. IV. Some effects of applied auxins. *Rev. Brasil. Biol.* **12,** 33–43.

Labouriau, L. G. (1958). Studies on the initiation of sporangia in ferns. *Arq. Mus. Nacional, Rio de Janero* **46,** 119–201.

Laird, S., and Sheffield, E. (1986). Antheridia and archegonia of the apogamous fern *Pteris cretica. Ann. Bot.* **57,** 139–43.

Laroche, J., Guervin C., and Le Coq, C. (1978). Le génome et l'expression sexuelle chez l'*Equisetum arvense* L. I. Comportement du gamétophyte cultivé sur milieu témoin. *Rev. Gen. Bot.*, **85,** 221–54.

Laurent, S., and Lefebvre, M. F. (1980). Etude de l'effet de différentes carences en éléments minéraux sur la teneur en tanoides de gamétophytes de Filicinées. *Bull. Soc. Bot. Fr. Lett. Bot.* **127,** 119–27.

Lawton, E. (1932). Regeneration and induced polyploidy in ferns. *Am. J. Bot.* **19,** 303–33.

Lawton, E. (1936). Regeneration and induced polyploidy in *Osmunda regalis* and *Cystopteris fragilis*. *Am. J. Bot.* **23,** 107–14.

Le Coq, C., Laroche, J., and Guervin, C. (1980), Le génome et l'expression sexuelle chez l'*Equisetum arvense* L. III. Comportement du gamétophyte cultivé sur milieu témoin additionné de testostérone. *Rev. Cytol. Biol. Veg.-Bot.* **3,** 147–56.

Leung, C., and Näf, U. (1979). On the cytology of antheridium formation in the fern species *Onoclea sensibilis*. I. Cytology of cell plate and cell wall formation. *Am. J. Bot.* **66,** 765–75.

Lever, M. (1971). The initial metabolism of germinating fern spores. *Phytochemistry* **10,** 2995–6.

Lintilhac, P. M. (1974). Differentiation, organogenesis, and the tectonics of cell wall orientation. III. Theoretical considerations of cell wall mechanics. *Am. J. Bot.* **61,** 230–7.

Lloyd, R. M. (1974). Mating systems and genetic load in pioneer and non-pioneer Hawaiian Pteridophyta. *Bot. J. Linn. Soc.* **69,** 23–35.

Lloyd, R. M. (1980). Reproductive biology and gametophyte morphology of new world populations of *Acrostichum aureum*. *Am. Fern J.* **70,** 99–110.

Lloyd, R. M. and Buckley, D. P. (1986). Effects of salinity on gametophyte growth of *Acrostichum aureum* and *A. danaeifolium*. *Fern Gaz.* **13,** 97–102.

Lloyd, R. M., and Gregg, T. L. (1975). Reproductive biology and gametophyte morphology of *Acrostichum danaeifolium* from Mexico. *Am. Fern J.* **65,** 105–20.

Lloyd, R. M., and Klekowski, E. J., Jr. (1970). Spore germination and viability in Pteridophyta: Evolutionary significance of chlorophyllous spores. *Biotropica* **2,** 129–37.

Lloyd, R. M., and Warne, T. R. (1978). The absence of genetic load in a morphologically variable sexual species, *Ceratopteris thalictroides* (Parkeriaceae). *Syst. Bot.* **3,** 20–36.

Lovis, J. D. (1973). A biosystematic approach to phylogenetic problems and its applications to the Aspleniaceae. In: *The Phylogeny and Classification of the Ferns,* ed. A.C. Jermy, J.A. Crabb and B.A. Thomas, *Bot. J. Linn. Soc.* **67** (Suppl. 1), 211–27.

Lovis, J. D. (1977). Evolutionary patterns and processes in ferns. *Adv. Bot. Res.* **4,** 229–415.

Loyal, D. S., and Chopra, H. K. (1976). Apogamous response of megagametophyte of *Regnellidium diphyllum*. *Phytomorphology* **26,** 355–8.

Loyal, D. S., and Chopra, H. K. (1977). *In vitro* life-cycle, regeneration and apospory in *Ceratopteris pteridoides* (Hook.) Hieron. *Curr. Sci.* **46,** 89–90.

Loyal, D. S., and Ratra, R. (1979). The gametophyte of three thelypteroid ferns with particular reference to segmentation pattern during two-dimensional growth. *J. Indian Bot. Soc.* **58,** 233–40.

Lucas, R. C., and Duckett, J. G. (1980). A cytological study of the male and female sporocarps of the heterosporous fern *Azolla filiculoides* Lasm. *New Phytol.* **85,** 409–18.

Ludueña, R. F., Myles, D. G., and Pfeffer, T. A. (1980). Isolation and partial characterization of flagellar tubulin from *Marsilea vestita. Expt. Cell Res.* **130,** 455–9.

Lugardon, B. (1974). La structure fine de l'exospore et de la périspore des filicinées isosporées. II. Filicales. Commentaires. *Pollen et Spores* **16,** 161–226.

Maeda, M., and Ito, M. (1981). Isolation of protoplasts from fern prothallia and their regeneration to gametophytes. *Bot. Mag. (Tokyo)* **94,** 35–40.

Mahlberg, P. G., and Baldwin, M. (1975). Experimental studies on megaspore viability, parthenogenesis, and sporophyte formation in *Marsilea, Pilularia,* and *Regnellidium. Bot. Gaz.* **136,** 269–273.

Mahlberg, P. G., and Yarus, S. (1977). Effects of light, pH, temperature, and crowding on megaspore germination and sporophyte formation in *Marsilea. J. Expt. Bot.* **28,** 1137–46.

Maly, R. (1951). Cytomorphologische Studien an Strahleninduzierten, Konstant abweichenden Plastidenformen bei Farnprothallien. *Z. Indukt. Abst. Vererbungslehre* **83,** 447–78.

Manabe, K., Ibushi, N., Nakayama, A., Takaya, S., and Sugai, M. (1987). Spore germination and phytochrome biosynthesis in the fern *Lygodium japonicum* as affected by gabaculine and cycloheximide. *Physiol. Plantarum* **70,** 571–6.

Manton, I. (1932). Contributions to the cytology of apospory in ferns. I. A case of induced apospory in *Osmunda regalis.* L. *J. Genet.* **25,** 423–30.

Manton I. (1950). *Problems of Cytology and Evolution in the Pteridophyta.* Cambridge: University Press.

Manton, I. (1959). Observations on the microanatomy of the spermatozoid of the bracken fern *(Pteridium aquilinum). J. Biophys. Biochem. Cytol.* **6,** 413–18.

Manton, I., and Sledge, W. A. (1954). Observations on the cytology and taxonomy of the pteridophyte flora of Ceylon. *Phil. Trans. Roy. Soc. Lond.* **238B,** 127–85.

Marc, J., and Gunning, B. E. S. (1986). Immunofluorescent localization of cytoskeletal tubulin and actin during spermatogenesis in *Pteridium aquilinum* (L.) Kuhn. *Protoplasma* **134,** 163–77.

Marc, J., Gunning, B. E. S., Hardham, A. R., Perkin, J. L., and Wick, S. M. (1988). Monoclonal antibodies to surface and cytoskeletal components of the spermatozoid of *Pteridium aquilinum. Protoplasma* **142,** 5–14.

Marcondes-Ferreira, W., and Felippe, G. M. (1984). Effects of light and temperature on the germination of spores of *Cyathea delgadii. Revis. Brasil. Bot.* **7,** 53–6.

Marengo, N. P. (1949). A study of the cytoplasmic inclusions during sporogenesis in *Onoclea sensibilis. Am. J. Bot.,* **36,** 603–13.

Marengo, N.P. (1954). The relation of the cytoplasmic inclusions to the establishment of tetrahedral symmetry in the spore quartet of *Osmunda regalis. Bull. Torrey Bot. Cl.* **81,** 501–8.

Marengo, N. P. (1962). The cytokinetic basis of tetrahedral symmetry in the spore quartet of *Adiantum hispidulum. Bull. Torrey Bot. Cl.* **89,** 42–8.

Marengo, N. P. (1973). The fine structure of the dormant spore of *Matteuccia struthiopteris. Bull. Torrey Bot. Cl.* **100,** 147–50.

Marengo, N. P. (1977). Ultrastructural features of the dividing meiocyte of *Onoclea sensibilis. Am. J. Bot.* **64,** 600–1.

Marengo, N. P., and Badalamente, M. A. (1978). The fine structure of the newly formed spore of *Onoclea sensibilis. Am. Fern J.* **69,** 52–4.

Masuyama, S. (1972). The sequence of sex expression in the prothallia of *Adiantum pedatum* L. and *A. capillus-veneris* L. *J. Jap. Bot.* **47,** 97–106.

Masuyama, S. (1975*a*). The sequence of the gametangium formation in homosporous fern gametophytes. I. Patterns and their possible effect on the fertilization, with special reference to the gametophytes of *Athyrium*. *Sci. Rep. Tokyo Kyoiku Daigaku* Sec. B **16,** 47–69.

Masuyama, S. (1975*b*). The sequence of the gametangium formation in homosporous fern gametophytes. II. Types and their taxonomic distribution. *Sci. Rep. Tokyo Kyoiku Daigaku* Sec. B **16,** 71–86.

Masuyama, S. (1975*c*). The gametophyte of *Pleurosoriopsis makinoi* (Maxim.) Fomin. *J. Jap. Bot.* **50,** 105–14.

Masuyama, S. (1979). Reproductive biology of the fern *Phegopteris decursive-pinnata*. I. The dissimilar mating systems of diploids and tetraploids. *Bot. Mag. (Tokyo)* **92,** 275–89.

Masuyama, S. (1986). Reproductive biology of the fern *Phegopteris decursive-pinnata*. II. Genetic analyses of self-sterility in diploids. *Bot. Mag. (Tokyo)* **99,** 107–21.

McCauley, D. E., Whittier, D. P., and Reilly, L. M. (1985). Inbreeding and the rate of self-fertilization in a grape fern, *Botrychium dissectum*. *Am. J. Bot.* **72,** 1978–81.

Mehra, P. N. (1938*a*). Apogamy in *Adiantum lunulatum* Burm, Part I. (Morphological). *Proc. Indian Acad. Sci.* **8B,** 192–201.

Mehra, P. N. (1938*b*). Apogamy in *Pteris biaurita* Linn. *Proc. Indian Acad. Sci.* **8B,** 202–10.

Mehra, P. N. (1944). Cytological investigation of apogamy in *Adiantum lunulatum* Burm. *Proc. Natl. Acad. Sci. India* **13B,** 189–204.

Mehra, P. N. (1952). Colchicine effect and the production of abnormal spermatozoids in the prothalli of *Dryopteris subpubescens* (Bl.) C. Chr. and *Goniopteris prolifera*. Roxb. *Ann. Bot.* **16,** 49–56.

Mehra, P. N. (1961). Chromosome numbers in Himalayan ferns. *Res. Bull. Panjab Univ.* **12,** 139–64.

Mehra, P. N. (1972). Some aspects of differentiation in cryptogams. *Res. Bull. Panjab Univ.* **23,** 221–42.

Mehra, P. N. and Bir, S. S. (1960). Cytological observations on *Asplenium cheilosorum* Kunze. *Cytologia* **25,** 17–27.

Mehra, P. N., and Loyal, D. S. (1956). Colchicine effect on the prothalli of *Goniopteris multilineata* (Wall.) Bedd. and *G. prolifera* (Roxb.) with emphasis on abnormal spermatogenesis in polyploid prothalli. *Ann. Bot.* **20,** 544–52.

Mehra, P. N., and Palta, H. K. (1969). Radiobiological investigations on *Pteris vittata* L. I. X-ray induced plastid variations in the gametophytes. *Radiation Bot.* **9,** 93–103.

Mehra, P. N., and Palta, H. K. (1971). *In vitro*-controlled differentiation of the root callus of *Cyclosorus dentatus*. *Phytomorphology* **21,** 367–75.

Mehra, P. N., and Singh, G. (1957). Cytology of Hymenophyllaceae. *J. Genet.* **55,** 379–93.

Mehra, P. N., and Sulklyan, D. S. (1969). *In vitro* studies on apogamy, apospory and controlled differentiation of rhizome segments of the fern, *Ampelopteris prolifera* (Retz.) Copel. *Bot. J. Linn. Soc.* **62,** 431–43.

Mehra, P. N., and Verma, S. C. (1960). Cytotaxonomic observations on some west Himalayan Pteridaceae. *Caryologia* **13,** 619–50.

Menke, W., and Fricke, N. (1964). Beobachtungen über die Entwicklung der Archegonien von *Dryopteris filix mas*. *Z. Naturforsch.* **19b,** 520–4.

Meyer, D.E. (1953). Über das verhalten einzelner isolierter Prothalliumzellen und dessen Bedeutung für Korrelation und Regeneration. *Planta* **41,** 642–5.

Meyer zu Bentrup, F. W. (1963). Vergleichende Untersuchungen zur Polaritätsinduktion durch das Licht an der *Equisetum*-Spore und der *Fucus*-Zygote. *Planta* **59,** 472–91.

Meyer zu Bentrup, F. W. (1964). Zur Frage eines Photoinaktivierungs-Effektes bei der Polaritätsinduktion in *Equisetum*Sporen und *Fucus* zygoten. *Planta* **63,** 356–65.

Miller, J. H. (1961). The effect of auxin and guanine on cell expansion and cell division in the gametophyte of the fern, *Onoclea sensibilis. Am. J. Bot.* **48,** 816–19.

Miller, J. H. (1968*a*). Fern gametophytes as experimental material. *Bot. Rev.* **34,** 361–440.

Miller, J. H. (1968*b*). An evaluation of specific and non-specific inhibition of 2-dimensional growth in fern gametophytes. *Physiol. Plantarum* **21,** 699–710.

Miller, J. H. (1980*a*). Differences in the apparent permeability of spore walls and prothallial cell walls in *Onoclea sensibilis. Am. Fern J.* **70,** 119–23.

Miller, J. H. (1980*b*). Orientation of the plane of cell division in fern gametophytes: The roles of cell shape and stress. *Am. J. Bot.* **67,** 534–42.

Miller, J. H. (1987). Inhibition of fern spore germination by lipophilic solvents. *Am. J. Bot.* **74,** 1706–8.

Miller, J. H., and Bassel, A. R. (1980). Effects of caffeine on germination and differentiation in spores of the fern, *Onoclea sensibilis. Physiol. Plantarum* **50,** 213–20.

Miller, J. H., and Greany, R. H. (1974). Determination of rhizoid orientation by light and darkness in germinating spores of *Onoclea sensibilis. Am. J. Bot.* **61,** 296–302.

Miller, J. H., and Greany, R. H. (1976). Rhizoid differentiation in fern spores: Experimental manipulation. *Science* **193,** 687–9.

Miller, J. H., and Kotenko, J. L. (1987). The use of alizarin red S to detect and localize calcium in gametophyte cells of ferns. *Stain Tech.* **62,** 237–45.

Miller, J. H., and Miller, P. M. (1961). The effect of different light conditions and sucrose on the growth and development of the gametophyte of the fern, *Onoclea sensibilis. Am. J. Bot.* **48,** 154–9.

Miller, J. H., and Miller, P. M. (1963). Effects of red and far-red illumination on the elongation of fern protonemata and rhizoids. *Plant Cell Physiol.* **4,** 65–72.

Miller, J. H., and Miller, P. M. (1964*a*). Blue light in the development of fern gametophytes and its interaction with far-red and red light. *Am. J. Bot.* **51,** 329–34.

Miller J. H., and Miller, P. M. (1964*b*). Effects of auxin on cell enlargement on fern gametophytes: Developmental stage of the cell and response to auxin. *Am. J. Bot.* **51,** 431–6.

Miller, J. H., and Miller, P. M. (1965). The relationship between the promotion of elongation of fern protonemata by light and growth substances. *Am. J. Bot.* **52,** 871–6.

Miller, J. H., and Miller, P. M. (1967*a*). Action spectra for light-induced elongation in fern protonemata. *Physiol. Plantarum* **20,** 128–38.

Miller, J. H., and Miller, P. M. (1967*b*). Interaction of photomorphogenetic pigments in fern gametophytes: Phytochrome and a yellow-light-absorbing pigment. *Plant Cell Physiol.* **8,** 765–9.

Miller, J. H., and Miller, P. M. (1970). Unusual dark-growth and antheridial differentiation in some gametophytes of the fern *Onoclea sensibilis. Am. J. Bot.* **57,** 1245–8.

Miller, J. H., and Stephani, M. C. (1971). Effects of colchicine and light on cell form in fern gametophytes. Implications for a mechanism of light-induced cell elongation. *Physiol. Plantarum* **24,** 264–71.

Miller, J. H., Vogelmann, T. C., and Bassel, A. R. (1983). Promotion of fern rhizoid elongation by metal ions and the function of the spore coat as an ion reservoir. *Plant Physiol.* **71,** 828–34.

Miller, J. H., and Wagner, P. M. (1987). Co-requirement for calcium and potassium in the germination of spores of the fern *Onoclea sensibilis. Am. J. Bot.* **74,** 1585–9.

Miller, J. H., and Wright, D. R. (1961). An age-dependent change in the response of fern gametophytes to red light. *Science* **134,** 1629.

Miller, P. M., and Miller, J. H. (1966). Temperature dependence of the effects of light and auxin on the elongation of the fern protonemata. *Plant Cell Physiol.* **7,** 485–8.

Miller, P. M., Sweet, H. C., and Miller, J. H. (1970). Growth regulation by ethylene in fern gametophytes. I. Effects on protonemal and rhizoidal growth and interaction with auxin. *Am. J. Bot.* **57,** 212–7.

Minamikawa, T., Koshiba, T., and Wada, M. (1984). Compositional changes in germinating spores of *Adiantum capillus-veneris* L. *Bot. Mag. (Tokyo)* **97,** 313–22.

Minamikawa, T., Masuda, N., Kadota, A., and Wada, M. (1987). Effect of sulfite on compositional changes of cell constituents in germinating spores of *Adiantum capillus-veneris* L. *Bot. Mag. (Tokyo)* **100,** 1–8.

Mineyuki, Y., and Furuya, M. (1980). Effect of centrifugation on the development and timing of premitotic positioning of the nucleus in *Adiantum* protonemata. *Develop. Growth Diffn.* **22,** 867–74.

Mineyuki, Y., and Furuya, M. (1985). Involvement of microtubules on nuclear positioning during apical growth in *Adiantum* protonemata. *Plant Cell Physiol.* **26,** 627–34.

Mineyuki, Y., and Furuya, M. (1986). Involvement of colchicine-sensitive cytoplasmic element in premitotic nuclear positioning of *Adiantum* protonemata. *Protoplasma* **130,** 83–90.

Mineyuki, Y., Takagi, M., and Furuya, M. (1984). Changes in organelle movement in the nuclear region during the cell cycle of *Adiantum* protonema. *Plant Cell Physiol.* **25,** 297–308.

Mineyuki, Y., Yamada, M., Takagi, M., Wada, M., and Furuya, M. (1983). A digital image processing technique for the analysis of particle movements: Its application to organelle movements during mitosis in *Adiantum* protonemata. *Plant Cell Physiol.* **24,** 225–34.

Miura, M., Koshiba, T., and Minamikawa, T. (1986). Characterization and role of RNAs synthesized during early spore germination of the fern *Cyathea. J. Plant Physiol.* **123,** 487–95.

Miyata, M., Wada, M., and Furuya, M. (1979). Effects of phytochrome and blue-near ultraviolet light-absorbing pigment on duration of component phases of the cell cycle in *Adiantum* gametophytes. *Develop. Growth Diffn.* **21,** 577–84.

Mizukami, I., and Gall, J. G. (1966). Centriole replication. II. Sperm formation in the fern, *Marsilea,* and the cycad, *Zamia. J. Cell Biol.* **29,** 97–111.

Mohan Ram, H. Y., and Chatterjee, J. (1970). Gametophytes of *Equisetum ramosissimum* subsp. *ramosissimum.* II. Sexuality and its modification. *Phytomorphology* **20,** 151–72.

Mohr, H. (1956*a*). Die Beeinflussung der Keimung von Farnsporen durch Licht und andere Faktoren. *Planta* **46,** 534–51.

Mohr, H. (1956*b*). Die Abhängigkeit des Protonemawachstums und der Protonemapolarität bei Farnen vom Licht. *Planta* **47,** 127–58.
Mohr, H. (1965). Die Steuerung der Entwicklung durch Licht an Beispiel der Farngametophyten. *Ber. Deut. Bot. Ges.* **78,** 54–68.
Mohr, H., and Barth, C. (1962). Ein Vergleich der Photomorphogenese der Gametophyten von *Alsophilia australis* (Br) und *Dryopteris filix-mas* (L.) Schott. *Planta* **58,** 580–93.
Mohr, H., and Holl, G. (1964). Die Regulation der Zellaktivität bei Farnvorkeimen durch Licht. *Z. Bot.* **52,** 209–21.
Mohr, H., Meyer, U., and Hartmann, K. (1964). Die Beeinflussung der Farnsporen-Keimung [*Osmunda cinnamomea* (L.) und *O. claytoniana* (L.)] über das Phytochromsystem und die Photosynthese. *Planta* **60,** 483–96.
Mohr, H., and Ohlenroth, K. (1962). Photosynthese und Photomorphogenese bei Farnvorkeimen von *Dryopteris filix-mas. Planta* **57,** 656–64.
Montardy-Pausader, J. (1982). Cytomorphogénèse du jeune gamétophyte de l'*Anemia phyllitidis* (L.) Sw. *Rev. Cytol. Biol. Veg.-Bot.* **5,** 217–54.
Morel, G. (1950). Sur la culture des tissus d'*Osmunda cinnamomea. Compt. Rend. Acad. Sci. Paris* **230,** 2318–20.
Morel, G. (1963). Leaf regeneration in *Adiantum pedatum. J. Linn. Soc. (Bot.)* **58,** 381–3.
Morel, G., and Wetmore, R. H. (1951). Fern callus tissue culture. *Am. J. Bot.* **38,** 141–3.
Morlang, C., Jr. (1967). Hybridization, polyploidy and adventitious growth in the genus *Asplenium. Am. J. Bot.* **54,** 887–97.
Morzenti, V. M. (1962). A first report on pseudomeiotic sporogenesis, a type of spore reproduction by which "sterile" ferns produce gametophytes. *Am. Fern J.* **52,** 69–78.
Mosebach, G. (1943). Über die Polarisierung der *Equisetum*-Spore durch das Licht. *Planta* **33,** 340-87.
Mottier, D. M. (1927). Behavior of certain fern prothallia under prolonged cultivation. *Bot. Gaz.* **83,** 244–66.
Munroe, M. H., and Bell, P. R. (1970). The fine structure of fern root cells showing apospory. *Develop. Biol.* **23,** 550–62.
Munroe, M. H., and Sussex, I. M. (1969). Gametophyte formation in bracken fern root cultures. *Can. J. Bot.* **47,** 617–21.
Munther, W. E., and Fairbrothers, D. E. (1980). Allelopathy and autotoxicity in three eastern North American ferns. *Am. Fern J.* **70,** 124–35.
Murata, T., Kadota, A., Hogetsu, T., and Wada, M. (1987). Circular arrangement of cortical microtubules around the subapical part of a tip-growing fern protonema. *Protoplasma* **141,** 135–8.
Muromtsev, G. S., Agnistikova, V. N., Lupova, L. M., Dubovaya, L. P., and Lekareva, T. A. (1964). Gibberellin-like substances in ferns and mosses. *Izvest. Akad. Nauk. SSSR (Ser. Biol.)* **5,** 727–34.
Myles, D. G. (1978). The fine structure of fertilization in the fern *Marsilea vestita. J. Cell Sci.* **30,** 265–81.
Myles, D. G., and Bell, P. R. (1975). An ultrastructural study of the spermatozoid of the fern, *Marsilea vestita. J. Cell Sci.* **17,** 633–45.
Myles, D. G., and Hepler, P. K. (1977). Spermiogenesis in the fern *Marsilea:* Microtubules, nuclear shaping, and cytomorphogenesis. *J. Cell Sci.* **23,** 57–83.
Myles, D. G., and Hepler, P. K. (1982). Shaping of the sperm nucleus in *Marsilea:* A distinction between factors responsible for shape generation and shape determination. *Develop. Biol.* **90,** 238–52.

Myles, D. G., Southworth, D., and Hepler, P. K. (1978*a*). Cell surface topography during *Marsilea* spermiogenesis: Flagellar reorientation and membrane particle arrays. *Protoplasma* **93,** 405–17.

Myles, D. G., Southworth, D., and Hepler, P. K. (1978*b*). A freeze-fracture study of the nuclear envelope during spermiogenesis in *Marsilea.* Formation of a pore-free zone associated with the microtubule ribbon. *Protoplasma* **93,** 419–31.

Näf, U. (1956). The demonstration of a factor concerned with the initiation of antheridia in polypodiaceous ferns. *Growth* **20,** 91–105.

Näf, U. (1958). On the physiology of antheridium formation in the bracken fern [*Pteridium aquilinum* (L.) Kuhn.]. *Physiol. Plantarum* **11,** 728–46.

Näf, U. (1959). Control of antheridium formation in the fern species *Anemia phyllitidis. Nature (London)* **184,** 798–800.

Näf, U. (1960). On the control of antheridium formation in the fern species *Lygodium japonicum. Proc. Soc. Expt. Biol. Med.* **105,** 82–6.

Näf, U. (1961*a*). On the physiology of antheridium formation in ferns. In: *Plant Growth Regulation. Fourth International Conference on Plant Growth Regulation,* ed. R.M. Klein, pp. 709–23. Ames: Iowa State University Press.

Näf, U. (1961*b*). Mode of action of an antheridium-inducing substance in ferns. *Nature (London)* **189,** 900–03.

Näf, U. (1962*a*). Loss of sensitivity to the antheridial factor in maturing gametophytes of the fern *Onoclea sensibilis. Phyton* **18,** 173–82.

Näf, U. (1962*b*). Developmental physiology of lower archegoniates. *Ann. Rev. Plant Physiol.* **13,** 507–32.

Näf, U. (1963). Antheridium formation in ferns – A model for the study of developmental change. *J. Linn. Soc. (Bot.)* **58,** 321–31.

Näf, U. (1965). On antheridial metabolism in the fern species *Onoclea sensibilis* L. *Plant Physiol.* **40,** 888–90.

Näf, U. (1966). On dark-germination and antheridium formation in *Anemia phyllitidis. Physiol. Plantarum* **19,** 1079–88.

Näf, U. (1967*a*). *Anemia phyllitidis*: Inducibility of physiological state antagonistic to antheridium formation. *Science* **156,** 1117–19.

Näf, U. (1967*b*). On the induction of a phase inhibitory to antheridium formation in the juvenile prothallus of the fern species *Anemia phyllitidis. Z. Pflanzenphysiol.* **56,** 353–65.

Näf, U. (1968). On separation and identity of fern antheridiogens. *Plant Cell Physiol.* **9,** 27–33.

Näf, U. (1969). On the control of antheridium formation in ferns. In: *Current Topics in Plant Science,* ed. J. E. Gunckel, pp. 77–93. New York: Academic Press.

Näf, U. (1979). Antheridiogens and antheridia development. In: *The Experimental Biology of Ferns,* ed. A. F. Dyer, pp. 435–70. London: Academic Press.

Näf, U., Nakanishi, K., and Endo, M. (1975). On the physiology and chemistry of fern antheridiogens. *Bot. Rev.* **41,** 315–59.

Näf, U., Sullivan, J., and Cummins, M. (1969). New antheridiogen from the fern *Onoclea sensibilis. Science* **163,** 1357–8.

Näf, U., Sullivan, J., and Cummins, M. (1974). Fern antheridiogen: Cancellation of a light-dependent block to antheridium formation. *Develop. Biol.* **40,** 355–65.

Nagatani, A., Suzuki, H., and Furuya, M. (1983). Protein synthesis during photocontrolled progression of the cell cycle in single-celled protonemata of the fern *Adiantum capillus-veneris. Develop. Growth Diffn.* **25,** 217–26.

Nagmani, R., and Raghavan, V. (1983). Origin of the rhizoid and protonemal cell during germination of spores of *Drymoglossum, Platycerium* and *Pyrrosia* (Polypodiaceae). *Bot. Gaz.* **144,** 67–72.

Nagy, A. H., Bocsi, J., Paless, G., and Vida, G. (1984). Red and far-red effect on the isoenzyme composition of young fern gametophytes. *Biochem. Physiol. Pflanzen* **179,** 335–7.

Nagy, A. H., Paless, G., and Vida, G. (1978). Differential protein synthesis after red light illuminations in germinating fern spores. *Biol. Plantarum* **20,** 193–200.

Nakanishi, K., Endo, M., Näf, U., and Johnson, L. F. (1971). Structure of the antheridium-inducing factor of the fern *Anemia phyllitidis*. *J. Am. Chem. Soc.* **93,** 5579–81.

Nakazawa, S. (1952). Studies on the polarity of *Equisetum arvense* L. *Bull. Yamagata Univ. Nat. Sci.* **2,** 125–52.

Nakazawa, S. (1956). The latent polarity in *Equisetum* spores. *Bot. Mag. (Tokyo)* **69,** 506–9.

Nakazawa, S. (1957). Centrifugacion de esporas de *Equisetum*. *Anal. Inst. Biol. Univ. Mexico* **28,** 11–5.

Nakazawa, S. (1958). The rhizoid point as the basophilic center of *Equisetum* spores. *Phyton* **10,** 1–6.

Nakazawa, S. (1959). Morphogenesis of the fern protonema. I. Polar susceptibility to colchicine in *Dryopteris varia*. *Phyton* **12,** 59–64.

Nakazawa, S. (1960*a*). Morphogenesis of the fern protonema. II. Modification of the apical differentiation in *Dryopteris* affected by IAA. *Protoplasma* **52,** 1–4.

Nakazawa, S. (1960*b*). Morphogenesis of the fern protonema. III. Differentiation of the 'metallophilic cytoplasm' in *Equisetum*. *Phyton* **14,** 37–41.

Nakazawa, S. (1960*c*). Cytodifferentiation patterns of *Dryopteris* protonema modified by some chemical agents. *Cytologia* **25,** 352–61.

Nakazawa, S. (1961). The polarity theory of morphogenetic fields. *Sci. Rep. Tohoku Univ.* Ser. IV. Biol. **27,** 57–92.

Nakazawa, S. (1963). Role of the protoplasmic connections in the morphogenesis of fern gametophytes. *Sci. Rep. Tohoku Univ.* Ser. IV. Biol. **29,** 247–55.

Nakazawa, S., and Kimura, S. (1964). Reduction of TTC and Janus green B in fern gametophytes in relation to polarity. *Bot. Mag. (Tokyo)* **77,** 222–7.

Nakazawa, S., and Ootaki, T. (1961). Polarity reversal in *Dryopteris* protonema. *Naturwissenschaften* **48,** 557–8.

Nakazawa, S., and Ootaki, T. (1962). A prepattern to the papilla differentiation in *Dryopteris* gametophyte. *Phyton* **18,** 113–20.

Nakazawa, S., and Tanno, N. (1965). Concentration gradients of RNA in fern protonema in relation to m-RNA. *Naturwissenschaften* **52,** 457.

Nakazawa, S., and Tanno, N. (1967). Estimation of messenger RNA in fern gametophytes. *Cytologia* **32,** 216–23.

Nakazawa, S., and Tsusaka, A. (1959*a*). Special cytoplasm detectable in fern rhizoids. *Naturwissenschaften* **46,** 609–10.

Nakazawa, S., and Tsusaka, A. (1959*b*). Appearance of 'metallophilic cytoplasm' as a prepattern to the differentiation of rhizoid in fern protonema. *Cytologia* **24,** 378–88.

Nayar, B. K. (1963*a*). Contributions to the morphology of some species of *Microsorium*. *Ann. Bot.* **27,** 89–100.

Nayar, B. K. (1963*b*). Contributions to the morphology of *Leptochilus* and *Paraleptochilus*. *Am. J. Bot.* **50,** 301–8.

Nayar, B. K., and Bajpai, N. (1964). Morphology of the gametophyte of some species of *Pellaea* and *Notholaena*. *J. Linn. Soc. (Bot.)*. **59,** 63–76.

Nayar, B. K., and Chandra, P. (1963). Observations on the morphology of the gametophyte of *Cyclosorus*. *J. Indian Bot. Soc.* **42,** 392–400.

Nayar, B. K., and Chandra, P. (1965). The gametophytes of some species of *Lastrea* Bory. *J. Indian Bot. Soc.* **44,** 84–94.

Nayar, B. K., and Kaur, S. (1968). Spore germination in homosporous ferns. *J. Palynol.* **4,** 1–14.

Nayar, B. K., and Kaur, S. (1971). Gametophytes of homosporous ferns. *Bot. Rev.* **37,** 295–396

Nester, J. E. (1985). Spore germination and early gametophyte development in *Anemia mexicana* Klotzsch. *Bot. Gaz.* **146,** 510–6.

Nester, J. E., and Coolbaugh, R. C. (1986). Factors influencing spore germination and early gametophyte development in *Anemia mexicana* and *Anemia phyllitidis*. *Plant Physiol.* **82,** 230–5.

Nester, J. E., and Schedlbauer, M. D. (1981). Gametophyte development in *Anemia mexicana* Klotzsch. *Bot. Gaz.* **142,** 242–50.

Nester, J. E., and Schedlbauer, M. D. (1982). Antheridiogen activity of *Anemia mexicana*. *Can. J. Bot.* **60,** 1606–10.

Nester, J. E., Veysey, S., and Coolbaugh, R. C. (1987). Partial characterization of an antheridiogen of *Anemia mexicana:* Comparison with the antheridiogen of *A. phyllitidis*. *Planta* **170.** 26–33.

Nienburg, W. (1924). Die Wirkung des Lichtes auf die Keimung der *Equisetum*spore. *Ber. Deut. Bot. Ges.* **42,** 95–99.

Nishida, M., and Sakuma, T. (1961). Abnormalities in the archegonia of the prothallia and its bearing to systematics. *J. Jap. Bot.* **36,** 142–52.

Ohlenroth, K., and Mohr, H. (1963). Die Steuerung der Proteinsynthese und der Morphogenese bei Farnvorkeimen durch Licht. *Planta* **59,** 427–41.

Ohlenroth, K., and Mohr, H. (1964). Die Steuerung der Proteinsynthese durch Blaulicht und Hellrot in den Vorkeimen von *Dryopteris filix-mas* (L.) Schott. *Planta* **62,** 160–70.

Okada, Y. (1929). Notes on the germination of the spores of some pteridophytes with special regard to their viability. *Sci. Rep. Tohoku Imp. Univ.* Ser. IV. Biol. **4,** 127–82.

Olsen, L. T., and Gullvåg, B. M. (1971). The presence of acid phosphatases in swelling spores of *Matteuccia struthiopteris* (L.) Tod. *Grana* **11,** 111–16.

Olsen, L. T., and Gullvåg, B. M. (1973). A fine-structural and cytochemical study of mature and germinating spores of *Equisetum arvense* L. *Grana* **13,** 113–18.

Ootaki, T. (1963). Modification of the developmental axis by centrifugation in *Pteris vittata*. *Cytologia* **28,** 21–9.

Ootaki, T. (1965). Branching in fern gametophyte, *Pteris vittata,* under various light conditions. *Cytologia* **30,** 182–93.

Ootaki, T. (1967). Branchings and regeneration patterns of isolated single cells of a fern protonema. *Bot. Mag. (Tokyo)* **80,** 1–10.

Ootaki, T. (1968). Polarity in the branching of *Pteris vittata* protonemata. *Embryologia* **10,** 126–37.

Ootaki, T., and Furuya, M. (1969). Experimentally induced apical dominance in protonemata of *Pteris vittata*. *Embryologia* **10,** 284–96.

Otto, E. A., Crow, J. H., and Kirby, E. G. (1984). Effects of acidic growth conditions on spore germination and reproductive development in *Dryopteris marginalis* (L.). *Ann. Bot.* **53,** 439–42.

Padhya, M. A., and Mehta, A. R. (1981). Induced apogamy in *Adiantum trapeziforme* L. *Curr. Sci.* **50,** 19–20.
Page, C. N. (1979*a*). The diversity of ferns. An ecological perspective. In: *The Experimental Biology of Ferns,* ed. A. F. Dyer, pp. 9–56. London: Academic Press.
Page, C. N. (1979*b*). Experimental aspects of fern ecology. In: *The Experimental Biology of Ferns,* ed. A. F. Dyer, pp. 551–89. London: Academic Press.
Paless, G., Vida, G., and Nagy, A. H. (1984). Phytochrome-mediated control of protein synthesis during germination of fern spores. *Acta Bot. Hung.* **30,** 191–200.
Palta, H. K., and Mehra, P. N. (1973). Radiobiological investigations on *Pteris vittata* L. II. X-ray effects on the gametophytic generation. *Radiation Bot.* **13,** 155–64.
Palta, H. K., and Mehra, P. N. (1986). *In vitro* induction of polyhaploid and octoploid *Pteris vittata* L. and their meiosis. *Caryologia* **36,** 325–32.
Panigrahi, G. (1962). Cytogenetics of apogamy in *Aleuritopteris farinosa* (Forsk.) Fee complex. *Nucleus* **5,** 53–64.
Parès, Y. (1958). Etude expérimentale de la morphogenèse du gamétophyte de quelques filicinées. *Ann. Sci. Nat. Bot.* (Ser. 11) **19,** 1–120.
Partanen, C. R. (1956). Comparative microphotometric determinations of deoxyribonucleic acid in normal and tumorous growth of fern prothalli. *Cancer Res.* **16,** 300–5.
Partanen, C. R. (1958). Quantitative technique for analysis of radiation-induced tumorization in fern prothalli. *Science* **128,** 1006–7.
Partanen, C. R. (1959). Quantitative chromosomal changes and differentiation in plant cells. In: *Developmental Cytology,* ed. D. Rudnick, pp. 21–45. New York: Ronald Press Co.
Partanen, C. R. (1960*a*). Suppression of radiation-induced tumorization in fern prothalli. *Science* **131,** 926–7.
Partanen, C. R. (1960*b*). Amino acid suppression of radiation-induced tumorization of fern prothalli. *Proc. Natl. Acad. Sci. USA* **46,** 1206–10.
Partanen, C. R. (1961). Endomitosis in a polyploid series of fern prothalli. *J. Hered.* **52,** 139–44.
Partanen, C. R. (1965). On the chromosomal basis for cellular differentiation. *Am. J. Bot.* **52,** 204–9.
Partanen, C. R. (1972). Comparison of gametophytic callus and tumor tissues of *Pteridium aquilinum. Bot. Gaz.* **133,** 287–92.
Partanen, C. R., and Nelson, J. (1961). Induction of plant tumors by ultraviolet radiation. *Proc. Natl. Acad. Sci. USA* **47,** 1165–9.
Partanen, C. R., Power J. B., and Cocking, E. C. (1980). Isolation and division of protoplasts of *Pteridium aquilinum. Plant Sci. Lett.* **17,** 333–8.
Partanen, C. R., and Steeves, T. A. (1956). The production of tumorous abnormalities in fern prothalli by ionizing radiations. *Proc. Natl. Acad. Sci. USA* **42,** 906–9.
Partanen, C. R., Sussex, I. M., and Steeves, T. A. (1955). Nuclear behavior in relation to abnormal growth in fern prothalli. *Am. J. Bot.* **42,** 245–56.
Partanen, J. N., and Partanen, C. R. (1963). Observations on the culture of roots of the bracken fern. *Can. J. Bot.* **41,** 1657–61.
Patterson, M. T., Sr. (1942). Cytology of apogamy in *Polystichum tsussemense. Bot. Gaz.* **104,** 107–14.
Payer, H. D. (1969). Untersuchungen zur Kompartimentierung der freien

Aminosaure Alanin in den Farnvorkeimen von *Dryopteris filix-mas* (L.) Schott im Rotlicht und im Blaulicht. *Planta* **86,** 103–15.

Payer, H. D., and Mohr, H. (1969). Ein spezifischer Einfluss von Blaulicht auf den Einbau von photosynthetisch assimiliertem ^{14}C in das Protein von Farnvorkeimen [*Dryopteris filix-mas* (L.) Schott]. *Planta* **86,** 286–94.

Payer, H. D., Sotriffer, U., and Mohr, H. (1969). Die Aufnahme von $^{14}CO_2$ und die Verteilung des ^{14}C auf freie Aminosäuren und auf Proteinaminosaüren im Hellrot und im Blaulicht [Objekt: Farnvorkeimen von *Dryopteris filix-mas* (L.) Schott]. *Planta* **85,** 270–83.

Petersen, R. L., Arnold, D., Lynch, D. G., and Price, S. A. (1980). A heavy metal bioassay based on percent spore germination of the sensitive fern, *Onoclea sensibilis. Bull. Environ. Contam. Toxicol.* **24,** 489–95.

Petersen, R. L., and Fairbrothers, D. E. (1980). Reciprocal allelopathy between the gametophytes of *Osmunda cinnamomea* and *Dryopteris intermedia. Am. Fern J.* **70,** 73–8.

Petersen, R. L., and Francis, P. C. (1980). Differential germination of fern and moss spores in response to mercuric chloride. *Am. Fern J.* **70,** 115–18.

Pettitt, J. M. (1966). Exine structures in some fossil and recent spores and pollen as revealed by light and electron microscopy. *Bull. Br. Mus. Nat. Hist. Geol.* **13,** 223–57.

Pettitt, J. M. (1979*a*). Developmental mechanisms in heterospory: Cytochemical demonstration of spore-wall enzymes associated with β-lectins, polysaccharides and lipids in water ferns. *J. Cell Sci.* **38,** 61–82.

Pettitt, J. M. (1979*b*). Ultrastructure and cytochemistry of spore wall morphogenesis. In: *The Experimental Biology of Ferns,* ed. A. F. Dyer, pp. 213–52. London: Academic Press.

Pickett-Heaps, J. D., and Northcote, D. H. (1966). Organization of microtubules and endoplasmic reticulum during mitosis and cytokinesis in wheat meristems. *J. Cell Sci.* **1.** 109–20.

Pietrykowska, J. (1962*a*). Investigations on the germination of spores of the fern *Matteuccia struthiopteris* (L.) Tod. *Acta Soc. Bot. Polon.* **31,** 437–47,

Pietrykowska, J. (1962*b*). Investigations on the differentiation of developmental stages of the fern, *Matteuccia struthiopteris* (L.) Tod. *Acta Soc. Bot. Polon.* **31,** 449–59.

Pietrykowska, J. (1963). Investigations on the action of light on the germination polarity of fern spores. *Acta Soc. Bot. Polon.* **32,** 677–91.

Pray, T. R. (1968). The gametophytes of *Pellaea* section *Pellaea:* Dark-stiped series. *Phytomorphology* **18,** 113–43.

Pray, T. R. (1971). The gametophyte of *Anemia colimensis. Am. J. Bot.* **58,** 323–8.

Pringle, R. B. (1961). Chemical nature of antheridogen-A, a specific inducer of the male sex organ in certain fern species. *Science* **133,** 284.

Pringle, R. B. (1970). Interaction between antheridogens and fatty acids in fern spore germination. *Plant Physiol.* **45,** 315–17.

Pringle, R. B., Näf, U., and Braun, A. C. (1960). Purification of a specific inducer of the male sex organ in certain fern species. *Nature (London)* **186,** 1066–7.

Quirk, H., and Chambers, T. C. (1981). Drought tolerance in *Cheilanthes* with special reference to the gametophyte. *Fern Gaz.* **12,** 121–9.

Racusen, R. H., and Cooke, T. J. (1982). Electrical changes in the apical cell of the fern gametophyte during irradiation with photomorphogenetically active light. *Plant Physiol.* **70,** 331–4.

Raghavan, V. (1964). Differentiation in fern gametophytes treated with purine and pyrimidine analogs. *Science* **146,** 1690–1.
Raghavan, V. (1965*a*). Actinomycin D: Its effects on two-dimensional growth in fern gametophytes. *Expt. Cell Res.* **39,** 689–92.
Raghavan, V. (1965*b*). Action of purine and pyrimidine analogs on the growth and differentiation of the gametophytes of the fern *Asplenium nidus. Am. J. Bot.* **52,** 900–10.
Raghavan, V. (1968*a*). RNA and protein metabolism in the particulate fractions of the gametophytes of bracken fern during growth in red and blue light. *Planta* **81,** 38–48.
Raghavan, V. (1968*b*). Ribonucleic acid and protein changes in the subcellular components of the gametophytes of *Pteridium aquilinum* during growth in red and blue light. *Physiol. Plantarum* **21,** 1020–8.
Raghavan, V. (1968*c*). Actinomycin D-induced changes in growth and ribonucleic metabolism in the gametophytes of bracken fern. *Am. J. Bot.* **55,** 767–72.
Raghavan, V. (1968*d*). A role for purines and pyrimidines of ribonucleic acid in the induction of two-dimensional growth in the gametophytes of the fern, *Asplenium nidus. J. Expt. Bot.* **19,** 553–66.
Raghavan, V. (1969*a*). Photocontrol of growth pattern in a tissue isolated from the gametophytes of bracken fern. *Plant Cell Physiol.* **10,** 481–4.
Raghavan, V. (1969*b*). Interaction of light quality and nucleases in the growth of the gametophytes of *Asplenium nidus. Am. J. Bot.* **56,** 871–9.
Raghavan, V. (1970). Germination of bracken fern spores. Regulation of protein and RNA synthesis during initiation and growth of the rhizoid. *Expt. Cell Res.* **63,** 341–52.
Raghavan, V. (1971*a*). Phytochrome control of germination of the spores of *Asplenium nidus. Plant Physiol.* **48,** 100–2.
Raghavan, V. (1971*b*). Synthesis of protein and RNA for initiation and growth of the protonema during germination of bracken fern spore. *Expt. Cell Res.* **65,** 401–7.
Raghavan, V. (1973*a*). Blue light interference in the phytochrome-controlled germination of the spores of *Cheilanthes farinosa. Plant Physiol.* **51,** 306–11.
Raghavan, V. (1973*b*). Photomorphogenesis of the gametophytes of *Lygodium japonicum. Am. J. Bot.* **60,** 313–21.
Raghavan, V. (1974*a*). Induction of biplanar growth in fern gametophytes. I. The relationship between cell division and biplanar growth. *Physiol. Plantarum* **30,** 132–6.
Raghavan, V. (1974*b*). Induction of biplanar growth in fern gametophytes. II. The inhibition of biplanar growth by 5-fluorouracil. *Physiol. Plantarum* **30,** 137–42.
Raghavan, V. (1974*c*). Control of differentiation in the fern gametophyte. *Am. Sci.* **62,** 465–75.
Raghavan, V. (1976). Gibberellic acid-induced germination of spores of *Anemia phyllitidis:* Nucleic acid and protein synthesis during germination. *Am. J. Bot.* **63,** 960–72.
Raghavan, V. (1977*a*). Gibberellic acid-induced germination of spores of *Anemia phyllitidis:* Autoradiographic study of the timing and regulation of nucleic acid and protein synthesis in relation to cell morphogenesis. *J. Cell Sci.* **23,** 85–100.
Raghavan, V. (1977*b*). Cell morphogenesis and macromolecule synthesis during phytochrome-controlled germination of spores of the fern, *Pteris vittata. J. Expt. Bot.* **28,** 439–56.

Raghavan, V. (1980). Cytology, physiology, and biochemistry of germination of fern spores. *Int. Rev. Cytol.* **62,** 69–118.

Raghavan, V. (1987). Changes in poly(A)+RNA concentrations during germination of spores of the fern, *Onoclea sensibilis. Protoplasma* **140,** 55–66.

Raghavan, V., and DeMaggio, A. E. (1971*a*). Enhancement of protein synthesis in isolated chloroplasts by irradiation of fern gametophytes with blue light. *Plant Physiol.* **48,** 82–5.

Raghavan, V., and DeMaggio, A. E. (1971*b*). Protein synthesis by chloroplasts isolated from gametophytes of the fern, *Todea barbara. Phytochemistry,* **10,** 2583–91.

Raghavan, V., and Huckaby, C. S. (1980). A comparative study of cell division patterns during germination of spores of *Anemia, Lygodium* and *Mohria* (Schizaeaceae). *Am. J. Bot.* **67,** 653–63.

Raghavan V., and Tung, H. F. (1967). Inhibition of two-dimensional growth and suppression of ribonucleic acid and protein synthesis in the gametophytes of the fern, *Asplenium nidus,* by chloramphenicol, puromycin and actinomycin D. *Am. J. Bot.* **54,** 198–204.

Rashid, A. (1970). *In vitro* studies on sex-expression in *Lygodium flexuosum. Phytomorphology* **20,** 255–61.

Rashid, A. (1972). *In vitro* sporogenesis in a fern. *Z. Pflanzenphysiol.* **66,** 277–9.

Reinert, J. (1952). Der Einfluss von 3,4-benzpyren auf das Flächenwachstum und den Auxinspiegel der Prothallien von *Stenochlaena palustris. Z. Bot.* **40,** 187–92.

Reuter, L. (1953). A contribution to the cell-physiologic analysis of growth and morphogenesis in fern prothallia. *Protoplasma* **42,** 1–29.

Reynolds, T. L. (1979). Apical dominance in *Anemia phyllitidis* gametophytes. *Am. Fern J.* **69,** 92–4.

Reynolds, T. L. (1981). Effects of auxin and abscisic acid on adventitious gametophyte formation by *Anemia phyllitidis.* (L.) Sw. *Z. Pflanzenphysiol.* **103,** 9–13.

Reynolds, T. L. (1982). Effects of cyanide, salicylhydroxamic acid, and temperature on respiration and germination of spores of the fern *Sphaeropteris cooperi. Physiol. Plantarum* **54,** 52–7.

Reynolds, T. L., and Corson, G. E. (1979). Apical dominance: The effects of growth regulators on the gametophyte of *Anemia phyllitidis. Am. J. Bot.* **66,** 1261–3.

Reynolds, T. L., and Raghavan, V. (1982). Photoinduction of spore germination in a fern, *Mohria caffrorum. Ann. Bot.* **49,** 227–33.

Rice, H. V., and Laetsch, W. M. (1967). Observations on the morphology and physiology of *Marsilea* sperm. *Am. J. Bot.* **54,** 856–66.

Robinson, A. I., Miller, J. H., Helfrich, R., and Downing, M. (1984). Metal-binding sites in germinating fern spores (*Onoclea sensibilis*). *Protoplasma* **120,** 1–11.

Robinson, P. M., Smith, D. L., Safford, R., and Nichols, B. W. (1973). Lipid metabolism in the fern *Polypodium vulgare. Phytochemistry* **12,** 1377–81.

Rosendahl, G. (1940). Versuche zur Erzeugung von Polyploidie bei Farnen durch Colchicinbehandlung sowie Beobachtungen an polyploiden Farnprothallien. *Planta* **31,** 597–637.

Rottmann, W. (1939). Versuche zur Gewinnung abweichender Formen mit Farnsporen und Gametophyten. *Beitr. Biol. Pflanzen* **26,** 1–80.

Rubin, G., and Paolillo, D. J., Jr. (1983). Sexual development of *Onoclea sensibilis* on agar and soil media without the addition of antheridiogen. *Am. J. Bot.* **70,** 811–15.

Rubin, G., and Paolillo, D. J., Jr. (1984). Obtaining a sterilized soil for the growth of *Onoclea* gametophytes. *New Phytol.* **97,** 621–8.

Rubin, G., Robson, D. S., and Paolillo, D. J., Jr. (1984). The spontaneous formation of antheridia in *Onoclea* is not blocked by light. *Ann. Bot.* **54,** 213–21.

Rubin, G., Robson, D. S., and Paolillo, D. J., Jr. (1985). Effects of population density on sex expression in *Onoclea sensibilis* L. on agar and ashed soil. *Ann. Bot.* **55,** 201–15.

Rutter, M. R., and Raghavan, V. (1978). DNA synthesis and cell division during spore germination in *Lygodium japonicum*. *Ann. Bot.* **42,** 957–65.

Sarbadhikari, P. C. (1936). Apospory in *Osmunda javanica* Bl. *Ceylon J. Sci.* **12,** 137–43.

Sarbadhikari, P. C. (1939). Cytology of apogamy and apospory in *Osmunda javanica* Bl. *Ann. Bot.* **3,** 137–45.

Sato, T., and Sakai, A. (1980). Freezing resistance of gametophytes of the temperate fern, *Polystichum retroso-paleaceum*. *Can. J. Bot.* **58,** 1144–8.

Sato, T., and Sakai, A. (1981). Cold tolerance of gametophytes and sporophytes of some cool temperate ferns native to Hokkaido. *Can. J. Bot.* **59,** 604–8.

Saus, G. L., and Lloyd, R. M. (1976). Experimental studies on mating systems and genetic load in *Onoclea sensibilis* L. (Aspleniaceae: Athyrioideae). *Bot. J. Linn. Soc.* **72,** 101–13.

Schedlbauer, M. D. (1974). Biological specificity of the antheridogen from *Ceratopteris thalictroides* (L.) Brongn. *Planta* **116,** 39–43.

Schedlbauer, M. D. (1976*a*). Specificity of the antheridogen from *Ceratopteris thalictroides* (L.) Brongn. *Plant Physiol.* **57,** 666–9.

Schedlbauer, M. D. (1976*b*). Fern gametophyte development: Controls of dimorphism in *Ceratopteris thalictroides*. *Am. J. Bot.* **63,** 1080–7.

Schedlbauer, M. D. (1978). Effects of benlate on fern gametophyte development. *Am. J. Bot.* **65,** 864–8.

Schedlbauer, M. D., Cave, C. F., and Bell, P. R. (1973). The incorporation of DL-[3-^{14}C]cysteine during spermatogenesis in *Ceratopteris thalictroides*. *J. Cell Sci.* **12,** 765–79.

Schedlbauer, M. D., and Klekowski, E. J., Jr. (1972). Antheridogen activity in the fern *Ceratopteris thalictroides* (L.) Brongn. *Bot. J. Linn. Soc.* **65,** 399–413.

Schnarrenberger, C., and Mohr, H. (1967). Die Wechselwirkung von Hellrot, Dunkelrot und Blaulicht bei der Photomorphogenese von Farngametophyten [*Dryopteris filix-mas* (L.) Schott]. *Planta* **75,** 114–24.

Schneller, J. J. (1979). Biosystematic investigations on the lady fern (*Athyrium filix-femina*). *Plant Syst. Evol.* **132,** 255–77.

Schraudolf, H. (1962). Die Wirkung von Phytohormonen auf Keimung und Entwicklung von Farnprothallien. I. Auslösung der Antheridienbildung und Dunkelkeimung bei Schizaeaceen durch Gibberellinsäure. *Biol. Zentralbl.* **81,** 731–40.

Schraudolf, H. (1963). Einige Beobachtungen zur Entwicklung der Antheridien von *Anemia phyllitidis*. *Flora* **153,** 282–90.

Schraudolf, H. (1964). Relative activity of the gibberellins in the antheridium induction in *Anemia phyllitidis*. *Nature (London)* **201,** 98–9.

Schraudolf, H. (1965). Einfluss von DNS-, RNS- und Proteinantimetaboliten auf die Antheridienbildung in Farnen. *Ber. Deut. Bot. Ges.* **78,** 73–5.

Schraudolf, H. (1966*a*). Die Wirkung von Phytohormonen auf Keimung und

Entwicklung von Farnprothallien. IV. Die Wirkung von unterschiedlichen Gibberelline und von Allo-Gibberinsäure auf die Auslösung der Antheridienbildung bei *Anemia phyllitidis* L. und einigen Polypodiaceen. *Plant Cell Physiol.* **7,** 277–89.

Schraudolf, H. (1966*b*). Die Wirkung von Phytohormonen auf Keimung und Entwicklung von Farnprothallien. II. Mitteilung. Analyse der Wechselbeziehung zwischen Gibberellinkonzentration, Antheridienbildung und physiologischem Alter der Prothalliumzellen in *Anemia phyllitidis*. *Planta* **68,** 335–52.

Schraudolf, H. (1966*c*). Die Wirkung von Phytohormonen auf Keimung und Entwicklung von Farnprothallien. III. Einfluss von Plasmolyse und Exstirpation auf die Auslösung der Antheridienbildung durch Gibberelline bei *Anemia phyllitidis* L. *Biol. Zentralbl.* **85,** 349–60.

Schraudolf, H. (1966*d*). Nachweis von Gibberellin in Gametophyten von *Anemia phyllitidis*. *Naturwissenschaften* **53,** 412.

Schraudolf, H. (1967*a*). Wirkung von Hemmstoffen der DNS-, RNS- und Proteinsynthese auf Wachstum und Antheridienbildung in Prothallien von *Anemia phyllitidis* L. *Planta* **74,** 123–47.

Schraudolf, H. (1967*b*). Die Steuerung der Antheridienbildung in *Polypodium crassifolium* L. (*Pessopteris crassifolia* Underw. and Maxon) durch Licht. *Planta* **76,** 37–46.

Schraudolf, H. (1967*c*). Die Wirkung von IES, Coumarin, und sogenannten "Antigibberellinen" auf die Auslösung der Antheridienbildung in *Anemia phyllitidis* L. durch Gibberelline. *Z. Pflanzenphysiol.* **56,** 375–86.

Schraudolf, H. (1967*d*). Wirkung von Terpenderivaten (Helminthosporol, Helminthosporsäure, Didydrohelminthosporsäure und Steviol) auf die Antheridienbildung in *Anemia phyllitidis*. *Planta* **74,** 188–93.

Schraudolf, H. (1968). Einige Beobachtungen zur Ausbildung des Antheridiums von Polypodiaceen. *Flora* **157,** 379–85.

Schraudolf, H. (1972). Chromatographischer Nachweis eines zweiten Antheridiogens im Nährmedium der Prothallien von *Anemia phyllitidis* L. Sw. *Z. Pflanzenphysiol.* **66,** 189–91.

Schraudolf, H. (1977*a*). Wirkungen von cycl.3',5'-AMP, Mononucleotiden und Theophyllin auf Antheridienbildung und Dunkelkeimung von *Anemia phyllitidis* L. Sw. *Z. Pflanzenphysiol.* **84,** 49–59.

Schraudolf, H. (1977*b*). Effects of 5-bromodeoxyuridine and 2-aminopurine on antheridium differentiation in *Anemia phyllitidis* L. *Experientia* **33,** 1161–2.

Schraudolf, H. (1980*a*). Induction of antheridium differentiation in the fern *Anemia phyllitidis* L. Sw. by halogenated base analogues. *Z. Pflanzenphysiol.* **96,** 67–75.

Schraudolf, H. (1980*b*). Cell division patterns during spore germination of *Anemia phyllitidis* L. Sw. *Beitr. Biol. Pflanzen* **55,** 285–8.

Schraudolf, H. (1982*a*). Effects of metronidazole on growth, chloroplast structure and differentiation in gametophytes of *Anemia phyllitidis* L. Sw. *Protoplasma* **113,** 144–9.

Schraudolf, H. (1982*b*). Activity of 2,2-dimethyl-gibberellin A_4 in the *Anemia phyllitidis* antheridiogen assay. *Naturwissenschaften* **69,** 286.

Schraudolf, H. (1983*a*). Nuclear morphology in rhizoidal cells of *Anemia phyllitidis* (L.) Sw. (Filicinae). *Flora* **173,** 359–61.

Schraudolf, H. (1983*b*). Antheridiogen-like activity of AC-94,377, a substituted phthalimide. *Z. Pflanzenphysiol.* **109,** 469–72.

Schraudolf, H. (1984). Ultrastructural events during sporogenesis of *Anemia phyllitidis* (L.) Sw. II. Spore wall formation. *Beitr. Biol. Pflanzen.* **59,** 237–60.

Schraudolf, H. (1987*a*). The effect of gabaculine on germination and gametophyte morphogenesis of *Anemia phyllitidis* L. Sw. *Plant Cell Physiol.* **28,** 53–60.

Schraudolf, H. (1987*b*). Parental predetermination of dark germination in spores of *Anemia phyllitidis. Naturwissenschaften* **74,** 138.

Schraudolf, H., and Fischer, A. (1979). Evidence for cytokinins in gametophytes of the fern *Anemia phylltidis* L. Sw. *Plant Sci. Lett.* **14,** 199–203.

Schraudolf, H., and Legler, K. (1969). Wechselwirkung von Riboflavin, Isoriboflavin und 2-Thiouracil auf die Induktion zweidimensionalen Wachstums von Farnprothallien. *Physiol. Plantarum* **22,** 312–8.

Schraudolf, H., and Richter, U. (1978). Elektronenmikroskopische Analyse der Musterbildung im Antheridium der Polypodiaceae. *Plant Syst. Evol.* **129,** 291–7.

Schraudolf, H., and Šonka, J. (1979). Effects of 5-bromo-deoxyuridine on chloroplast structure in gametophytes of *Anemia phyllitidis* L. Sw. *Eur. J. Cell Biol.* **19,** 135–8.

Schwabe, W. W. (1951). Physiological studies in plant nutrition. XVI. The mineral nutrition of bracken. Part I. Prothallial culture and the effects of phosphorus and potassium supply on leaf production in the sporophyte. *Ann. Bot.* **15,** 417–46.

Scott, R. J., and Hickok, L. G. (1987). Genetic analysis of antheridiogen sensitivity in *Ceratopteris richardii. Am. J. Bot.* **74,** 1872–7.

Seilheimer, A. V. (1978). Chlorophyll and lipid changes on germination in the non-green spores of *Thelypteris dentata. Am. Fern J.* **68,** 67–70.

Sharma, B. D., and Singh, R. (1983). Effects of morphactin on spore germination and differentiation in *Actinopteris radiata. Phytomorphology* **33,** 13–6.

Sharp, L. W. (1914). Spermatogenesis in *Marsilia. Bot. Gaz.* **58,** 419–31.

Sharp, P. B., Keitt, G. W., Jr., Clum, H. H., and Näf, U. (1975). Activity of antheridiogen from *Anemia phyllitidis* in three flowering plant bioassays. *Physiol. Plantarum* **34,** 101–5.

Shattuck, C. H. (1910). The origin of heterospory in *Marsilia. Bot. Gaz.* **49,** 19–40.

Sheffield, E. (1984). Apospory in the fern *Pteridium aquilinum* (L.) Kuhn. 1. Low temperature scanning electron microscopy. *Cytobios* **39,** 171–6.

Sheffield, E., and Bell, P. R. (1978). Phytoferritin in the reproductive cells of a fern, *Pteridium aquilinum* (L.) Kuhn. *Proc. Roy. Soc. Lond.* **202B,** 297–306.

Sheffield, E., and Bell, P. R. (1979). Ultrastructural aspects of sporogenesis in a fern, *Pteridium aquilinum* (L.) Kuhn. *Ann. Bot.* **44,** 393–405.

Sheffield, E., and Bell, P. R. (1981*a*). Experimental studies of apospory in ferns. *Ann. Bot.* **47,** 187–95.

Sheffield, E., and Bell, P. R. (1981*b*). Cessation of vascular activity correlated with aposporous development in *Pteridium aquilinum* (L.) Kuhn. *New Phytol.* **88,** 533–8.

Sheffield, E., Bell, P. R., and Laird, S. (1982). Tracheary occlusion in fronds of *Pteridium aquilinum* (L.) Kuhn showing apospory. *New Phytol.* **90,** 321–5.

Sheffield, E., Laird, S., and Bell, P. R. (1983). Ultrastructural aspects of sporogenesis in the apogamous fern *Dryopteris borreri. J. Cell Sci.* **63,** 125–34.

Sigee, D., and Bell, P. R. (1968). Deoxyribonucleic acid in the cytoplasm of the female reproductive cells of *Pteridium aquilinum. Expt. Cell Res.* **49,** 105–15.

Sigee, D. C. (1972). The origin of cytoplasmic DNA in the mature egg cell of *Pteridium aquilinum. Protoplasma* **75,** 323–34.

Sigee, D. C., and Bell, P. R. (1971). The cytoplasmic incorporation of tritiated thymidine during oogenesis in *Pteridium aquilinum*. *J. Cell Sci.* **8,** 467–87.
Singh, I. P., and Roy, S. K. (1987). Methyl isocyanate and fern gametophyte. *Curr. Sci.* **56,** 679–81.
Singh, S. P., and Roy, S. K. (1984*a*). Effects of colchicine on the gametophyte of *Lygodium flexuosum*. *Phytomorphology* **34,** 60–4.
Singh, S. P., and Roy, S. K. (1984*b*). Effect of chloral hydrate on the gametophyte of *Drynaria quercifolia* (L.) J. Sm. *Acta Bot. Indica* **12,** 185–9.
Singh, V. P., and Roy, S. K. (1977). Mating systems and distribution in some tropical ferns. *Ann. Bot.* **41,** 1055–60.
Sinnott, E. W. (1960). *Plant Morphogenesis*. New York: McGraw-Hill Book Co.
Smith, D. L. (1972*a*). Localization of phosphatases in young gametophytes of *Polypodium vulgare* L. *Protoplasma* **74,** 133–48.
Smith, D. L. (1972*b*). Staining and osmotic properties of young gametophytes of *Polypodium vulgare* L. and their bearing on rhizoid function. *Protoplasma* **74,** 465–79.
Smith D. L. (1973). Phosphatase and amylase activity of intact fern gametophytes. *Z. Pflanzenphysiol.* **69,** 447–55.
Smith, D. L., and Robinson, P. M. (1969). The effects of fungi on morphogenesis of gametophytes of *Polypodium vulgare* L. *New Phytol.* **68,** 112–22.
Smith, D. L., and Robinson, P. M. (1971). Growth factors produced by germinating spores of *Polypodium vulgare* L. *New Phytol.* **70,** 1043–52.
Smith, D. L., and Robinson, P. M. (1975). The effects of spore age on germination and gametophyte development in *Polypodium vulgare* L. *New Phytol.* **74,** 101–8.
Smith, D. L., Robinson P. M., and Govier, R. N. (1973). Growth factors produced by gametophytes of *Polypodium vulgare* L. grown under red and blue light. *New Phytol.* **72,** 1261–8.
Smith, D. L., and Rogan, P. G. (1970). The effects of population density on gametophyte morphogenesis in *Polypodium vulgare* L. *New Phytol.* **69,** 1039–51.
Smith, H. M., and Smith, D. S. (1969). An electron-microscopic study of the starch-containing plastids in the fern *Todea barbara*. *J. Cell Sci.* **4,** 211–21.
Sobota, A. E. (1970). Incompatibility of meristematic and filamentous growth in the fern gametophyte. *Am. J. Bot.* **57,** 530–4.
Sobota, A. E. (1972). Ribonucleic acid synthesis associated with a developmental change in the gametophyte of *Pteridium aquilinum*. *Plant Physiol.* **49,** 914–8.
Sobota, A. E., and Partanen, C. R. (1966). The growth and division of cells in relation to morphogenesis in fern gametophytes. I. Photomorphogenetic studies in *Pteridium aquilinum*. *Can. J. Bot.* **44,** 498–506.
Sobota, A. E., and Partanen, C. R. (1967). The growth and division of cells in relation to morphogenesis in fern gametophytes. II. The effect of biochemical agents on the growth and development of *Pteridium aquilinum*. *Can. J. Bot.* **45,** 595–603.
Soltis, D. E., and Soltis, P. S. (1986). Electrophoretic evidence for inbreeding in the fern *Botrychium virginianum* (Ophioglossaceae). *Am. J. Bot.* **73,** 588–92.
Soltis, D. E., and Soltis, P. S. (1987*a*). Breeding system of the fern *Dryopteris expansa:* Evidence for mixed mating. *Am. J. Bot.* **74,** 504–9.
Soltis, P. S., and Soltis, D. E. (1987*b*). Population structure and estimates of gene flow in the homosporous fern *Polystichum munitum*. *Evolution* **41,** 620–9.

Šonka, J., and Schraudolf, H. (1979). Incorporation of 5-bromo-2-deoxyuridine into the chloroplast DNA of *Anemia phyllitidis*. *Z. Naturforsch.* **34c,** 449–51.
Sossountzov, I. (1953). Action de l'hydrazide maléique sur la germination *in vitro* des spores de *Gymnogramme calomelanos*. *Compt. Rend. Soc. Biol.* **147,** 287–91.
Sossountzov, I. (1954). Le développement *in vitro* des colonies prothalliennes de *Gymnogramme calomelanos* en présence des acides aspartique et glutamique et de leurs amides (asparagine et glutamine). *Physiol. Plantarum* **7,** 726–42.
Sossountzov, I. (1955). Sexualité et dimensions des prothalles de *Gymnogramme calomelanos* (Filicinée Polypodiacée) cultivés sur des milieux à concentrations variables en alanine et en phenylalanine. *Compt. Rend. Soc. Biol.* **149,** 1374–7.
Sossountzov, I. (1956). Action du potassium sur le gamétophyte de la fougère *Gymnogramme calomelanos.* La germination des spores en présence de con-concentrations variables en potassium. *Compt. Rend. Soc. Biol.* **150,** 35–8.
Sossountzov, I. (1957*a*). Le développement des colonies prothalliennes de *Gymnogramme calomelanos* (Filicinée Polypodiacée) en présence de méthylamine comme seule source d'azote. *Phyton* **8,** 79–96.
Sossountzov, I. (1957*b*). Le développement des prothalles de *Gymnogramme calomelanos* (Filiciné Polypodiacée) en présence de concentrations variables en phosphore. *Compt. Rend. Soc. Biol.* **151,** 1713–16.
Sossountzov, I. (1957*c*). Action de trois auxines de synthèse sur le développement *in vitro* des germinations de fougères. *Compt. Rend. Soc. Biol.* **151,** 531–6.
Sossountzov, I. (1958). La croissance *in vitro* de colonies prothalliennes de *Gymnogramme calomelanos,* Filicinée Polypodiacée en présence de trois acides aminés aliphatiques. *Compt. Rend. Soc. Biol.* **152,** 1353–6.
Sossountzov, I. (1959). Le développement *in vitro* des colonies prothalliennes de *Gymnogramme calomelanos* (Filicinée Polypodiacée) en présence de mélanges binaires d'acides aminés aliphatiques. II. Sexualité et dimensions des prothalles cordiformes. *Phyton* **13,** 21–8.
Sossountzov, I. (1961). La germination *in vitro* des spores de *Gymnogramme calomelanos* (Filicinée) en présence de coumarine. *Compt. Rend. Soc. Biol.* **155,** 1006–10.
Soyerman, A. (1963). Action de la pénicilline sur la régénération des prothalles de *Gymnogramme calomelanos* (variété *argentea*). *Compt. Rend. Acad. Sci. Paris* **257,** 957–60.
Soyerman, A. (1964). Action de la pénicilline sur la taille des colonies prothalliennes de *Gymnogramme calomelanos* variété *argentea* et sur les dimensions des parois prothalles régéneres. *Compt. Rend. Acad. Sci. Paris* **258,** 2157–9.
Spiess, L. D., and Krouk, M. G. (1977). Photocontrol of germination of spores of the fern *Polypodium aureum*. *Bot. Gaz.* **138,** 428–33.
Srinivasan, J. C., and Kaufman, P. B. (1978). Sex expression in *Equisetum scirpoides in vitro*. *Phytomorphology* **28,** 331–5.
Stansfield, F. W. (1899). On the production of apospory by environment in *Athyrium filix-foemina,* var. *unco-glomeratum,* an apparently barren fern. *J. Linn. Soc. (Bot.).* **34,** 262–8.
Stebbins, G. L. (1967). Gene action, mitotic frequency, and morphogenesis in higher plants. In: *Control Mechanisms in Developmental Processes,* ed. M. Locke, pp. 113–35. New York: Academic Press.
Steeves, T. A., Sussex, I. M., and Partanen, C. R. (1955). *In vitro* studies on abnormal growth of prothalli of the bracken fern. *Am. J. Bot.* **42,** 232–45.

Steeves, T. A., and Wetmore, R. H. (1953). Morphogenetic studies on *Osmunda cinnamomea* L.: Some aspects of the general morphology. *Phytomorphology* **3,** 339–54.

Steil, W. N. (1919*a*). A study of apogamy in *Nephrodium hirtipes,* Hk. *Ann. Bot.* **33,** 109–32.

Steil, W. N. (1919*b*). Apospory in *Pteris sulcata* L. *Bot. Gaz.* **67,** 469–82.

Steil, W. N. (1939). Apogamy, apospory, and parthenogenesis in pteridophytes. *Bot. Rev.* **5,** 433–53.

Steil, W. N. (1944). Apospory and apogamy in species of *Tectaria. Bot. Gaz.* **105,** 369–73.

Steil, W. N. (1951). Apogamy, apospory and parthenogenesis in the pteridophytes. II. *Bot. Rev.* **17,** 90–104.

Stein, D. B. (1971). Gibberellin-induced fertility in the fern *Ceratopteris thalictroides* (L.) Brongn. *Plant Physiol.* **48,** 416–8.

Steiner, A. M. (1967*a*). Dose-response curves for polarotropism in germlings of a fern and a liverwort. *Naturwissenschaften* **54,** 497.

Steiner, A. M. (1967*b*). Action spectra for polarotropism in germlings of a fern and a liverwort. *Naturwissenschaften* **54,** 497–8.

Steiner, A. M. (1969*a*). Dose response behaviour for polarotropism of the chloronema of the fern *Dryopteris filix-mas* (L.) Schott. *Photochem. Photobiol.* **9,** 493–506.

Steiner, A. M. (1969*b*). Action spectrum for polarotropism in the chloronema of the fern *Dryopteris filix-mas* (L.) Schott. *Photochem. Photobiol.* **9,** 507–13.

Steiner, A. M. (1970). Red light interactions with blue and ultraviolet light in polarotropism of germlings of a fern and a liverwort. *Photochem. Photobiol.* **12,** 169–74.

Stetler, D. A., and DeMaggio, A. E. (1972). An ultrastructural study of fern gametophytes during one- to two-dimensional development. *Am. J. Bot.* **59,** 1011–7.

Stockwell, C. R., and Miller, J. H. (1974). Regions of cell wall expansion in the protonema of a fern. *Am. J. Bot.* **61,** 375–8.

Stokey, A. G. (1940). Spore germination and vegetative stages of the gametophytes of *Hymenophyllum* and *Trichomanes. Bot. Gaz.* **101,** 759–90.

Stokey, A. G. (1948). Reproductive structures of the gametophytes of *Hymenophyllum* and *Trichomanes. Bot. Gaz.* **109,** 363–80.

Stokey, A. G. (1951*a*). The contribution by the gametophyte to classification of the homosporous ferns. *Phytomorphology* **1,** 39–58.

Stokey A. G. (1951*b*). Duration of viability of spores of the Osmundaceae. *Am. Fern J.* **41,** 111–5.

Stokey, A. G., and Atkinson, L. R. (1956*a*). The gametophyte of the Osmundaceae. *Phytomorphology* **6,** 19–40.

Stokey, A. G., and Atkinson, L. R. (1956*b*). The gametophytes of *Plagiogyria glauca* (Bl.) Mett. and *P. semicordata* (Pr.) Christ. *Phytomorphology* **6,** 239–49.

Stokey, A. G., and Atkinson, L. R. (1957). The gametophyte of some American species of *Elaphoglossum* and *Rhipidopteris. Phytomorphology* **7,** 275–92.

Stokey, A. G., and Atkinson, L. R. (1958). The gametophyte of the Grammitidaceae. *Phytomorphology* **8,** 391–403.

Stone, I. G. (1958). The gametophyte and embryo of *Polyphlebium venosum* (R.Br) Copeland (Hymenophyllaceae). *Aust. J. Bot.* **6,** 183–203.

Stone, I. G. (1961). The gametophytes of the Victorian Blechnaceae. I. *Blechnum nudum* (Labill.) Luerss. *Aust. J. Bot.* **9,** 20–36.

Stone, I. G. (1962). The ontogeny of the antheridium in some leptosporangiate ferns with particular reference to the funnel-shaped wall. *Aust. J. Bot.* **10,** 76–92.

Stone, I. G. (1965). The gametophytes of the Victorian Hymenophyllaceae. *Aust. J. Bot.* **13,** 195–224.

Sugai, M. (1968). The changes of respiratory quotient during the early development in fern gametophytes. *Embryologia* **10,** 164–72.

Sugai, M. (1970). Photomorphogenesis in *Pteris vittata.* III. Protective action of ethanol on blue-light-induced inhibition of spore germination. *Develop. Growth Diffn.* **12,** 13–20.

Sugai, M. (1971). Photomorphogenesis in *Pteris vittata.* IV. Action spectra for inhibition of phytochrome-dependent spore germination. *Plant Cell Physiol.* **12,** 103–9.

Sugai, M. (1982). Effects of anaerobiosis and respiratory inhibitors on blue-light inhibition of spore germination in *Pteris vittata. Plant Cell Physiol.* **23,** 1155–60.

Sugai, M., and Furuya, M. (1967). Photomorphogenesis in *Pteris vittata.* I. Phytochrome-mediated spore germination and blue light interaction. *Plant Cell Physiol.* **8,** 737–48.

Sugai, M., and Furuya, M. (1968). Photomorphogenesis in *Pteris vittata.* II. Recovery from blue-light-induced inhibition of spore germination. *Plant Cell Physiol.* **9,** 671–80.

Sugai, M., and Furuya, M. (1985). Action spectrum in ultraviolet and blue light region for the inhibition of red-light-induced spore germination in *Adiantum capillus-veneris* L. *Plant Cell Physiol.* **26,** 953–6.

Sugai, M., Nakamura, K., Yamane, H., Sato, Y., and Takahashi, N. (1987). Effects of gibberellins and their methyl esters on dark germination and antheridium formation in *Lygodium japonicum* and *Anemia phyllitidis. Plant Cell Physiol.* **28,** 199–202.

Sugai, M., Takeno, K., and Furuya, M. (1977). Diverse responses of spores in the light-dependent germination of *Lygodium japonicum. Plant Sci. Lett.* **8,** 333–8.

Sugai, M., Tomizawa, K., Watanabe, M., and Furuya, M. (1984). Action spectrum between 250 and 800 nanometers for the photoinduced inhibition of spore germination in *Pteris vittata. Plant Cell Physiol.* **25,** 205–12.

Sulklyan, D. S., and Mehra, P. N. (1977). *In vitro* morphogenetic studies in *Nephrolepis cordifolia. Phytomorphology* **27,** 396–407.

Surova, T. D. (1981). The development of the spores in *Anemia phyllitidis* (Schizaeaceae). The membrane junctions during the exine formation. *Bot. Zhurn.* **66,** 371–9.

Sussex, I. M. (1966). The origin and development of heterospory in vascular plants. In: *Trends in Plant Morphogenesis,* ed. E. G. Cutter, pp. 140–52. London: Longmans, Green.

Sussex, I. M., and Steeves, T. A. (1958). Experiments on the control of fertility of fern leaves in sterile culture. *Bot. Gaz.* **119,** 203–8.

Sussman, A. E. (1965). Physiology of dormancy and germination in the propagules of cryptogamic plants. *Encyl. Plant Physiol.* **15/2,** 932–1025.

Swami, P., and Raghavan, V. (1980). Control of morphogenesis in the gametophyte of a fern by light and growth hormones. *Can. J. Bot.* **58,** 1464–73.

Takahashi, C. (1961). The growth of protonema cells and rhizoids in bracken. *Cytologia* **26,** 62–6.

Takahashi, C. (1962). Cytological study on induced apospory in ferns. *Cytologia* **27,** 79–96.
Takahashi, C. (1969). Studies on the regeneration of detached organs in pteridophytes (1). *Pteris vittata, Lygodium japonicum* and *Equisetum arvense. J. Jap. Bot.* **44,** 149–58.
Takeno, K., and Furuya, M. (1975). Bioassay of antheridiogen in *Lygodium japonicum. Develop. Growth Diffn.* **17,** 9–18.
Takeno, K., and Furuya, M. (1977). Inhibitory effect of gibberellins on archegonial differentiation in *Lygodium japonicum. Physiol. Plantarum* **39,** 135–8.
Takeno, K., and Furuya, M. (1980). Sexual differentiation in population of prothallia in *Lygodium japonicum. Bot. Mag. (Tokyo)* **93,** 67–76.
Takeno, K., and Furuya, M. (1987). Sporophyte formation in experimentally-induced unisexual female and bisexual gametophytes of *Lygodium japonicum. Bot. Mag. (Tokyo)* **100,** 37–41.
Takeno, K., Furuya, M., Yamane, H., and Takahashi, N. (1979). Evidence for naturally occurring inhibitors of archegonial differentiation in *Lygodium japonicum. Physiol. Plantarum* **45,** 305–10.
Takeno, K., Yamane, H., Nohara, K., Takahashi, N., Corey, E. J., Myers, A. G., and Schraudolf, H. (1987). Biological activity of antheridic acid, an antheridiogen of *Anemia phyllitidis. Phytochemistry* **26,** 1855–7.
Templeman, T. S., DeMaggio, A. E., and Stetler, D. A. (1987). Biochemistry of fern spore germination: Globulin storage proteins in *Matteuccia struthiopteris* L. *Plant Physiol.* **85,** 343–9.
Tilquin, J. P. (1981). Note on apomixis in ferns. *Acta Soc. Bot. Polon.* **50,** 217–22.
Tittle, F. L. (1987). Auxin-stimulated ethylene production in fern gametophytes and sporophytes. *Physiol. Plantarum* **70,** 499–502.
Toia, R. E., Jr., Marsh, B. H., Perkins, S. K., McDonald, J. W., and Peters, G. A. (1985). Sporopollenin content of the spore apparatus of *Azolla. Am. Fern J.* **75,** 38–43.
Tomizawa, K., Manabe, K., and Sugai, M. (1982). Changes in phytochrome content during imbibition in spores of the fern *Lygodium japonicum. Plant Cell Physiol.* **23,** 1305–8.
Tomizawa, K., Sugai, M., and Manabe, K. (1983). Relationship between germination and P_{FR} level in spores of the fern *Lygodium japonicum. Plant Cell Physiol.* **24,** 1043–8.
Tourte, Y. (1967). Contribution a l'étude de la sexualité d'une fougère Leptosporangiée en culture pure. *Rev. Gen. Bot.* **74,** 197–223.
Tourte, Y. (1968). Observations sur le comportement du noyau, des plastes et des mitochondries au cours de la maturation de l'oosphère du *Pteridium aquilinum* L. *Compt. Rend. Acad. Sci. Paris* **266D,** 2324–6.
Tourte, Y. (1970). Nature, origine et évolution d'enclaves cytoplasmiques particulières au cours de l'oogénèse chez le *Pteridium aquilinum* (L.) *Rev. Cytol. Biol. Veg.* **33,** 311–24.
Tourte, Y. (1975). Étude ultrastructurale de l'oogenèse chez une ptéridophyte: Le *Pteridium aquilinum* (L.) Kuhn. I. Évolution des structures nucléaires. *J. Micros. Biol. Cell.* **22,** 87–108.
Tourte, Y., and Hurel-Py, G. (1967). Ontogénie et ultrastructure de l'appareil cinétique des spermatozoides de *Pteridium aquilinum* L. *Compt. Rend. Acad. Sci. Paris* **265D,** 1289–92.
Tourte, Y., Kuligowski-Andres J., and Barbier-Ramond, C. (1980). Comporte-

ment différentiel des chromatines paternelles et maternelles au cours de l'embryogenèse d'une fougère: Le *Marsilea. Eur. J. Cell Biol.* **21,** 28–36.

Towill, L. R. (1978). Temperature and photocontrol of *Onoclea* spore germination. *Plant Physiol.* **62,** 116–9.

Towill, L. R., (1980). Analysis of starch accumulation and germination in *Onoclea* spores. *Am. J. Bot.* **67,** 88–94.

Towill, L. R., and Ikuma, H. (1973). Photocontrol of the germination of *Onoclea* spores. I. Action spectrum. *Plant Physiol.* **51,** 973–8.

Towill, L. R. and Ikuma, H. (1975*a*). Photocontrol of the germination of *Onoclea* spores. II. Analysis of germination processes by means of anaerobiosis. *Plant Physiol.* **55,** 150–4.

Towill, L. R., and Ikuma, H. (1975*b*). Photocontrol of the germination of *Onoclea* spores. III. Analysis of germination processes by means of cycloheximide. *Plant Physiol.* **55,** 803–8.

Towill, L. R., and Ikuma, H. (1975*c*). Photocontrol of the germination of *Onoclea* spores. IV. Metabolic changes during germination. *Plant Physiol.* **56,** 468–73.

Trivedi, B. S., and Bajpai, U. (1977). Ultrastructural study on effects of colchicine on gametophytes of *Lygodium flexuosum. Phytomorphology* **27,** 337–43

Tryon, A. F. (1964). *Platyzoma* – A Queensland fern with incipient heterospory. *Am. J. Bot.* **51,** 939–42.

Tryon, A. F. (1968). Comparisons of sexual and apogamous races in the fern genus *Pellaea. Rhodora* **70,** 1–24.

Tryon, A. F., and Britton, D. M. (1958). Cytotaxonomic studies on the fern genus *Pellaea. Evolution* **12,** 137–45.

Tryon, R. M., and Vitale, G. (1977). Evidence for antheridogen production and its mediation of a mating system in natural populations of fern gametophytes. *Bot. J. Linn. Soc.* **74,** 243–9.

Vaudois, B. (1963). Modifications cellulaires provoquées par le repiquage dans des prothalles de *Gymnogramme calomelanos* (Filicinées) en culture aseptque. *Compt. Rend. Acad. Sci. Paris* **256,** 251–3.

Vaudois, B. (1964). Modifications des noyaux et des nucléoles provoquées par le repiquage, dans le prothalle de *Lygodium scandens* (Filicinées). *Rev. Gen. Bot.* **71,** 592–9.

Vaudois, B. (1969). Contribution a l'étude de la régénération du prothalle de *Lygodium japonicum* (Filicinées). I. Etude en microscopie photonique. *Rev. Gen. Bot.* **76,** 361–415.

Vaudois, B. (1980). Intergeneric hybridization and speciation between two leptosporangiate ferns. *Cytobios* **29,** 69–79.

Vaudois, B. (1983). Transitory coenocyte during spermiogenesis of a leptosporangiate fern. *Cytobios* **38,** 21–31.

Vaudois, B., and Tourte, Y. (1979). Spermatogenesis in a pteridophyte. 1. First stages of the motile apparatus. *Cytobios* **24,** 143–56.

Vazart, J. (1964). Données sur le développement et l'infrastructure de l'appareil cinétique dans les spermatides du *Polypodium vulgare. Compt. Rend. Acad. Sci. Paris* **259,** 631–4.

Verma, S. C., and Khullar, S. P. (1966). Ontogeny of the polypodiaceous fern antheridium with particular reference to some Adiantaceae. *Phytomorphology* **16,** 302–14.

Voeller, B. R. (1964*a*). Antheridogens in ferns. In *Régulateurs Naturels de la Croissance Végétale,* Colloq. Internat. CNRS, Paris, No. 12, pp. 665–84.

Voeller, B. R. (1964*b*). Gibberellins: Their effect on antheridium formation in fern gametophytes. *Science* **143,** 373–5.

Voeller, B. R., and Weinberg, E. S. (1967). Antheridium induction and the number of sperms per antheridium in *Anemia phyllitidis. Am. Fern J.* **57,** 107–12.

Voeller, B. R., and Weinberg, E. S. (1969). Evolutionary and physiological aspects of antheridium induction in ferns. In: *Current Topics in Plant Sciences,* ed. J. E. Gunckel, pp. 77–93. New York: Academic Press.

Vogelmann, T. C., Bassel, A. R., and Miller, J. H. (1981). Effects of microtubule-inhibitors on nuclear migration and rhizoid differentiation in germinating fern spores (*Onoclea sensibilis*). *Protoplasma* **109,** 295–316.

Vogelmann, T. C., and Miller, J. H. (1980). Nuclear migration in germinating spores of *Onoclea sensibilis:* The path and kinetics of movement. *Am. J. Bot.* **67,** 648–52.

Vogelmann, T. C., and Miller, J. H. (1981). The effect of methanol on spore germination and rhizoid differentiation in *Onoclea sensibilis. Am. J. Bot.* **68,** 1177–83.

von Aderkas, P. (1984). Promotion of apogamy in *Matteuccia struthiopteris,* the ostrich fern. *Am. Fern J.* **74,** 1–6.

von Aderkas, P. (1986). Enhancement of apospory in liquid culture of *Matteuccia struthiopteris. Ann. Bot.* **57,** 505–10.

von Aderkas, and Cutter, E. G. (1983*a*). The role of the meristem in gametophyte development of the osmundaceous fern *Todea barbara* (L.) Moore. *Bot. Gaz.* **144,** 519–24.

von Aderkas, P., and Cutter, E. G. (1983*b*). Gametophyte plasticity and its bearing on sex expression in *Todea barbara* (L.) Moore. *Bot. Gaz.* **144,** 525–32.

von Aderkas, P., and Raghavan, V. (1985). Spore germination and early development of the gametophyte of *Schizaea pusilla. Am. J. Bot.* **72,** 1067–73.

von Witsch, H., and Rintelen, J. (1962). Die Entwicklung von Farnprothallien unter Gibberellineinfluss. *Planta* **59,** 115–8.

Wada, M., and Furuya, M. (1970). Photocontrol of the orientation of cell division in *Adiantum.* I. Effects of the dark and red periods in the apical cell of gametophyte. *Develop. Growth Diffn.* **12,** 109–18.

Wada, M., and Furuya, M. (1971). Photocontrol of the orientation of cell division in *Adiantum.* II. Effects of the direction of white light on the apical cell of gametophytes. *Planta* **98,** 177–85.

Wada, M., and Furuya, M. (1972). Phytochrome action on the timing of cell division in *Adiantum* gametophytes. *Plant Physiol.* **49,** 110–3.

Wada, M., and Furuya, M. (1973). Photocontrol of the orientation of cell division in *Adiantum.* III. Effects of metabolic inhibitors. *Develop. Growth Diffn.* **15,** 73–80.

Wada, M., and Furuya, M. (1974). Action spectrum for the timing of photo-induced cell division in *Adiantum* gametophytes. *Physiol. Plantarum* **32,** 377–81.

Wada, M., and Furuya, M. (1978). Effects of narrow-beam irradiations with blue and far-red light on the timing of cell division in *Adiantum* gametophytes. *Planta* **138,** 85–90.

Wada, M., Hayami, J., and Kadota, A. (1984). Returning dark-induced cell cycle to the beginning of G1 phase by red light irradiation in fern *Adiantum* protonemata. *Plant Cell Physiol.* **25,** 1053–8.

Wada, M., Kadota, A., and Furuya, M. (1978). Apical growth of protonemata in *Adiantum capillus-veneris.* II. Action spectra for the induction of apical swelling and the intracellular photoreceptive site. *Bot. Mag. (Tokyo)* **91,** 113–20.

Wada, M., Kadota, A., and Furuya, M. (1981). Intracellular photoreceptive site for polarotropism in protonema of the fern *Adiantum capillus-veneris* L. *Plant Cell Physiol.* **22,** 1481–8.

Wada, M., Kadota, A., and Furuya, M. (1983). Intracellular localization and dichroic orientation of phytochrome in plasma membrane and/or ectoplasm of a centrifuged protonema of fern *Adiantum capillus-veneris* L. *Plant Cell Physiol.* **24,** 1441–7.

Wada, M., Mineyuki, Y., and Furuya, M. (1982). Change in the rate of organelle movement during progression of the cell cycle in *Adiantum* protonemata. *Protoplasma* **113,** 132–6.

Wada, M., Mineyuki, Y., Kadota, A., and Furuya, M. (1980). The changes of nuclear position and distribution of circumferentially aligned microtubules during the progression of cell cycle in *Adiantum* protonemata. *Bot. Mag. (Tokyo)* **93,** 237–45.

Wada, M., and O'Brien, T. P. (1975). Observations on the structure of the protonema of *Adiantum capillus-veneris.* L. undergoing cell division following white-light irradiation. *Planta* **126,** 213–27.

Wada, M., Shimizu, H., Abe, H., Kadota, A., and Kondo, N. (1986). A model system to study the effect of SO_2 on plant cells. I. Experimental conditions in the case of fern gametophytes. *Environ. Cont. Biol.* **24,** 95–102.

Wada, M., Shimizu, H., and Kondo, N. (1987). A model system to study the effect of SO_2 on plant cells. II. Effect of sulfite on fern spore germination and rhizoid development. *Bot. Mag. (Tokyo)* **100,** 51–62.

Wada, M., and Staehelin, L. A. (1981). Freeze-fracture observations on the plasma membrane, the cell wall and the cuticle of growing protonemata of *Adiantum capillus-veneris* L. *Planta* **151,** 462–8.

Wadhwani, C., and Bhardwaja, T. N. (1981). Effect of *Lantana camara* L. extract on fern spore germination. *Experientia* **37,** 245–6.

Wadhwani, C., Bhardwaja, T. N., and Mahna, S. K. (1981). Growth and morphogenetic responses of fern gametophytes treated with extract of *Lantana camara. Phytomorphology* **31,** 51–5.

Wagner, W. H., Jr., and Sharp, A. J. (1963). A remarkably reduced vascular plant in the United States. *Science* **142,** 1483–4.

Walker, T. G. (1962). Cytology and evolution in the fern genus *Pteris* L. *Evolution* **16,** 27–43.

Walker, T. G. (1966). A cytotaxonomic survey of the pteridophytes of Jamaica. *Trans. Roy. Soc. Edin.* **66,** 169–237.

Walker, T. G. (1979). The cytogenetics of ferns. In: *The Experimental Biology of Ferns,* ed. A. F. Dyer, pp. 87–132. London: Academic Press.

Walker, T. G. (1985). Some aspects of agamospory in ferns – The Braithwaite system. *Proc. Roy. Soc. Edin.* **86B,** 59–66.

Ward, M. (1954). Fertilization in *Phlebodium aureum* J. Sm. *Phytomorphology* **4,** 1–17.

Ward, M. (1963). Developmental patterns of adventitious sporophytes in *Phlebodium aureum* J. Sm. *J. Linn. Soc. (Bot.).* **58,** 377–80.

Wardlaw, C. W., and Sharma, D. N. (1963). Experimental and analytical studies of pteridophytes. XL. Factors in the formation and distribution of sori in leptosporangiate ferns. *Ann. Bot.* **27,** 101–21.

Warne, T. R., and Hickok, L. G. (1987*a*). Single gene mutants tolerant to NaCl in the fern *Ceratopteris:* Characterization and genetic analysis. *Plant Sci.* **52,** 49–55.

Warne, T. R., and Hickok, L. G. (1987*b*). (2-Chloroethyl)phosphonic acid promotes germination of immature spores of *Ceratopteris richardii* Brongn. *Plant Physiol.* **83,** 723–5.
Warne, T. R., and Lloyd, R. M. (1980). The role of spore germination and gametophyte development in habitat selection: Temperature responses in certain temperate and tropical ferns. *Bull. Torrey Bot. Cl.* **107,** 57–64.
Warne, T. R., and Lloyd, R. M. (1987). Gametophyte density and sex expression in *Ceratopteris. Can. J. Bot.* **65,** 362–5.
Wayne, R., and Hepler, P. K. (1984). The role of calcium ions in phytochrome-mediated germination of spores of *Onoclea sensibilis* L. *Planta* **160,** 12–20.
Wayne, R., and Hepler, P. K. (1985*a*). Red light stimulates an increase in intracellular calcium in the spores of *Onoclea sensibilis. Plant Physiol.* **77,** 8–11.
Wayne, R., and Hepler, P. K. (1985*b*). The atomic composition of *Onoclea sensibilis* spores. *Am. Fern J.* **75,** 12–8.
Wayne, R., Rice, D., and Hepler, P. K. (1986). Intracellular pH does not change during phytochrome-mediated spore germination in *Onoclea. Develop. Biol.* **113,** 97–103.
Weinberg, E. S., and Voeller, B. R. (1969*a*). External factors inducing germination of fern spores. *Am. Fern J.* **59,** 153–67.
Weinberg, E. S., and Voeller, B. R. (1969*b*). Induction of fern spore germination. *Proc. Natl. Acad. Sci. USA* **64,** 835–42.
Werth, C. R., Guttman, S. I., and Eshbaugh, W. H. (1985). Electrophoretic evidence of reticulate evolution in the Appalachian *Asplenium* complex. *Syst. Bot.* **10,** 184–92.
Wetmore, R. H., DeMaggio, A. E., and Morel, G. (1963). A morphogenetic look at the alternation of generations. *J. Indian Bot. Soc.* **42A,** 306–20.
Whittier, D. P. (1962). The origin and development of apogamous structures in the gametophyte of *Pteridium* in sterile culture. *Phytomorphology* **12,** 10–20.
Whittier, D. P. (1964*a*). The effect of sucrose on apogamy in *Cyrtomium falcatum* Presl. *Am. Fern J.* **54,** 20–5.
Whittier, D. P. (1964*b*). The influence of cultural conditions on the induction of apogamy in *Pteridium* gametophytes. *Am. J. Bot.* **51,** 730–6.
Whittier, D. P. (1965). Obligate apogamy in *Cheilanthes tomentosa* and *C. alabamensis. Bot. Gaz.* **126,** 275–81.
Whittier, D. P. (1966*a*). The influence of growth substances on the induction of apogamy in *Pteridium* gametophytes. *Am. J. Bot.* **53,** 882–6.
Whittier, D. P. (1966*b*). Induced apogamy in diploid gametophytes of *Pteridium. Can. J. Bot.* **44,** 1717–21.
Whittier, D. P. (1970). The initiation of sporophytes by obligate apogamy in *Cheilanthes castanea. Am. J. Bot.* **57,** 1249–54.
Whittier, D. P. (1972). Gametophytes of *Botrychium dissectum* as grown in sterile culture. *Bot. Gaz.* **133,** 363–9.
Whittier, D. P. (1973). The effect of light and other factors on spore germination in *Botrychium dissectum. Can. J. Bot.* **51,** 1971–4.
Whittier, D. P. (1975). The influence of osmotic conditions on induced apogamy in *Pteridium* gametophytes. *Phytomorphology* **25,** 246–9.
Whittier, D. P. (1976). Tracheids, apogamous leaves, and sporophytes in gametophytes of *Botrychium dissectum. Bot. Gaz.* **137,** 237–41.
Whittier, D. P. (1978). Apospory in haploid leaves of *Botrychium. Phytomorphology* **28,** 215–9.
Whittier, D. P. (1981). Spore germination and young gametophyte development of *Botrychium* and *Ophioglossum* in axenic culture. *Am. Fern J.* **71,** 13–19.

Whittier, D. P. (1983). Gametophytes of *Ophioglossum engelmannii*. *Can. J. Bot.* **61,** 2369–73.

Whittier, D. P. (1984). The organic nutrition of *Botrychium* gametophytes. *Am. Fern J.* **74,** 77–86.

Whittier, D. P. (1987). Germination of *Helminthostachys* spores. *Am. Fern J.* **77,** 95–9.

Whittier, D. P., and Pratt, L. H. (1971). The effect of light quality on the induction of apogamy in prothalli of *Pteridium aquilinum*. *Planta* **99,** 174–8.

Whittier, D. P., and Steeves, T. A. (1960). The induction of apogamy in the bracken fern. *Can. J. Bot.* **38,** 925–30.

Whittier, D. P., and Steeves, T. A. (1962). Further studies on induced apogamy in ferns. *Can. J. Bot.* **40,** 1525–31.

Wiebe, D. H., Towill, L. R., and Campbell-Domsky, C. (1987). Soluble-starch-synthesizing and sucrose-degrading enzymes in *Onoclea* spores. *Phytochemistry* **26,** 949–54.

Wilkie, D. (1954). The movements of spermatozoa of bracken (*Pteridium aquilinum*). *Expt. Cell Res.* **6,** 384–91.

Wilkie, D. (1956). Incompatibility in bracken. *Heredity* **10,** 247–56.

Wilkie, D. (1963). Genetic analysis of variation in the bracken prothallus. *J. Linn. Soc. (Bot).* **58,** 333–6.

Willson, M. F. (1981). Sex expression in fern gametophytes: Some evolutionary possibilities. *J. Theor. Biol.* **93,** 403–9.

Wilson, K. A. (1958). Ontogeny of the sporangium of *Phlebodium* (*Polypodium*) *aureum*. *Am. J. Bot.* **45,** 483–91.

Windham, M. D., Wolf, P. G., and Ranker, T. A. (1986). Factors affecting prolonged spore viability in herbarium collections of three species of *Pellaea*. *Am. Fern J.* **76,** 141–8.

Wolf, P. G., Haufler, C. H., and Sheffield, E. (1987). Electrophoretic evidence for genetic diploidy in the bracken fern (*Pteridium aquilinum*). *Science* **236,** 947–9.

Yamane, H., Nohara, K., Takahashi, N., and Schraudolf, H. (1987). Identification of antheridic acid as an antheridiogen in *Anemia rotundifolia* and *Anemia flexuosa*. *Plant Cell Physiol.* **28,** 1203–7.

Yamane, H., Sato, Y., Takahashi, N., Takeno, K., and Furuya, M. (1980). Endogenous inhibitors for spore germination in *Lygodium japonicum* and their inhibitory effects on pollen germinations in *Camellia japonica* and *Camellia sinensis*. *Agr. Biol. Chem.* **44,** 1697–9.

Yamane, H., Takahashi, N., Takeno, K., and Furuya, M. (1979). Identification of gibberellin A_9 methyl ester as a natural substance regulating formation of reproductive organs in *Lygodium japonicum*. *Planta* **147,** 251–6.

Yamane, H., Yamaguchi, I., Kobayashi, M., Takahashi, M., Sato, Y., Takahashi, N., Iwatsuki, K., Phinney, B. O., Spray, C.R., Gaskin, P., and MacMillan, J. (1985). Identification of ten gibberellins from sporophytes of the tree fern, *Cyathea australis*. *Plant Physiol.* **78,** 899–903.

Yamasaki, N. (1954). Über den Einfluss von Colchicin auf Farnpflanzen. I. Die jungen Prothallien von *Polystichum craspedosorum* Diels. *Cytologia* **19,** 249–54.

Yatsuhashi, Y., Kadota, A., and Wada, M. (1985). Blue- and red-light action in photoorientation of chloroplasts in *Adiantum* protonemata. *Planta* **165,** 43–50.

Yatsuhashi, Y., Hashimoto, T., and Wada, M. (1987). Dichroic orientation of photoreceptors for chloroplast movement in *Adiantum* protonemata. Non-helical orientation. *Plant Sci.* **51,** 165–70.

Yeoh, O. C., and Raghavan, V. (1966). Riboflavin as photoreceptor in the induction of two-dimensional growth in fern gametophytes. *Plant Physiol.* **41,** 1739–42.

Zanno, P. R., Endo, M., Nakanishi, K., Näf, U., and Stein, C. (1972). On the structural diversity of fern antheridiogens. *Naturwissenschaften* **59,** 512.

Zilberstein, A., Arzee, T., and Gressel, J. (1984). Early morphogenetic changes during phytochrome-induced fern spore germination. I. The existence of a pre-photoinduction phase and the accumulation of chlorophyll. *Z. Pflanzenphysiol.* **114,** 97–107.

Zilberstein, A., Gressel, J., Arzee, T., and Edelman, M. (1984). Early morphogenetic changes during phytochrome-induced fern spore germination. II. Transcriptional and translational events. *Z. Pflanzenphysiol.* **114,** 109–22.

Zirkle, R. E. (1932). Some effects of alpha radiation upon plant cells. *J. Cell. Comp. Physiol.* **2,** 251–74.

Zirkle, R. E. (1936). Modification of radiosensitivity by means of readily penetrating acids and bases. *Am. J. Roentgenol.* **35,** 230–7.

Author index

Subject index